LONDON MATHEMATICAL SOCIETY LECTURE NOTE SERIES

Managing Editor: Professor J.W.S. Cassels,
Department of Pure Mathematics and Mathematical Statistics,
16 Mill Lane, Cambridge CB2 1SB.

1. General cohomology theory and K-theory, P.HILTON
4. Algebraic topology, J.F.ADAMS
5. Commutative algebra, J.T.KNIGHT
8. Integration and harmonic analysis on compact groups, R.E.EDWARDS
9. Elliptic functions and elliptic curves, P.DU VAL
10. Numerical ranges II, F.F.BONSALL & J.DUNCAN
11. New developments in topology, G.SEGAL (ed.)
12. Symposium on complex analysis, Canterbury, 1973, J.CLUNIE & W.K.HAYMAN (eds.)
13. Combinatorics: Proceedings of the British Combinatorial Conference 1973, T.P.McDONOUGH & V.C.MAVRON (eds.)
15. An introduction to topological groups, P.J.HIGGINS
16. Topics in finite groups, T.M.GAGEN
17. Differential germs and catastrophes, Th.BROCKER & L.LANDER
18. A geometric approach to homology theory, S.BUONCRISTIANO, C.P. ROURKE & B.J.SANDERSON
20. Sheaf theory, B.R.TENNISON
21. Automatic continuity of linear operators, A.M.SINCLAIR
23. Parallelisms of complete designs, P.J.CAMERON
24. The topology of Stiefel manifolds, I.M.JAMES
25. Lie groups and compact groups, J.F.PRICE
26. Transformation groups: Proceedings of the conference in the University of Newcastle-upon-Tyne, August 1976, C.KOSNIOWSKI
27. Skew field constructions, P.M.COHN
28. Brownian motion, Hardy spaces and bounded mean oscillations, K.E.PETERSEN
29. Pontryagin duality and the structure of locally compact Abelian groups, S.A.MORRIS
30. Interaction models, N.L.BIGGS
31. Continuous crossed products and type III von Neumann algebras, A.VAN DAELE
32. Uniform algebras and Jensen measures, T.W.GAMELIN
33. Permutation groups and combinatorial structures, N.L.BIGGS & A.T.WHITE
34. Representation theory of Lie groups, M.F. ATIYAH et al.
35. Trace ideals and their applications, B.SIMON
36. Homological group theory, C.T.C.WALL (ed.)
37. Partially ordered rings and semi-algebraic geometry, G.W.BRUMFIEL
38. Surveys in combinatorics, B.BOLLOBAS (ed.)
39. Affine sets and affine groups, D.G.NORTHCOTT
40. Introduction to Hp spaces, P.J.KOOSIS
41. Theory and applications of Hopf bifurcation, B.D.HASSARD, N.D.KAZARINOFF & Y-H.WAN
42. Topics in the theory of group presentations, D.L.JOHNSON
43. Graphs, codes and designs, P.J.CAMERON & J.H.VAN LINT
44. Z/2-homotopy theory, M.C.CRABB
45. Recursion theory: its generalisations and applications, F.R.DRAKE & S.S. WAINER (eds.)
46. p-adic analysis: a short course on recent work, N.KOBLITZ
47. Coding the Universe, A.BELLER, R.JENSEN & P.WELCH
48. Low-dimensional topology, R.BROWN & T.L.THICKSTUN (eds.)
49. Finite geometries and designs, P.CAMERON, J.W.P.Hirschfield & D.R. Hughes (eds.)

50. Commutator calculus and groups of homotopy classes, H.J.BAUES
51. Synthetic differential geometry, A.KOCK
52. Combinatorics, H.N.V.TEMPERLEY (ed.)
53. Singularity theory, V.I.ARNOLD
54. Markov processes and related problems of analysis, E.B.DYNKIN
55. Ordered permutation groups, A.M.W.GLASS
56. Journëes arithmëtiques 1980, J.V.ARMITAGE (ed.)
57. Techniques of geometric topology, R.A.FENN
58. Singularities of smooth functions and maps, J.MARTINET
59. Applicable differential geometry, M.CRAMPIN & F.A.E.PIRANI
60. Integrable systems, S.P.NOVIKOV et al.
61. The core model, A.DODD
62. Economics for mathematicians, J.W.S.CASSELS
63. Continuous semigroups in Banach algebras, A.M.SINCLAIR
64. Basic concepts of enriched category theory, G.M.KELLY
65. Several complex variables and complex manifolds I, M.J.FIELD
66. Several complex variables and complex manifolds II, M.J.FIELD
67. Classification problems in ergodic theory, W.PARRY & S.TUNCEL
68. Complex algebraic surfaces, A.BEAUVILLE
69. Representation theory, I.M.GELFAND et al.
70. Stochastic differential equations on manifolds, K.D.ELWORTHY
71. Groups - St Andrews 1981, C.M.CAMPBELL & E.F.ROBERTSON (eds.)
72. Commutative algebra: Durham 1981, R.Y.SHARP (ed.)
73. Riemann surfaces: a view towards several complex variables, A.T.HUCKLEBERRY
74. Symmetric designs: an algebraic approach, E.S.LANDER
75. New geometric splittings of classical knots (algebraic knots), L.SIEBENMANN & F.BONAHON
76. Linear differential operators, H.O.CORDES
77. Isolated singular points on complete intersections, E.J.N.LOOIJENGA
78. A primer on Riemann surfaces, A.F.BEARDON
79. Probability, statistics and analysis, J.F.C.KINGMAN & G.E.H.REUTER (eds.)
80. Introduction to the representation theory of compact and locally compact groups, A.ROBERT
81. Skew fields, P.K.DRAXL
82. Surveys in combinatorics: Invited papers for the ninth British Combinatorial Conference 1983, E.K.LLOYD (ed.)
83. Homogeneous structures on Riemannian manifolds, F.TRICERRI & L.VANHECKE
84. Finite group algebras and their modules, P.LANDROCK
85. Solitons, P.G.DRAZIN
86. Topological topics, I.M.JAMES (ed.)
87. Surveys in set theory, A.R.D.MATHIAS (ed.)
88. FPF ring theory, C.FAITH & S.PAGE
89. An F-space sampler, N.J.KALTON, N.T.PECK & J.W.ROBERTS
90. Polytopes and symmetry, S.A.ROBERTSON
91. Classgroups of group rings, M.J.TAYLOR
92. Representation of rings over skew fields, A.H. SCHOFIELD
93. Aspects of topology, I.M.JAMES & E.H.KRONHEIMER (eds.)
94. Representations of general linear groups, G.D.JAMES
95. Low dimensional topology 1982: Proceedings of the Sussex Conference, 2-6 August 1982, R.A.FENN (ed.)
96. Diophantine equations over function fields, R.C.MASON
97. Varieties of constructive mathematics, D.S.BRIDGES & F.RICHMAN
98. Localization in Noetherian rings, A.V.JATEGAONKAR
99. Methods of differential geometry in algebraic topology, M.KAROUBI & C.LERUSTE
100. Stopping time techniques for analysts and probabilists, L.EGGHE

Aspects of topology

London Mathematical Society Lecture Note Series: 93

Aspects of topology

In memory of Hugh Dowker 1912 - 1982

Edited by

I.M.JAMES

Savilian Professor of Geometry,Oxford University

and

E.H.KRONHEIMER

Birkbeck College,University of London

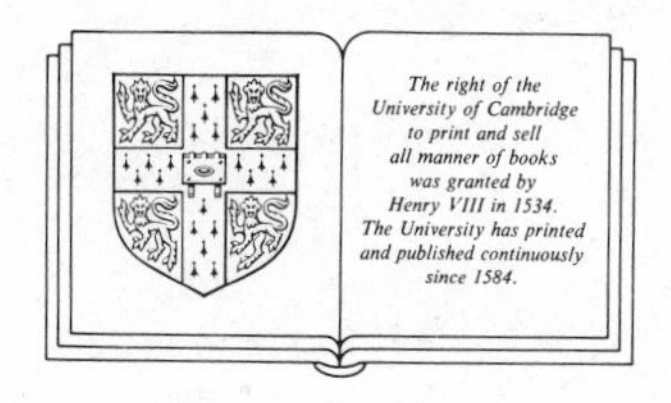

CAMBRIDGE UNIVERSITY PRESS

Cambridge

London New York New Rochelle

Melbourne Sydney

Published by the Press Syndicate of the University of Cambridge
The Pitt Building, Trumpington Street, Cambridge CB2 1RP
32 East 57th Street, New York, NY 10022, USA
10 Stamford Road, Oakleigh, Melbourne 3166, Australia

First published 1985

Printed in Great Britain at the University Press, Cambridge

Library of Congress catalogue card number:

ISBN 0 521 27815 5

British Library cataloguing in publication data

Aspects of topology.— (London Mathematical
Society lecture note series; 93)
1. Topology
I. James, I.M. II. Kronheimer, E.H.
III. Series
514 QA611

ISBN 0 521 27815 5

CONTENTS

CONTRIBUTORS

Professor S. Antonian, Department of Mathematics, State University of Yerevan, 375049 Yerevan-49, U.S.S.R.

Professor J. Dugundji, Department of Mathematics, University of Southern California, DRB 306, University Park, Los Angeles, California 90089-1113, U.S.A.

Professor P. Erdős, Mathematical Institute of Hungarian Academy of Sciences, Budapest, Reáltanoda u. 13-15, H-1053, Hungary.

Professor S. Halperin, Department of Mathematics, University of Toronto, Toronto, Canada M5S 1A1.

Professor P. Hilton, Division of Science and Mathematics, State University of New York at Binghamton, Binghamton, New York 13901, U.S.A.

Professor J. Isbell, Department of Mathematics, Wesleyan University, Middletown, Connecticut 06457, U.S.A.

Professor I.M. James, Mathematical Institute, University of Oxford, 24-29 St. Giles, Oxford OX1 3LB, England.

Dr P. Johnstone, Department of Pure Mathematics and Mathematical Statistics, University of Cambridge, 16 Mill Lane, Cambridge CB2 1SB, England.

Professor K. Morita, Department of Mathematics, Sophia University, Kioicho Chiyoda-ku, Tokyo 102, Japan.

Professor J. Nagata, Department of Mathematics, Osaka Kyoiku University, Tennoiji, Osaka 543, Japan.

Professor J. Pach, Mathematical Institute of Hungarian Academy of Sciences, Budapest, Reáltanoda u. 13-15, H-1053, Hungary.

Professor B.A. Pasynkov, Moskowskii Univ., Mehmat, Moscow 117234, U.S.S.R.

Professor M.E. Rudin, Mathematics Department, University of Wisconsin-Madison, Van Vleck Hall, 480 Lincoln Drive, Madison, Wisconsin 53706, U.S.A.

Professor A.H. Stone, Department of Mathematics, Ray P. Hylan Building, University of Rochester, New York 14627, U.S.A.

Dr D. Strauss, Department of Pure Mathematics, University of Hull, 22-24 Newland Park, Hull HU5 2RD.

Dr M. Thistlethwaite, Polytechnic of the South Bank, London SE1 0AA, England.

Dr M. Warner, Department of Mathematics, City University, Northampton Square, London EC1V 0HB, England.

NOTES

1. The photograph of Hugh Dowker was kindly provided by Yael Dowker.

2. The obituary notice, by Dona Strauss, is reprinted from the Bulletin of the London Mathematical Society, Volume 16, 1984.

3. The editors have made some minor alterations to the text of contributions by authors whose native language is other than english.

4. The final article was written by Dona Strauss on the basis of joint work by Hugh Dowker and herself.

OBITUARY

CLIFFORD HUGH DOWKER

Clifford Hugh Dowker was born in 1912 in Western Ontario, and grew up in a rural community, where his family owned a small farm. His ancestors on his father's side were of Yorkshire origin, while his mother was a McGregor of Scottish descent.

This rural background might appear unexpected for an important mathematician. Indeed, Hugh was the first Dowker to go to High School. Neither of his brothers had academic careers. His elder brother, Gordon, left school at thirteen and worked in the Canadian forests. His younger brother, Arthur, followed the family tradition of working as a farmer.

Hugh Dowker's first school was a one-room country school to which he had to walk a couple of miles. His next school was the High School in Parkhill, where the mathematics teacher appears to have had little understanding of the subject. Hugh was paid to stay in after school in order to teach mathematics to his teacher!

There was one teacher in Parkhill—not a mathematician—who seems to have had an important influence on Dowker. This was a teacher with a deep knowledge of wildlife, botany and geology, who took him and other pupils to the Muskoka Lakes and the Bruce Peninsula. This experience probably had a lasting effect on Dowker, who, throughout his life, displayed a keen interest in the countryside around him.

When Dowker was seventeen, he went to the University of Western Ontario, having been awarded a scholarship on the basis of his excellent examination results. He intended to be a school teacher. However, unexpectedly for himself and his family, his talents were to lead him into mathematics. This was a period of some penury for Dowker; the room that he shared with another student was heated by a chicken-coop heater, and he lived largely on tinned salmon and carrots—the cheapest foods available.

He studied a variety of subjects, including physics and economics, and received his BA degree in 1933. Then, because of his evident brilliance in mathematics, he was persuaded to continue his studies at the University of Toronto. After obtaining an MA there in the following year, he was advised to go to Princeton University to study under Lefschetz.

It was at Princeton that Dowker became fully aware of the power and beauty of mathematics, and that he became an active topologist, running one of Lefschetz' seminars. He obtained his Ph.D. there in 1938. Apart from Lefschetz, the mathematicians who were to have an important influence on Dowker included Alexandrov, Fox, Hurewicz and Steenrod.

He subsequently held a position as instructor at the University of Western Ontario, and was then an assistant to Von Neumann at the Princeton Institute for Advanced Study. He then went to Johns Hopkins University as an instructor. It was here that he met Yael Naim, whom he was to marry in 1944. Yael, at the time, was a young graduate student who had come to Johns Hopkins from Israel. She is herself a highly gifted mathematician, who was to become well-known for her work in ergodic

Received 22 February, 1984.

theory. Dowker once remarked that all measure theorists marry topologists—this law being exemplified by the Dowkers, Rudins and Stones!

In 1943 Dowker was seconded to the United States Air Force as a civilian adviser, and carried out work on gunnery and the trajectories of projectiles, which took him to Libya and Egypt. Then, from 1943 to 1946, he and Yael both worked at the M.I.T. Radiation Laboratory. After the war he was an associate professor at Tufts, and then a visiting lecturer at Princeton and at Harvard.

This was the period of McCarthyism, when the atmosphere in North American Universities was very difficult. Several of Dowker's friends in the mathematical community were severely harassed, and one had been arrested. Hugh and Yael decided to leave North America. They came to England in 1950, where Yael had a post at the University of Manchester, and where Hugh was soon appointed to a readership in Applied Mathematics at Birkbeck College.

Although Dowker is famous for his work in the purest and most abstract branches of mathematics, it is a mark of his versatility that he deserved to hold a post as an applied mathematician. Indeed, he has made a real contribution to applied mathematics, through his work on projectiles and on servo-mechanisms [**5**].

In 1962 Dowker was appointed to a personal chair at Birkbeck College. He remained at Birkbeck until his retirement in 1979. He died in London in 1982, after a long and difficult illness against which he had struggled for seven years. He had been a member of the Society since 1951.

In manner, Dowker was reserved and gentle, with an innate dignity and a penetrating wit. He possessed a high degree of integrity and moral strength which enabled him to endure seven years of illness uncomplainingly. Although supremely tolerant towards others, he had only the highest standards of behaviour for himself. He was totally without ostentation or pretention and totally disinterested in wealth, honours or managerial power.

Hugh Dowker was unfailingly kind and generous and was always ready to spend time in aiding others. Over the course of some thirteen years, he and Yael did a great deal of work with children who were sent to them by the National Association for Gifted Children. The Dowkers were very committed to this work and were highly successful in it. They wrote an interesting joint paper [**32**], in which they describe how they had helped more than thirty gifted children—many of whom had difficulties in school—to experience the delight of mathematical discovery. It should not be thought that it was only with gifted children that Dowker was concerned. He had an affection for all young people and was known among his students for his helpfulness and patience. Even children in his neighbourhood would come to him for help with their homework.

Hugh and Yael were well known for their kindness and hospitality, which earned them many deep and lasting friendships in the mathematical community.

Dowker was widely travelled. In his early twenties he had twice "hoboed" across the United States and Canada, jumping on and off freight trains at suitable points. Later, as a mathematician, he held posts as a visiting professor in Russia, Israel, India and Canada. He also spent some time working on a kibbutz in Israel. He was able to speak Russian and knew some Georgian and some Hebrew. He loved the countryside, and often went walking or mountain-climbing in the national parks and in Switzerland.

That Dowker was impressively knowledgeable in mathematics was widely known; his letters and papers contain a wealth of answers to other mathematicians' questions. What might be less widely known, is that he had a deep knowledge of many subjects, including Georgian culture, the early history of Christian religions, the geography of many lands and localities and the history and anthropology of arithmetic in different societies. He always had a thorough knowledge of the history and culture of whatever area he happened to live in.

He had a deep love of mathematics, and continued to work even when he was weak and in pain. I visited him during the last days of his life, and was very moved by his determination to discuss mathematics even though he scarcely had the strength to speak.

He was a man who was respected and loved by his students, his many friends and the world-wide mathematical community.

Mathematical work

Dowker's mathematical work lay mainly in the field of topology. Although the number of his published papers is not large, they have been remarkably influential. His name is very widely quoted among topologists—indeed, it has been said that his best known paper [9] is one of the most frequently quoted in the whole of mathematics.

Dowker's papers are always striking, not only for their fundamental importance, but also for the elegance and clarity with which they are written. They contain a wealth of ingenious examples, often answering difficult problems posed by other mathematicians. He was constantly concerned to find the "right" basic definitions and axioms, and this led to his proving very general results under very few assumptions.

The results of his Ph.D. thesis were announced in [**1**] and presented, with additional material, in [**3**]. The theme of these papers was the extension of basic theorems in homotopy theory from compact metric to normal or paracompact spaces, a key role being played by the concept of uniform homotopy. [**3**] contains a pioneering exposition of Čech cohomology from a geometric point of view, and includes a proof of the surprising fact that the first Betti number of the real line is c. It also provides a proof of the important fact that the same covering dimension of a normal space is obtained whether finite, star-finite or locally finite covers are used. The techniques employed, involving the use of canonical maps to nerves of covers, led naturally to [**6**], in which it was shown that a canonical map of a space into the nerve of an arbitrary open covering exists if and only if the space is paracompact and normal. Although there are two different ways of topologising the nerve of a covering (called "geometric" and "natural"), they are shown to be equivalent in this context.

This observation led to a systematic study of metrisable topologies on infinite complexes in [**10**], in which the important fact is established that all the various reasonable topologies have the same homotopy type. In this paper, a question of J. H. C. Whitehead is answered by an example of two CW complexes whose product is not CW. The study of infinite complexes was continued in subsequent papers ([**18**, **21**]). An interesting corollary of theorems proved in [**21**] is that isomorphic Euclidean complexes are homeomorphic.

Dowker's interest in locally finite covers led to [**4**], in which it was shown that paracompact metric spaces are precisely the spaces that can be embedded in Hilbert spaces. (It was not yet known that all metric spaces are paracompact.) This paper now provides a key step in the famous Bing–Nagata–Smirnov metrisation theorem.

[9] is probably Dowker's best known paper. This is the paper which introduced the important concept of countable paracompactness. In homotopy theory, the properties of a product space $X \times I$, where I denotes the closed unit interval, are fundamentally important. In **[9]**, Hugh showed that $X \times I$ is normal precisely in the case in which X is countably paracompact. He gave several striking and useful characterisations of countable paracompactness, and asked whether there were any normal spaces which failed to be countably paracompact. This question turned out to be one of the most challenging in general topology. It was finally answered twenty years later by Mary Ellen Rudin, who constructed an ingenious and intricate example of a normal space which was not countably paracompact. Spaces of this kind—which play a significant role in the study of non-metric spaces—are now known as "Dowker" spaces.

Dowker was constantly concerned with the fundamental definitions of homology and cohomology groups in general spaces. In **[7]** he showed how the Čech construction of direct and inverse limits can be used to deal with singular homology and cohomology. This work showed the identity of the Alexander and Čech cohomology groups in a wide class of spaces. In **[8]** he showed that the Eilenberg–Steenrod axioms were satisfied by Čech cohomology theory based on infinite open coverings. (It was known that the homotopy axiom failed to hold if finite covers were used.) The identity of the Čech and Vietoris homology groups and the Čech and Alexander cohomology groups was established in **[11]** for *arbitrary* spaces. A corollary of this is the fact that the Eilenberg–Steenrod axioms are satisfied by Alexander cohomology theory.

In **[20, 22, 23]**, Dowker developed results in excision theory. In **[20]** he gave very general conditions for the strong excision property to hold for Čech cohomology. He also showed that some conditions, going beyond normality, are needed, thus proving a conjecture due to A. D. Wallace.

An interest in the extension of maps was implicit in **[2, 3]** and developed further in **[12, 17, 18]**. Dowker's main contribution was given in **[17]**, where he gave very general conditions sufficient for extensibility, and gave criteria for spaces to be ANR and NES (neighbourhood extension) spaces. In an earlier paper **[12]**, he had improved and generalised Hanner's characterisation of the separable metric spaces which are ANRs for normal spaces, and had extended it to non-separable spaces.

Dimension theory was a recurring theme in Dowker's work. **[14]** contains a thorough study of large inductive dimension (Ind) in completely normal spaces, and introduces the slightly more restrictive class of totally normal spaces, to which many of the principal theorems of classical dimension theory are extended. This study was continued in **[15]**, where the subset theorem is extended to totally normal spaces. In this paper, Dowker shows that local dimension and dimension coincide in a wide class of spaces, and gives an example of a normal space in which they differ. In **[19]** it is shown that the dimension of a metric space can be defined in terms of a sequence of open covers satisfying the condition that the closures of the sets in the $(i+1)$-st cover form a refinement of the i-th cover. This paper also contains a new proof of the Katetov–Morita theorem which states that dim and Ind coincide for metric spaces.

[24] contains a generalisation of the concepts of proximity and uniformity, in which symmetry is not required. By discarding the property of symmetry, it becomes possible to define quotient, open and closed maps. Product structures are also defined and investigated. A question of Smirnov is answered by an example of a (symmetric) proximity space which has no finest consistent (symmetric) uniformity.

In joint papers with myself, published from 1966 onwards, Dowker did pioneering work in the study of frames. A frame is a complete lattice which satisfies the infinite distributivity condition: $x \wedge \bigvee_{\alpha} x_a = \bigvee_{a} (x \wedge x_a)$. Lattices of this kind are significant in topology, as the lattice formed by the open subsets of a topological space is an example of a frame. They are also significant in other fields; for example, they furnish models for intuitionist logic. In recent years, they have aroused interest because of their applications in sheaf theory and in the new field of topos theory. Dowker was responsible for establishing some of the basic features of the category of frames, including the properties of quotients and co-products. (It should be noted that some of these ideas were developed independently by other mathematicians, including J. R. Isbell.)

[**34**] gives an indication of the versatility of Dowker's work. In this paper, written with M. Thistlethwaite, a complete classification of all 12,765 knots with at most thirteen crossings is given. The authors pioneered the use of the computer for the purpose of tabulating knots, and developed some strikingly ingenious new techniques.

Dowker was also interested in the general theory of categories. In [**26**] he showed how the connecting morphism of homology theory arises from a functor definable in any abelian category. In his final paper [**35**], written during the last weeks of his life, he proved the interesting fact that two categories must be isomorphic if there are functors between them which are injective on objects and have composites naturally equivalent to the identity functors.

Finally Dowker's influential lectures on sheaf theory [**16**] should be mentioned among his publications. They provide a very clear and careful exposition of the subject, with a wealth of illustrative examples, and were for some years the only source for several important results about sheaves and sheaf cohomology. Mathematicians, who are now experts in sheaf theory, have told me that it was these lecture notes which first awakened their interest in the subject.

This description of Dowker's mathematical work is not exhaustive. His command of mathematics was so wide that he has contributed to fields of which I am unaware. For example, I have recently learned of a theorem in pure geometry which bears his name.

Acknowledgements

I wish to express my gratitude to the many friends who have written to me about Hugh Dowker's life and work. I am particularly grateful to Yael Dowker, who has spent many hours in talking to me about him, and to Gordon Dowker, who sent me a very long and very interesting letter about his brother's childhood. I am also most grateful to Arthur Stone, who sent me an invaluable assessment of Dowker's mathematical work. A great deal of the account that I have given above, is, in fact, due to him, as I often felt unable to improve on his formulation.

Bibliography

1. 'Hopf's Theorem for non-compact spaces', *Proc. Nat. Acad. Sci. U.S.A.*, 23 (1937), 293–294.

2. 'On minimum circumscribed polygons', *Bull. Amer. Math. Soc.*, 50 (1944), 120–122 [*Math. Rev.* 5, #153 (John)].

3. 'Mapping theorems for non-compact spaces', *Amer. J. Math.*, 69 (1947), 200–242 [*Math. Rev.* 8, #594 (Freudenthal)].
4. 'An imbedding theorem for paracompact metric spaces', *Duke Math. J.*, 14 (1947), 639–645 [*Math. Rev.* 9, #196 (Dieudonné)].
5. (with R. Phillips) Chapter VIII, *Theory of servo-mechanisms* (McGraw Hill, 1947), 340–360. (Publication of the M.I.T. Radiation Laboratory.)
6. 'An extension of Alexandroff's mapping theorem', *Bull. Amer. Math. Soc.*, 54 (1948), 386–391 [*Math. Rev.* 9, #523 (Dieudonné)].
7. (with W. Hurewicz and J. Dugundji) 'Connectivity groups in terms of limit groups', *Ann. of Math.*, 49 (1948), 391–406 [*Math. Rev.* 9, #606 (Freudenthal)].
8. Čech cohomology theory and the axioms', *Ann. of Math.* (2), 51 (1950), 278–292 [*Math. Rev.* 11, #450 (Freudenthal)].
9. 'On countably paracompact spaces', *Canad. J. Math.*, 3 (1951), 219–224 [*Math. Rev.* 13, #264 (A. H. Stone)].
10. 'Topology of metric complexes', *Amer. J. Math.*, 74 (1952), 555–577 [*Math. Rev.* 13, #965 (Spanier)].
11. 'Homology groups of relations', *Ann. of Math.* (2), 56 (1952), 84–95 [*Math. Rev.* 13, #967 (Spanier)].
12. 'On a theorem of Hanner', *Ark. Mat.*, 2 (1952), 307–313 [*Math. Rev.* 14, #296 (Michael)].
13. 'A problem in set theory', *J. London Math. Soc.*, 27 (1952), 371–374 [*Math. Rev.* 13, #924 (Gustin)].
14. 'Inductive dimension of completely normal spaces', *Quart. J. Math., Oxford Ser.* (2), 4 (1953), 267–281 [*Math. Rev.* 16, #157 (Katětov)].
15. 'Local dimension of normal spaces', *Quart. J. Math. Oxford Ser.* (2), 6 (1955), 101–120 [*Math. Rev.* 19, #157 (M. Henriksen)].
16. *Lectures on sheaf theory.* Notes by S. V. Adavi and N. Ramabbadran (Tata Institute of Fundamental Research, Bombay, 1956, v+212+iv+iii pp. mimeographed) [*Math. Rev.* 19, #301 (M. F. Atiyah)].
17. 'Homotopy extension theorems', *Proc. London Math. Soc.* (3), (1956), 100–116 [*Math. Rev.* 17, #518 (J. Dugundji)].
18. 'Imbedding of metric complexes', *Algebraic geometry and topology*, A symposium in honour of S. Lefschetz (Princeton University Press, Princeton, N.J., 1957), pp. 239–242 [*Math. Rev.* 18, #920 (J. Dugundji)].
19. (with W. Hurewicz) 'Dimension of metric spaces', *Fund. Math.*, 43 (1956), 83–88 [*Math. Rev.* 18, #56 (Haskell Cohen)].
20. 'The excision theorem' (Russian), *Dokl. Akad. Nauk. SSSR*, 125 (1959), 1190–1192 [*Math. Rev.* 21, #3840 (Isbell)].
21. 'Affine and Euclidean complexes' (Russian), *Dokl. Akad. Nauk SSSR*, 128 (1959), 655–656 [*Math. Rev.* 22, #8483 (Kahn)].
22. 'The Kolmogorov–Aleksandrov duality theorem' (Russian), *Mat. Sb.* (*N.S.*), 50 (92) (1960), 247–255 [*Math. Rev.* 22, #12518 (Kahn)].
23. 'The map excision theorem' (Russian), *Soobshch. Akad. Nauk. Gruzin. SSR*, 24 (1960), 649–654 [*Math. Rev.* 24, #A1124 (Isbell)].
24. 'Mapping of proximity structures', *General topology and its relations to modern analysis and algebra*, Proc. Sympos., Prague, 1961 (Academic Press, New York), pp. 139–141; Publ. House Czech. Acad. Sci., Prague; 1962 [*Math. Rev.* 26, #4312 (Krishnan)].
25. (with Dona Papert Strauss) 'Quotient frames and subspaces', *Proc. London Math. Soc.* (3), 16 (1966), 275–296 [*Math. Rev.* 34, #2510 (D. F. Brown)].
26. 'Composite morphisms in abelian categories', *Quart. J. Math. Oxford Ser.* (2), 37 (1966), 98–105 [*Math. Rev.* 34, #2652 (A. Heller)].
27. (with D. P. Strauss) 'On Urysohn's Lemma', *General topology and its relations to modern analysis and algebra, II*, Proc. Second Prague Topological Sympos., 1966 (Academia, Prague, 1967), pp. 111–114 [*Math. Rev.* 39, #108 (O. Frink)].
28. (with D. P. Strauss) 'Separation axioms for frames', *Topics in topology*, Proc. Colloq., Keszthely, 1972, *Colloq. Math. Soc. Janos Bolyai*, 8 (North Holland, Amsterdam, 1974), pp. 223–240 [*Math. Rev.* 52, #15360, 54D10 (06A23) (E. P. Rozycki)].
29. (with D. P. Strauss) 'Paracompact frames and closed maps', *Symposia Mathematica, Vol. XVI*, Convegno Sulla Topologia Isiemistica e Generale, Indam, Rome, 1973 (Academic Press, London, 1975), pp. 93–116 [*Math. Rev.* 53, #14411 54D20 (54F05, 06A23) (Stephen Willard)].
30. (with D. Strauss) 'Products and sums in the category of frames', *Categorical topology*, Proc. Conf. Mannheim, 1975, Lecture Notes in Mathematics 540 (Springer, Berlin, 1976), pp. 208–219 [*Math. Rev.* 55, #11216, 54F05 (E. J. Braude)].
31. (with D. Strauss) 'Sums in the category of frames', *Houston J. Math.*, 3 (1977), no. 1, 17–32. [*Math. Rev.* 56, #1275, 54F05 (06A23) (Stephen Willard). See also Problems (in Topology) *Math. Rev.* 50, #8399].
32. (with Yael Dowker) 'Helping gifted children with mathematics', *Journal of the gifted child*, No. 1, I (1979), 52–66.
33. (with M. Thistlethwaite) 'On the classification of knots', *Comptes Rendus Mathématiques de l'Académie de Science* (*Royal Soc. of Canada*), vol. VI 2, (1982), 129–131.

34. (with M. Thistlethwaite) 'Classification of knot projections', *Topology and its applications*, 16 (1983), 19–31.
35. 'Isomorphism of categories', *J. Pure Appl. Algebra*, 27 (1983), 205–206.
36. (with D. Strauss) 'T_1- and T_2-axioms for frames', *Aspects of topology* (Cambridge University Press), to appear.

Department of Pure Mathematics,
The University of Hull,
22–24 Newland Park,
Hull HU5 2DW,
England.

D. STRAUSS

KNOT TABULATIONS AND RELATED TOPICS

Morwen B. Thistlethwaite
Polytechnic of the South Bank, London SE1 0AA, England

Contents

0 Introduction

The aim of this article is to examine some of the ideas connected with the problem of classifying 1-dimensional knots. As the title suggests, the flavour is intended to be rather pragmatic. We follow through the history of knot tabulations, from the pre-topological dark ages of the last century up to the present day.

Dowker developed an interest in knot tabulations in the 1960's, and collaborated recently with the present author in the classification of knots of up to 13 crossings.

It is hoped that the expository parts of sections 1, 3 and 4 will make the article reasonably self-contained. We pay only cursory attention to "abelian" knot theory, as this is dealt with extensively in Cameron Gordon's survey article **(Gor)**.

I would like to thank John Conway, Raymond Lickorish and Larry Siebenmann for valuable conversations.

1 Preliminaries

A *knot* is a submanifold of S^3 homeomorphic to S^1 . Below are drawings of three famous knots.

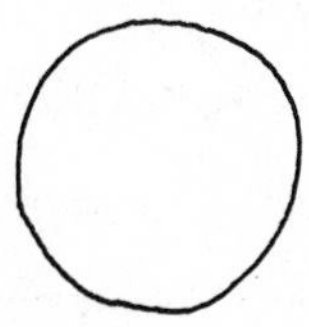

Fig. 1.1. The unknot The trefoil The figure-eight

Two knots K , L are *equivalent* if there exists an autohomeomorphism of S^3 mapping K onto L . The equivalence class of K is its *knot type*. We shall consider exclusively knot types with piecewise linear (or, equivalently, smooth) representatives, in order to steer clear of the virtually unexplored world of *wild knots*, i.e. knot types with no piecewise linear representative. The reader interested in wild knots can find information and references in **(Fox$_3$)**. We shall often use the word "knot" as an abbreviation for "knot type", when there is no danger of confusion. Thus the term "trefoil knot" really refers to the knot type of the knot depicted above, rather than the specific knot.

Choosing orientations of S^3 or K leads to stronger notions of knot equivalence. Let us suppose that an orientation of S^3 has been chosen. Then two knots K , L are equivalent via an orientation-preserving homeomorphism of S^3 if and only if they are *isotopic*, in the sense that there exists a homotopy $h_t : S^3 \to S^3$ $(0 \leqslant t \leqslant 1)$ such that $h_0 = 1$, each h_t is a homeomorphism, and $h_1(K) = L$.

This leads us to consideration of symmetries of knots. Let $\overline{K}$ be a mirror-image of K . If it happens that K can be isotoped to $\overline{K}$, then the above two notions of equivalence coincide for K , and K is said to be *amphicheiral*. The reader can check that the figure-eight knot is amphicheiral; the trefoil is not, though this of course requires proof. If, in addition, we orient the actual knots, an even stronger version of equivalence results: we can insist that the homeomorphism of S^3 mapping K to L preserves orientations of the knots, as well as that of S^3 . In this way, we also get finer notions of symmetry. Let K' be K with the orientation of the knot reversed. If K can be isotoped to K' , the knot K is said to be *invertible*. Both the trefoil and the figure-eight are readily seen to be invertible, by marking arrowheads on the knots and turning them over. For oriented knots, there are really two kinds of amphicheirality, depending on whether K can be isotoped to $\overline{K}$ or to $\overline{K}'$; Fox calls these + and - amphicheirality **(Fox$_3$)**. In practice, it is easy to find topological invariants which will detect non-amphicheirality, and it is exceedingly hard to detect non-invertibility. Significantly, Dehn **(De)** proved in 1914 that the trefoil was not amphicheiral, but the first proof of the existence of *any* non-invertible knots was due to Trotter **(Tro)** in 1964. For the important class of algebraic knots, these symmetry problems have been settled by Bonahon and Siebenmann.

Much analysis of knots is conveniently carried out in a more or less

2-dimensional setting. Let us regard S^3 as $\mathbb{R}^3 + \infty$. Then any knot is isotopic under an arbitrarily small deformation to a knot K in $\mathbb{R}^3$, whose image under the parallel projection $(x,y,z) \to (x,y,0)$ is a closed curve in the plane with no multiple points other than finitely many transverse double points. K is said to be in *regular position* with respect to this projection, and the image of K in the plane is called a *regular projection* of K . When drawing such a closed curve, if we leave suitable small gaps as in Fig. 1.1, we can indicate at each double point which of the two pre-image points on K has the larger z-coordinate, thus recovering completely the knot type, and indeed the isotopy class, of K . We can think of each "gap" as an open arc in the curve, coloured differently (white in this instance!). Such an embellished regular projection is called a *knot diagram* of K . Although knot diagrams obscure much of the 3-dimensional "feel" of knots, they are of undoubted use in computing knot invariants, and in tabulating knots.

The reader might well prefer a crisper approach to knot diagrams. We can define a knot diagram in the oriented plane to be a regular knot projection with signed crossing points. The intended interpretation of the signs is that a *positive* crossing conforms to the picture with respect to some orientation of the closed curve, and a *negative* crossing conforms to the picture . Note that the choice of orientation of the curve is immaterial here.

Any knot diagram drawn in the plane $z = 0$ can of course be converted into a corresponding knot in $\mathbb{R}^3$ by substituting, for each pair of gaps adjacent to a crossing, a semicircle pointing downwards:-

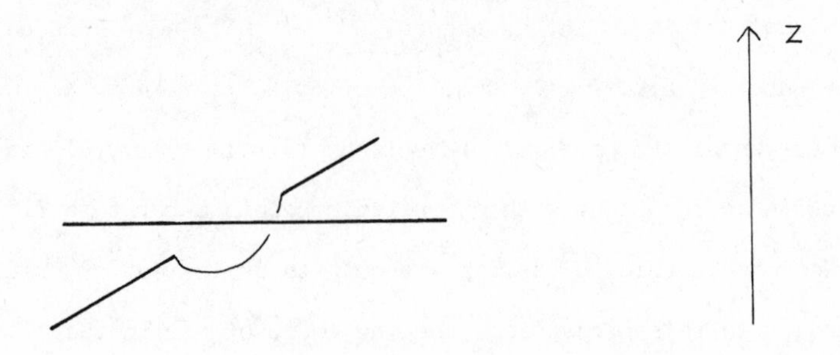

Fig. 1.2.

An important observation is that a diagram of a knot K can be deformed by an isotopy in the *extended* plane $\mathbb{R}^2 + \infty$ without losing the isotopy class of K . To see why this is so, observe that a diagram drawn on the 2-sphere of radius 1, centre (0,0,1) can be converted into a knot by replacing each adjacent pair of gaps by an outward-pointing semicircle, and that this realization is compatible with the realization for diagrams in the plane, via stereographic projection. For example, the following three diagrams represent isotopic knots.

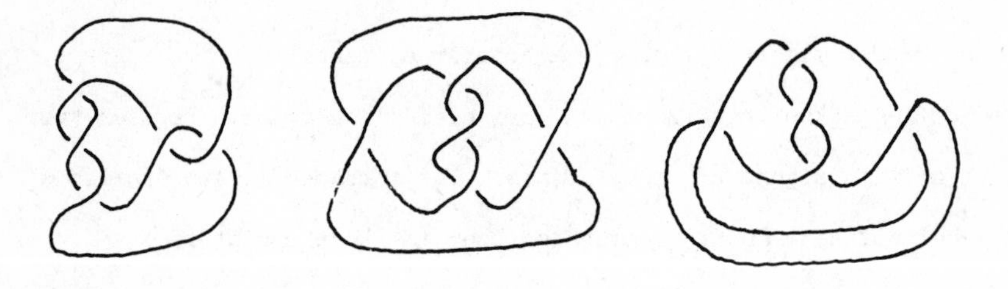

Fig. 1.3.

It now seems sensible to say that two diagrams in the plane are *equivalent* if there is a homeomorphism of the extended plane mapping one onto the other, preserving signs of crossings, and that they are *isotopic* if they are equivalent via an orientation-preserving homeomorphism of the extended plane. Equivalent (respectively isotopic) diagrams represent equivalent (respectively isotopic) knots.

The converse of this last statement is of course far from true. For instance, the crossings of any regular knot projection may be given signs so as to make a diagram of the unknot! Also, illustrated in Fig. 2.6 is a knot with no diagram of fewer than 13 crossings, yet with 769 inequivalent 13-crossing diagrams. However, if diagrams D_1 , D_2 represent isotopic knots, D_1 can be transformed to D_2 by means of a finite sequence of so-called *Reidemeister moves*:-

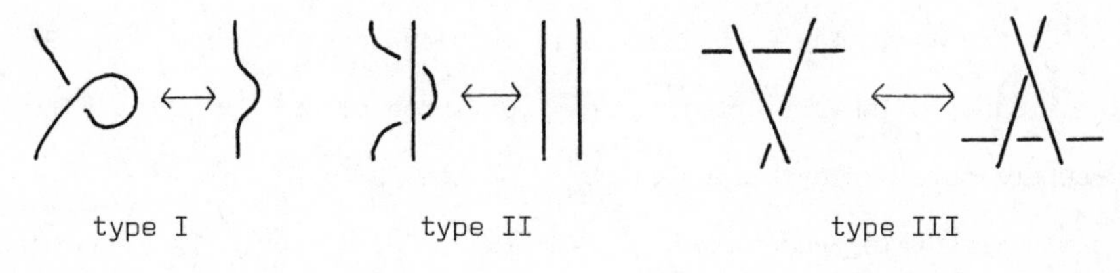

type I type II type III

Fig. 1.4.

Proof of this fact is elementary, if somewhat tedious, and relies on standard techniques in PL topology for adjusting maps so as to simplify sets of self-intersection. For this reason, a proof is not included here. Of course, the Reidemeister moves do not give us an effective procedure for deciding whether two diagrams represent equivalent knots.

We now turn our attention to the factorization of knots. A knot K in S^3 is *prime* if any 2-sphere which meets K transversely in two points bounds one and only one 3-ball intersecting K in an unknotted spanning arc. That is, the ball-arc pair in question is homeomorphic to the standard pair (B,A) , where $B = \{\mathbf{x} \in \mathbb{R}^3 : |\mathbf{x}| \leqslant 1\}$ and $A = \{(t,0,0) \in \mathbb{R}^3 : -1 \leqslant t \leqslant 1\}$. Thus, the unknot is not prime. Many knots are known to be prime, but proving that a given knot is prime can be a task which is not altogether trivial. A non-trivial knot which is not prime is *composite*; in layman's terms, a composite knot results from tying two knots separately in the same piece of string, but the validity of this interpretation depends on the theorem that two knots combined in this way cannot "cancel each other out", to produce the unknot. One of the many existing proofs of this "non-cancellation theorem" is given in §3.

In order to define the *connected sum*, or *composite* of two knots unambiguously, it is necessary to work in the oriented category. Let K , L be oriented knots in oriented S^3 . Let S^2_K , S^2_L bound 3-balls B^3_K , B^3_L which intersect K , L respectively in unknotted spanning arcs. Let S^2_K , S^2_L meet K , L respectively in 0-spheres denoted S^0_K , S^0_L . All these spheres inherit orientations from those of S^3 , K , L . Then the 3-sphere obtained by attaching $S^3 - \mathring{B}^3_K$ to $S^3 - \mathring{B}^3_L$ by an orientation-reversing homeomorphism $h: (S^2_L , S^0_L) \to (S^2_K , S^0_K)$ contains a knot $K \# L$, formed by joining $K \cap (S^3 - \mathring{B}^3_K)$ to $L \cap (S^3 - \mathring{B}^3_L)$. $K \# L$ is the *connected sum* of K and L , and is unique up to oriented equivalence. The reason for making h orientation-reversing is that $(S^3 , K \# L)$ then inherits an orientation naturally from the oriented pairs (S^3 , K) and (S^3 , L) .

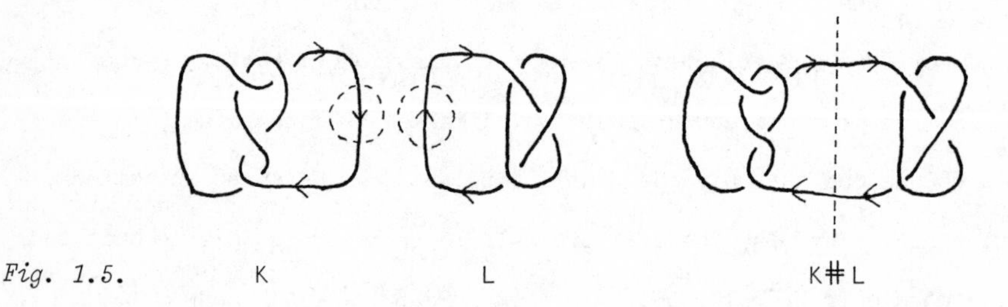

Fig. 1.5. K L K#L

Schubert proved in 1949 **(Schu$_1$)** that factorization of knots into primes is unique with respect to # , up to order of the factors. Since the trefoil T is not amphicheiral, it follows that the knots $T \# T$ (the *granny* knot), $T \# \overline{T}$ (the *reef* or *square* knot), and $\overline{T} \# \overline{T}$ are mutually inequivalent in the oriented category (there is no need to specify a particular orientation of T itself as the trefoil is invertible). Since $\overline{T\#T} = \overline{T}\#\overline{T}$, it follows that the granny knot and the reef knot are not equivalent even in the weak (unoriented) sense. This last fact was proved by Seifert in 1933 **(Sei$_1$)**. Using a similar argument, if K , L are neither amphicheiral nor invertible, the knots $K\#L$, $K\#\overline{L}$, $K\#L'$, $K\#\overline{L}'$ are pairwise inequivalent in the weak sense.

The formation of connected sums of knots is subsumed in a different, more general construction, namely that of the satellites of a knot K . Working, as usual, in the PL-category, let h be an embedding of the solid torus $V^3 = S^1 \times D^2$ in S^3 such that the image of the *core* $S^1 \times \{0\}$ of V^3 is K . Let C be any simple closed curve in V^3 , which is not contained in any 3-ball in V^3 . Then $h(C)$ is called a *satellite* of K , and K is a *companion* of $h(C)$. The study of companionship was initiated by Schubert in 1953 **(Schu$_2$)**, and recently Jaco, Shalen and Johannson **(JS, Joh)**, using some of Schubert's ideas, proved the important *characteristic variety theorem*, one of whose many consequences is that, essentially, knots factorize uniquely into companions. For further details, see **(BS)**. Illustrated in Fig. 1.6 are three satellites of the trefoil: a connected sum, a double and a cable (see **(Rol)** for details).

Fig. 1.6.

An important class of knots is that of alternating knots. Let us say that a diagram is *alternating* if (i) it has at least one crossing; (ii) the curve goes alternately over and under at successive crossing points; (iii) the diagram has no "nugatory" crossing  , where the shaded disc represents a portion of knot diagram which may or may not be trivial. Then an *alternating knot* is a knot admitting an alternating diagram. It is not difficult to see that every singular closed curve in the plane, which is a regular projection with at least one crossing of a knot, and in which there is no occurrence of  , is a regular projection of some alternating knot. Also, it is clear that such a curve must in fact have at least three crossing points. However, it is by no means obvious that there exist any knots which are not alternating. Bankwitz proved in 1930 that the *determinant* of a knot (see §4) is not exceeded by the number of crossings of any alternating diagram of the knot. It follows that the unknot, having determinant 1, is not alternating. Also, the 8-crossing torus knot, listed as 8_{19} in **(AB)** and denoted $T_{3,4}$ in §5 of this article, is a non-trivial non-alternating knot for the following reasons: its determinant is 3, there does not exist a non-trivial knot apart from the trefoil with a diagram of up to 3 crossings (the reader can quickly verify this), and $T_{3,4}$ is distinguished from the trefoil by its Alexander polynomial (again, see §5).

One of the most commonly used measures of complexity of a knot is its *crossing-number* $c(K)$. If D is a diagram of K , let $c(D)$ be the number of crossing points of D . Then $c(K) = \inf \{c(D) : D \text{ is a diagram of } K\}$. The advantage of this measure is that there are only finitely many diagrams, hence only finitely many knots of a given crossing-number. Standard practice when tabulating knots is to list all prime knots of up to a certain crossing-number. That tabulating knots is a precarious undertaking is evidenced by the numerous errors in published tables. For instance, a duplication in

C.N. Little's 1899 tabulation of (supposedly) non-alternating 10-crossing knots (**Lit$_3$**) remained undetected until K.A. Perko spotted it in 1974. As this duplication is somewhat notorious, we quote it here:-

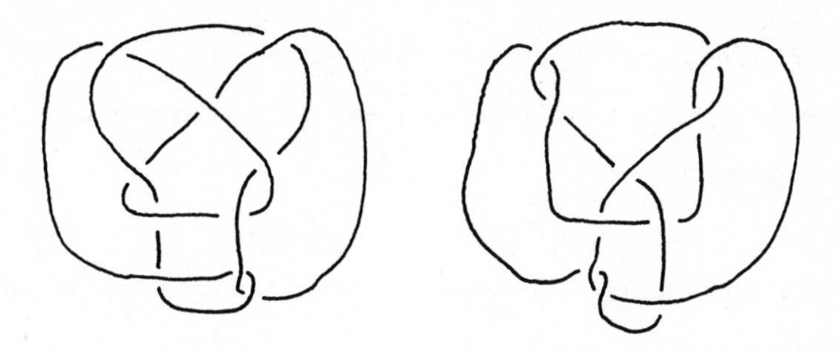

Fig. 1.7.

The first (presumably) correct list of 11-crossing knots was produced by Caudron in 1978 **(Cau)**, and the present author has continued the tabulation up to 12 crossings (Dec. 1981) and 13 crossings (Dec.1982). These latest tabulations use a notation for knot diagrams developed by Dowker, which is particularly well suited to computer methods, and which is described in §6. All 12965 knots of up to 13 crossings have been proved to be pairwise distinct, by means of topological invariants. However, it would be most unwise to claim categorically that this listing was correct, in the absence of independent verification. The numbers of prime knots with crossing-number up to 13 would seem to be as in the following table:-

Crossing-number	3	4	5	6	7	8	9	10	11	12	13
Number of knots	1	1	2	3	7	21	49	165	552	2176	9988

Perko, Bonahon and Siebenmann have proved that all the listed knots of up to 11 crossings are prime. I have not yet got round to checking primality of the 12- and 13-crossing knots, but this should be a fairly routine matter. Adequate techniques are not available for determining exactly how many of these knots are non-alternating.

When embarking upon the subject of knot theory, one is inclined to ask certain basic and apparently simple questions, such as

(i) If K , L are inequivalent knots, are their complements in S^3 non-homeomorphic?

(ii) Is crossing-number additive with respect to connected sum of knots?

(iii) Can a knot with crossing-number n have an alternating projection with more than n crossings?

The answer is not known to any of these questions, as of this writing. Gordon's article **(Gor)** contains a discussion of question (i). Our ignorance is further highlighted by the following. The *gordian number* or *unknotting number* **(Wen)** of a knot K is the minimal number of overcrossing-undercrossing changes required to transform K to the unknot (one is allowed to isotope the knot between changes). This is probably the most down-to-earth measure of complexity of a knot, and is the one naturally considered when attempting to untangle a pile of string. Well, the unknotting number is unknown for the 8-crossing knot 4 4 (Conway's notation; see §5) :-

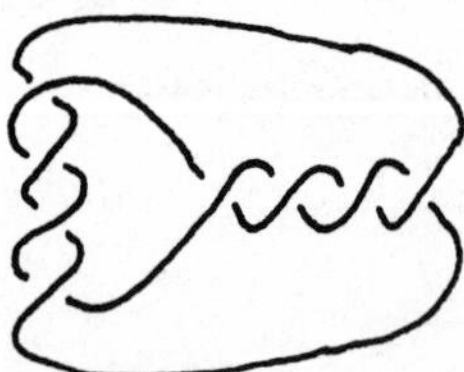

Fig. 1.8. A knot whose unknotting number is unknown.

We do, at least, have the following important (and very recent) result of M. Scharlemann: a knot whose unknotting number equals 1 is prime. This is perhaps a first step towards settling the famous conjecture that the unknotting number is additive with respect to connected sum of knots.

2 The nineteenth century literature

The three figures who dominated the science of knot tabulations in the nineteenth century were Kirkman, Tait and Little. However, some mention should be made in this context of Gauss, and his pupil Listing.

Gauss is recorded as showing an interest in knots at an early age. He wrote little on the subject, but did invent a notation for regular projections, which was re-invented in a slightly different form by Tait, and which was used in a further modified form by Dowker. To obtain a symbol in this notation, one first chooses a starting point on the curve, and a direction. One then travels round the curve, labelling the different crossing points A, B, C, ... in the order in which they occur, and also recording the sequence, complete with repetitions, of crossing points traversed. For example, 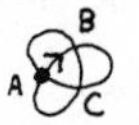

would be denoted ABCABC . Gauss mentioned the problem of finding necessary and sufficient conditions for a sequence of 2n letters, where each letter occurs twice, to be realizable in this way as a closed curve, but apparently did not address himself further to the matter. Algorithms for deciding realizability have been written by topologists **(De)**, and also by graph theorists **(RR)**. An algorithm described in **(DT)** has been implemented as a computer program of some 30 lines, and is crucial in the tabulations of Dowker-Thistlethwaite.

Listing wrote a substantial section on knots in **(Lis)**, in which he revealed an awareness of the amphicheirality of the figure-eight knot, and went some way towards describing a notation for alternating knot diagrams, of a contrasting nature from that of Gauss. This notation was to be refined into a usable form by Tait. Listing simply coloured the regions of the knot diagram black and white, and then listed the numbers of regions of each colour bounded by 2 arcs, by 3 arcs, and so on. There are two major drawbacks to this notation: firstly, symbols do not determine knot projections (or equivalently, alternating knot diagrams) uniquely; secondly, given a symbol, there is no natural way of constructing a corresponding diagram. Despite these shortcomings, there is no doubt that Listing's essay had a significant influence on the work of Tait and Little.

The first of our trio of tabulating titans was the Reverend Thomas P. Kirkman, F.R.S. Kirkman was a mathematician of international repute, claimed to be the inventor of "Hamiltonian circuits", and contributed to the subject of transitive permutation groups. He was also the acknowledged master of obscure terminology. His individual style manifests itself admirably in his definition of "knot", at the beginning of his article **(Kirk)**:-

"By a knot of n crossings, I understand a reticulation of any number of meshes of two or more edges, whose summits, all tessaraces (ὰκή), are each a single crossing, as when you cross your forefingers straight or slightly curved, so as not to link them, and such meshes that every thread is either seen, when the projection of the knot with its n crossings and no more is drawn in double lines, or conceived by the reader of its course when drawn in single line, to pass alternately under and over the threads to which it comes at successive crossings."

As far as one can make out, a knot was seen by Kirkman as being an alternating knot described by some particular alternating diagram. By his own admission [**(Kirk)**, p. 282], he was interested solely in the classification of knot projections; it was anathema to consider deforming a knot in space, or even to consider the different ways in which a specific simple closed curve in $\mathbb{R}^3$ could be projected onto a plane.

Later in the same article, we learn that a *flap* is a two-edged region in a polyhedral diagram; a *solid* knot is a diagram containing no flaps; a *subsolid* knot is a diagram which admits no simple closed curve in the plane cutting it transversely in just two points (whether crossing points or distinct edges), except through the two crossings of a flap; any other diagram is *unsolid*.

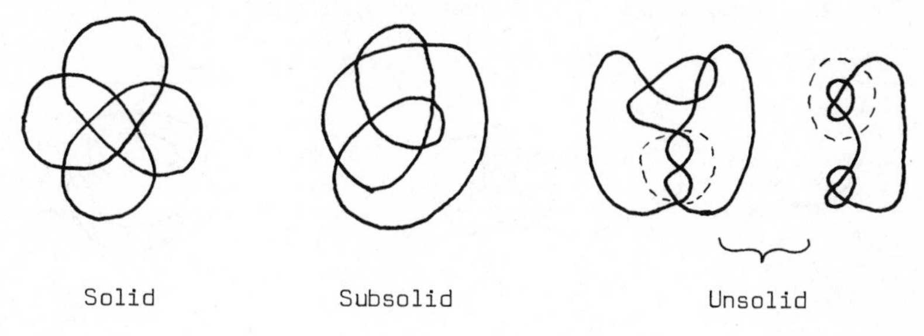

Fig. 2.1.

Kirkman used the novel idea of systematically reducing any such polyhedral diagram to an irreducible form. Then, to enumerate diagrams, one starts from the irreducible diagrams, which hopefully are few in number, and works backwards. The principal transformation used in Kirkman's reduction process was the removal of flaps (Fig. 2.2). For want of better terminology, let us say that a polyhedral diagram is a *diagram of n curves* if it consists of n closed curves meeting one another transversely. One feature of the transformation of Fig. 2.2 is that if both edges of a flap belong to the same curve, but are oriented oppositely, thus: , as in Fig. 2.2(c), the number of curves in the diagram is increased by 1. Consequently, if one wishes to tabulate n-crossing diagrams of 1 curve by this method, it is necessary to tabulate diagrams of more than one curve (i.e. diagrams of *links* of more than one component), up to n-2 crossings. Kirkman used his method to construct polyhedral diagrams (of 1 curve) up to 11 crossings, and these diagrams were duly used by Tait and Little to tabulate knots. Some omissions were found by Tait, others by Conway. Although Kirkman's use of "flaps" and "solid knots" may seem ad hoc to the reader (it probably was!), in fact these concepts have connections with hyperbolic structure (see the end of §5). Conway, keeping to the ideas of "flap" and "solid knot", used an improved method of reduction with great success in 1970 **(Con)**.

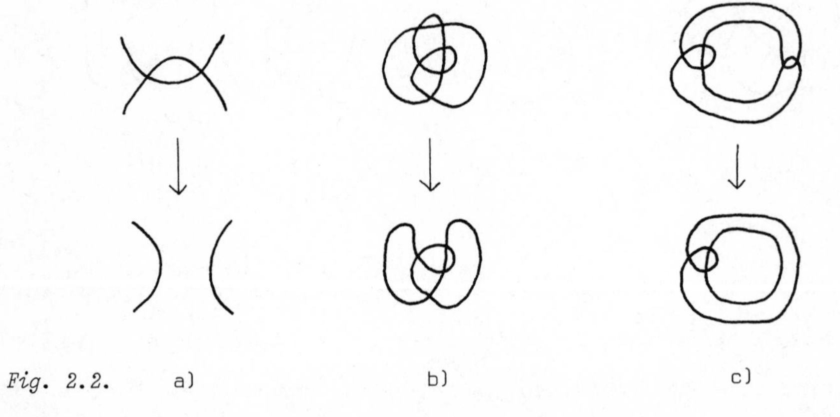

Fig. 2.2. a) b) c)

Peter Guthrie Tait, eminent Scottish physicist, was undoubtedly the central figure in the nineteenth century world of knots. He was originally drawn to the subject by Kelvin's theory of vortex atoms, but one suspects that at least in part this provided an excuse for doing some mathematics. Inequivalent knots were thought to correspond in some sense to different elements. In a series of three substantial and energetically written papers **(Tai)**, he discussed knots extensively, and included a tabulation of alternating knots up to 10 crossings. His concept of knot, in contrast to Kirkman's, was essentially the modern one, although he was still hampered to some extent by the 2-dimensional way of representing knots. It is intriguing to see him grappling with the problems caused by the complete absence of the subject of topology; he was definitely aware that techniques were not available for proving certain empirically-observed "facts". Thus, in connection with the problem of deciding whether two knot diagrams represent equivalent knots (**(Tai)**, p. 319): "... though I have grouped together many widely different but equivalent forms, I cannot be *absolutely* certain that all those groups are essentially different from one another." Also, in connection with "beknottedness", i.e. the unknotting number (**(Tai)**,p.308): "There must be some very simple method of determining the amount of beknottedness for any given knot; but I have not hit upon it." Tait would be amused to know that we still have not hit upon it.

Tait attacked the problem of enumerating knot projections by means of two separate notations. Firstly, he used the "alphabetic" notation of Gauss to enumerate diagrams as far as 7 crossings, abandoning it at that point because of the "combinatorial explosion" of numbers of possible sequences. As an independent check, he used an elaborated version of Listing's "compartmental" notation, which no longer had the ambiguity of Listing's prototype. To each polyhedral diagram of n crossings, with regions coloured alternately black and white, a graph is constructed by placing a node inside each white region, and then connecting the nodes by n arcs which pass through the crossing points of the diagram, and which avoid the interiors of black regions.

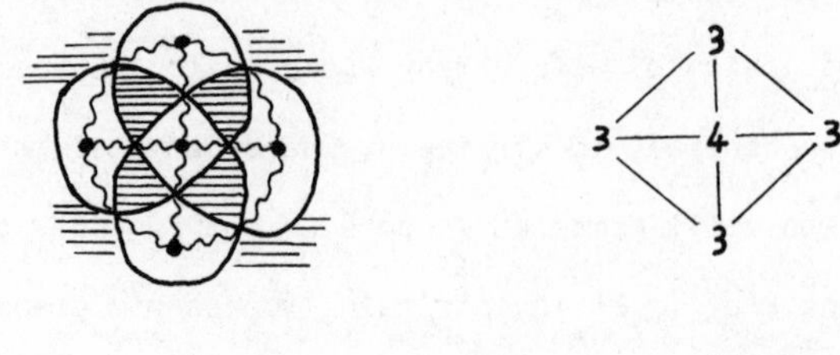

Fig. 2.3. a) b)

Tait actually wrote in the valencies of the nodes, as in b). From the black regions of the diagram, one obtains another graph which is the dual of the first. It is relatively easy to see that any embedded, connected graph in the plane determines a unique 4-valent polyhedral diagram, but Tait does not raise the question as to whether the alphabetic notation is ambiguous. It is shown in **(DT)** that in fact it is unambiguous for projections of prime knots. For his tabulations of knots beyond 7 crossings, Tait relied on Kirkman's enumeration of polyhedral diagrams.

For finding equivalences between alternating knot diagrams, Tait introduced the important idea of a *tangle*, that is a portion of knot diagram connected to the remainder of the diagram by just four arcs. (Here, of course, we mean that two diagrams are equivalent if they represent equivalent knots; lack of appropriate vocabulary has forced this duplicity of word usage.) Then a *flype* is a transformation of the following sort:-

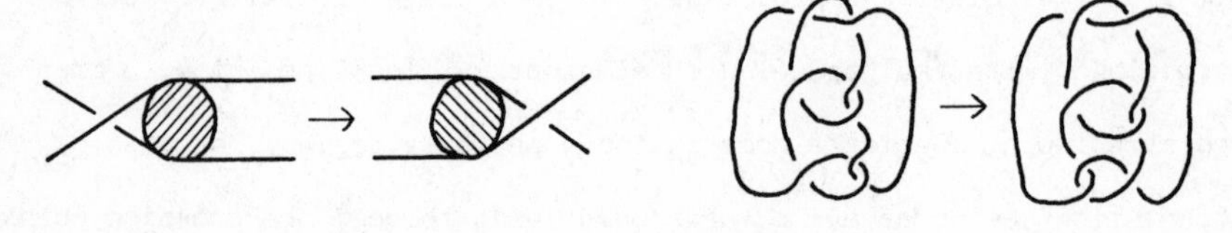

Fig. 2.4. a) b)

The shaded disc in Fig. 2.4a) represents an arbitrary tangle which is turned upside-down when the crossing on its left is removed by twisting. Fig. 2.4b) is a specific example of a flype. It is still not known whether flypes generate all equivalences of alternating knot diagrams. There does not seem to be a natural way of expressing a flype as a sequence of Reidemeister moves.

Tait's original census of knots with alternating diagrams of up to 10 crossings contained a duplication which was brought to his attention, just before going to print, by Little's independent census **(Lit$_1$)**. The final published version, however, has stood the test of time.

Mention should be made of Tait's work on amphicheiral knots. Amphicheirality seemed to be, for Tait, more a property of knot diagrams than of the knots themselves. Thus distinction was made between a diagram isotopic to its mirror-image via an orientation-preserving homeomorphism of the extended plane (first order amphicheirality), and one merely transformable to its mirror-image by means of flypes (second order amphicheirality). This had the strange effect that different diagrams of the same knot could have different orders of amphicheirality. At any rate, Tait constructed first order amphicheiral diagrams out of two symmetrically placed halves, and had other techniques for the second order amphicheirals. These techniques were adequate for his purposes, in that Perko has proved that Tait's list of amphicheirals of up to 10 crossings is complete **(Per$_1$)**.

Mary G. Haseman, for her doctoral dissertation of 1917 **(Has)**, extended Tait's list of amphicheirals up to 12 crossings, using the same empirical methods (neither Haseman nor Tait considered the possibility of the existence of amphicheirals of odd crossing-number). A check on her tabulations reveals that there are no omissions of "obviously" amphicheiral 12-crossing alternating knots, but that there are a few duplications. Specifically, 16) = 13) , 36) = 20) , 51) = 50) , 54) = 40) , 57) = 45) , and also 60) = 59) , 61) = 6) probably due to subdivision of amphicheirality into orders. Evidently, the influence of the infant subject of topology was not widespread in 1917, as no mention is made of it in Haseman's paper: recall that Dehn proved in 1914, using topological invariants, that the trefoil was not amphicheiral.

C.N. Little, Professor at the State University of Nebraska, rivalled Kirkman in his dedication to the cause of large-scale tabulations. Little undertook, for the first time, the enumeration of non-alternating knots. These do not materialize until 8 crossings, at which point there are three:-

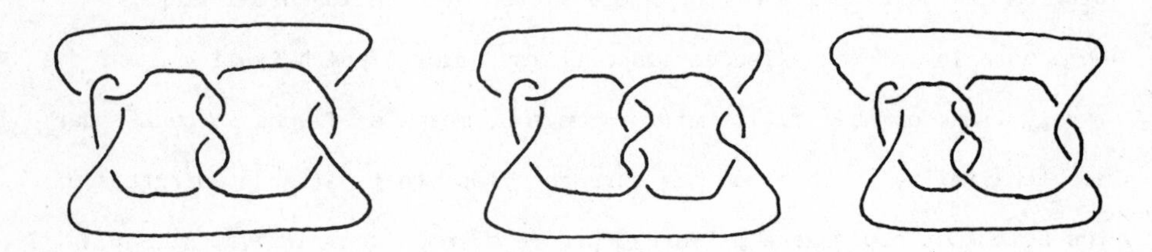

Fig. 2.5. 8-crossing non-alternating knots.

Non-alternating knots require much more work than their alternating counterparts, for two reasons. Firstly, whereas a given knot projection with n crossings corresponds essentially to only one alternating diagram, it corresponds potentially to $2^{n-1} - 1$ different non-alternating diagrams, up to homeomorphism of the extended plane. Secondly, a given non-alternating n-crossing knot often has huge numbers of n-crossing diagrams, all of which have to be considered. Flyping no longer suffices: it is necessary to consider other types of transformation. In addition to the flype, Little used the *2-pass*: , where the vertical string is simply passed over the tangle. The type III Reidemeister move is a special case of a 2-pass. Little might have been discouraged if he had known that flyping and 2-passing, when applied to the diagram of Fig. 2.6, generate a set of 769 13-crossing diagrams.

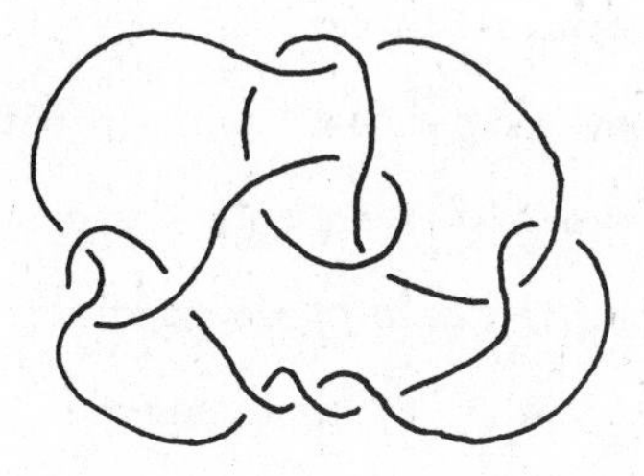

Fig. 2.6.

Little eventually published a list of 43 10-crossing non-alternating knots, complete with some 551 drawings of their various 10-crossing diagrams. He reported that this task took him six years, on and off, and remarkably there are very few errors. There are a few omissions in the lists of diagrams, but the only serious error was the duplication of Fig. 1.7, discovered by Perko in 1974. Thus, the actual number of knots is 42. To assist him in his computations, Little invented an "invariant", called the *twist* of the knot, equal to the absolute value of the sum of the signs of the crossings in any diagram with minimal number of crossings. For example, the twist of the trefoil is 3, and that of the figure-eight is 0. Little observed that the twist of a diagram is invariant under flypes and passes, and guessed (quite reasonably) that it was a genuine knot invariant. Alas, it is not: Fig. 1.7 was the first known counterexample. At any rate, Little grouped diagrams according to their twist, and this accounts partly for the duplication. Perhaps Little's most serious error was to proclaim the invariance of the twist as a theorem (**(Lit$_3$)**, §8), and then to give a "proof" based on the rash assumption that, for a given knot, all equivalences of diagrams of minimal crossing-number are generated by flypes and passes. Without wishing to dwell on such negative matters, there is also a fallacious "proof" of an incorrect statement regarding "beknottedness" of certain knots, in an earlier paper (**(Lit$_2$)**, §8).

It is not possible to convey here the spirit of enthusiasm which exudes copiously from the papers of Kirkman, Tait and Little. The modern knot tabulator must be indebted to these three, and also to Mary Haseman, who ventured courageously into the hitherto uncharted regions of 12-crossing knots.

3 The group of a knot

The topological invariant upon which almost all other knot invariants depend is the fundamental group of the complement of the knot, $\pi_1(\mathbb{R}^3 - K)$. Of course, the natural inclusion $\mathbb{R}^3 \to S^3 = \mathbb{R}^3 + \infty$ induces an isomorphism $\pi_1(\mathbb{R}^3 - K) \to \pi_1(S^3 - K)$.

Given any diagram of a knot K , we can read off a presentation for $\pi_1(\mathbb{R}^3 - K)$, called the *Wirtinger presentation*. First, let us choose orientations of $\mathbb{R}^3$ and K . Let us take $x_0 = (0,0,1)$ as the basepoint of $\mathbb{R}^3$. Let us consider K as being a realization of a diagram, as in Fig. 1.2. The part of K in the plane $z = 0$ consists of a collection of disjoint arcs $\alpha_1, \ldots, \alpha_n$ say; the closure of $K - \bigcup\{\alpha_i\}$ consists of semicircular arcs $\beta_1, \ldots, \beta_n$ say. We assume that the labelling of these arcs is such that the sequence $\alpha_1, \beta_1, \alpha_2, \beta_2, \ldots, \alpha_n, \beta_n$ is consistent with the orientation of K .

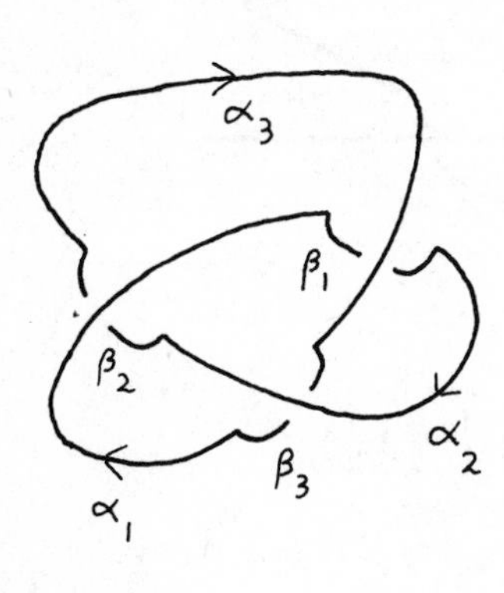

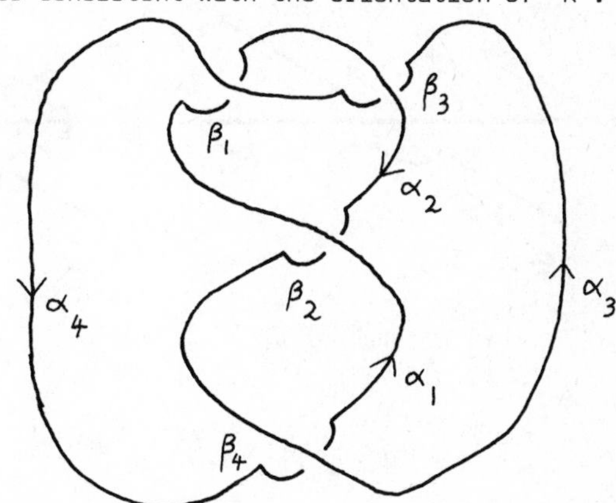

Fig. 3.1.

For each $i = 1,2, \ldots, n$, let μ_i be the element of $\pi_1(R^3 - K)$ represented by a loop of form $p_i * c_i * p_i^{-1}$, where p_i is a straight line path from x_0 to a point above, and suitably close to α_i, and c_i encircles α_i in the direction of a positive screw, as in Fig. 3.2.

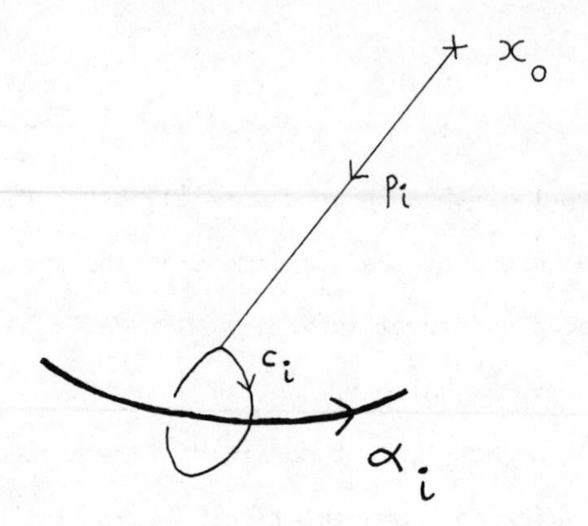

Fig. 3.2.

Corresponding to each crossing-point of the diagram of K, there is an obvious relation between the μ_i ; if α_k passes above β_i, we get $\mu_{i+1} = \mu_k^{-\varepsilon_i} \mu_i \mu_k^{\varepsilon_i}$, where ε_i is equal to the sign of the crossing, and suffixes are taken mod n.

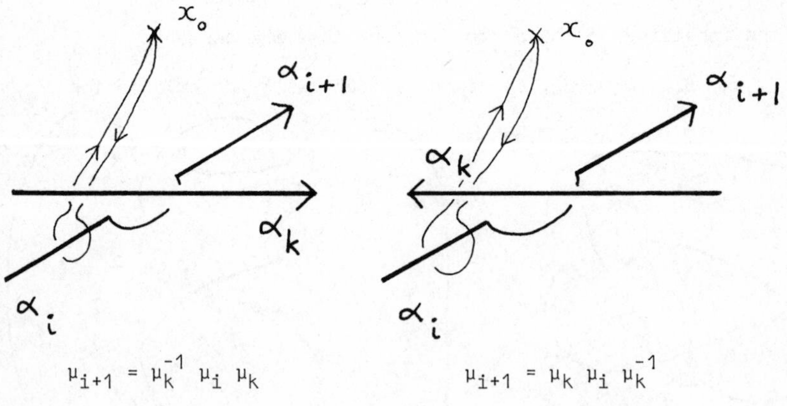

Fig. 3.3.

Let r_i be the word $\mu_k^{-\varepsilon_i} \mu_i \mu_k^{\varepsilon_i} \mu_{i+1}^{-1}$. A straightforward application of Van Kampen's theorem produces the result that $G = \pi_1(\mathbb{R}^3 - K)$ has a presentation $\langle \mu_1, \ldots, \mu_n \mid r_1, \ldots, r_n \rangle$, and that any one of the relators r_i may be omitted. For details, see **(Fox$_3$)**.

We can observe immediately that the μ_i are all in the same conjugacy class, and that the abelianized group of G , $G/[G,G] \cong H_1(\mathbb{R}^3 - K)$ is infinite cyclic (note also that $H_1(\mathbb{R}^3 - K) \cong Z$ by Alexander duality). As G is finitely presented, it is possible to search exhaustively for all homomorphisms of G onto any given finite group H ; an assignment $\mu_i \to h_i$ of elements h_i of H to the generators μ_i of G yields a unique homomorphism $G \to H$ precisely when the equations $h_{i+1} = h_k^{-\varepsilon_i} h_i h_k^{\varepsilon_i}$ all hold in H . Of course, the h_i must all lie in the same conjugacy class in H .

As an application of these elementary facts, let us prove that the unknot, the trefoil and the figure-eight are pairwise distinct knot types. Let G_1, G_2, G_3 be the groups of the unknot, the trefoil and the figure-eight respectively. Then $G_1 \cong Z$, and the diagrams of Fig. 3.1 yield Wirtinger presentations

$$G_2 = \langle \mu_1, \mu_2, \mu_3 \mid \mu_3 \mu_1 \mu_3^{-1} \mu_2^{-1} , \mu_1 \mu_2 \mu_1^{-1} \mu_3^{-1} \rangle ,$$
$$G_3 = \langle \nu_1, \nu_2, \nu_3, \nu_4 \mid \nu_4 \nu_1 \nu_4^{-1} \nu_2^{-1} , \nu_1^{-1} \nu_2 \nu_1 \nu_3^{-1} , \nu_2 \nu_3 \nu_2^{-1} \nu_4^{-1} \rangle .$$

From the relators, G_2 is generated by μ_1, μ_2 and G_3 by ν_1, ν_2 . The assignment $\mu_1 \to (12)$, $\mu_2 \to (23)$ determines a homomorphism of G_2 onto the symmetric group S_3 , and the assignment $\nu_1 \to (12)(34)$, $\nu_2 \to (13)(45)$ determines a homomorphism of G_3 onto the dihedral group D_5 , represented as a subgroup of S_5 . Hence G_2 , G_3 are not abelian. Also, the assignment $\nu_1 \to (12)$, $\nu_2 \to (23)$ is not consistent with the relations in G_3, so G_3 does not admit a homomorphism onto S_3 (it is fruitless to search further, since any homomorphism of G_3 onto S_3 would have to map ν_1, ν_2 to distinct transpositions). Since G_1, G_2, G_3 have been shown to be pairwise non-isomorphic, the result follows.

Given a homomorphism of a knot group onto S_3, the Wirtinger relators dictate that the transpositions associated with the three arcs $\alpha_i, \alpha_{i+1}, \alpha_k$ incident at a crossing point are either all equal or all different. This leads to the well-known pictorial interpretation of such a homomorphism, known as a *3-colouring* of a knot diagram. A 3-colouring of a diagram is a prescription for painting each arc α_i a certain colour, such that (i) three colours are used overall; (ii) the colours of three arcs incident at a crossing are either all the same or all different. In fact, it is easy to show that the property of 3-colourability is preserved by each of the Reidemeister moves, so 3-colourability can be used to prove that the trefoil is knotted, without bringing in any group theory.

More generally, a homomorphism of the group G of a knot K onto the dihedral group D_n (n odd) corresponds to a non-trivial assignment of an integer to each arc α_i of a diagram of K, such that at each crossing, if x, y, z are assigned to $\alpha_i, \alpha_{i+1}, \alpha_k$ respectively (k as above), then $x + y - 2z = 0 \pmod n$ **($\mathbf{Per_3}$)**. A knot group cannot admit a homomorphism onto D_n if n is even, as such a D_n is not generated by the elements of any single conjugacy class. Another explanation is that each such D_n has $Z_2 + Z_2$ as a homomorphic image, and $Z_2 + Z_2$ is not a homomorphic image of the abelianized group of G.

The group of a knot is indeed a phenomenally powerful invariant:

no example is known of distinct prime knots with isomorphic groups. However, the isomorphism class of the group is obviously independent of a choice of orientation of $\mathbb{R}^3$ or K ; this means that the isomorphism class alone of the group cannot distinguish between a knot and its mirror-image, or between a knot and its reverse. We can strengthen the invariant by introducing *peripheral elements* of the group. There are two sorts of peripheral elements: *meridians*, and *longitudes*. A *meridian* is an element of $\pi_1(\mathbb{R}^3 - K)$ represented by a loop of form $p * c * p^{-1}$, where c equals one of the small circular paths c_i used in the definition of μ_i above, and p is any path in $\mathbb{R}^3 - K$ from x_0 to the start of c . The μ_i of a Wirtinger presentation are, of course, meridians. A *longitude* is an element represented by a loop of form $p * \ell * p^{-1}$, where ℓ is a closed path which follows the knot round its course, remaining close to it and without linking it (i.e. ℓ is null-homologous in $\mathbb{R}^3 - K$), and p is any path in $\mathbb{R}^3 - K$ from x_0 to the start of ℓ .

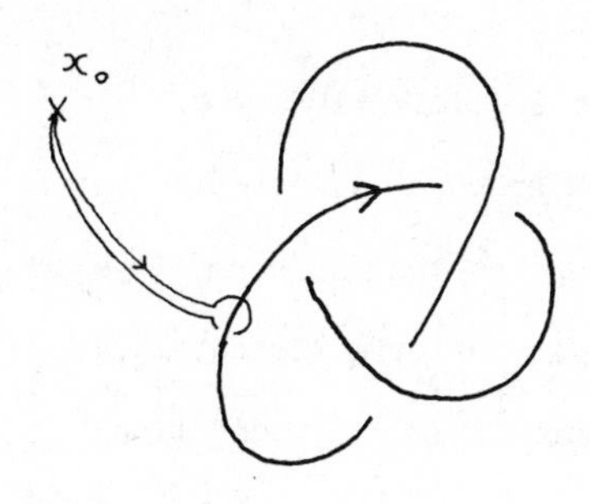

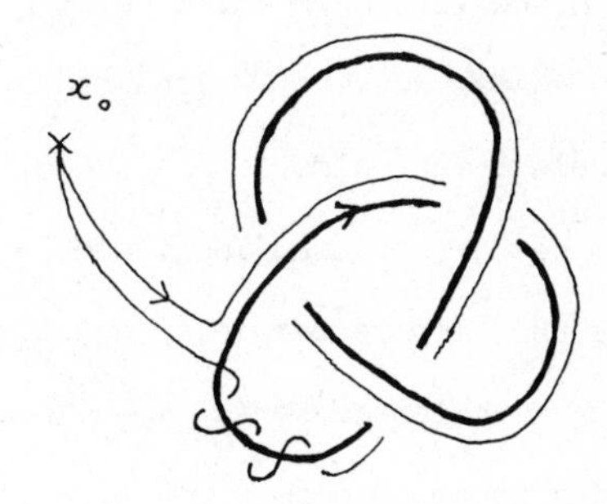

Fig. 3.4. A meridian A longitude

From the definition, it follows that the meridians form a conjugacy class in $\pi_1(\mathbb{R}^3 - K)$, as do the longitudes. Let us say that a meridian and a longitude form a *meridian-longitude pair* if they may be constructed using the same path p . It is easy to see geometrically that such a meridian and longitude commute. Under abelianization (the Hurewicz homomorphism), the meridians all map to a fixed generator of $H_1(\mathbb{R}^3 - K) \cong Z$, and the longitudes all map to 0 . In fact, each longitude belongs to the *second* commutator subgroup of G **(Fox$_3$)**.

Most importantly, if f is an orientation-preserving homeomorphism of $\mathbb{R}^3$ mapping K to L , preserving orientations of knots, then the induced isomorphism of fundamental groups f_* maps each meridian-longitude pair of $\pi_1(\mathbb{R}^3 - K)$ to a meridian-longitude pair of $\pi_1(\mathbb{R}^3 - L)$ (simply consider what happens under a corresponding isotopy of $\mathbb{R}^3$). If f preserves the orientation of $\mathbb{R}^3$, but reverses that of knots, then f_* maps a meridian-longitude pair to an (inverse meridian)-(inverse longitude) pair. Therefore, a knot cannot be invertible if its group does not admit an automorphism mapping meridians to inverses of meridians and longitudes to inverses thereof. Trotter used this to prove that certain "pretzel" knots were not invertible **(Tro)**. Similar properties hold for other types of homeomorphism f .

As a consequence of Waldhausen's work on sufficiently-large, irreducible 3-manifolds, it is known that the group of a knot together with the set of meridian-longitude pairs (known as the *peripheral system* of the group) form a complete knot invariant in the oriented category. That is, two oriented knots K, L are equivalent via an orientation-preserving homeomorphism of $\mathbb{R}^3$, which preserves the orientations of the knots, if and only if there exists an isomorphism of their groups mapping each meridian-longitude pair to a meridian-longitude pair. It is then a simple matter to write down weaker criteria for the various weaker types of knot equivalence.

A case in point is that of composite knots $K \# L$. Let G_K denote the group of K etc. From a diagram of $K \# L$, one sees immediately that if

$$G_K = \langle \mu_1, \ldots, \mu_m \mid r_1, \ldots, r_m \rangle \ , \quad G_L = \langle \nu_1, \ldots, \nu_n \mid s_1, \ldots, s_n \rangle \ ,$$

then $G_{K \# L}$ has a Wirtinger presentation

$$G_{K \# L} = \langle \mu_1, \ldots, \mu_m, \nu_1, \ldots, \nu_n \mid r_1, \ldots, r_m, s_1, \ldots, s_n, \ \mu_1 = \nu_1 \rangle \ ,$$

i.e. $G_{K \# L}$ is the free product of G_K with G_L , amalgamated along the subgroup $\langle \mu_1 \rangle = \langle \nu_1 \rangle$. Thus the group alone cannot distinguish between the various connected sums $K \# L$, $K \# \overline{L}$, $K \# L'$, $K \# \overline{L}'$, even though these may be inequivalent in the weak (unoriented) sense. In particular, the groups of the granny knot and the reef knot are isomorphic. Fox distinguished these knots in **(Fox$_1$)** by showing that no isomorphism of their groups preserved meridian-longitude pairs, even up to inverses of peripheral elements. From his analysis follows the stronger result that the complements of the knots are not homeomorphic (see Gordon's comment at the foot of p.6 of **(Gor)**).

There are many other points of interest concerning the group of a knot, a few of which we indicate here. It follows immediately from Dehn's lemma **(Papa)** that a longitude is trivial if and only if the knot is trivial. This gives us the celebrated *unknotting theorem*, that $\pi_1(\mathbb{R}^3 - K)$ is infinite cyclic if and only if K is trivial. If a composite $K \# L$ is trivial, the amalgamated-free-product structure of $\pi_1(\mathbb{R}^3 - K \# L)$ gives us that the groups of K and L are both isomorphic to Z . This, together with the unknotting theorem, gives us the *non-cancellation theorem*: if $K \# L$ is unknotted, then K and L are unknotted. For another proof, which does not use the big guns of Dehn's lemma, but which uses the fact that the *genus* of a knot, i.e. the minimal genus of an orientable surface in $\mathbb{R}^3$ spanning the knot, is additive with respect to connected sum, see **(Rol)**. For yet another, using an infinite construction due to Mazur, see **(Fox$_3$)**.

As a consequence of the sphere theorem **(Papa)**, the complement of a knot in S^3 is *aspherical*, i.e. all its higher homotopy groups are trivial. Thus the complement of a knot in S^3 is an *Eilenberg-Maclane space* $K(G_K,1)$. It follows that the homotopy type of $S^3 - K$ is determined by the isomorphism class of the group G . Also it is known that if a group G contains non-trivial elements of finite order, then $K(G,1)$ is infinite-dimensional (**(Hem)**, p.76). Hence we have the interesting result that a knot group is *torsion-free*. Hence each meridian-longitude pair in the group of a non-trivial knot generates a subgroup isomorphic to $Z + Z$. A fairly recent result **(Sha)** is that a knot group cannot contain an infinitely divisible element other than the identity (an element g is infinitely divisible if it is an nth power for infinitely many n).

An outcome of Thurston's work on hyperbolic metrics on 3-manifolds is the theorem that knot groups are *residually finite*: given any non-identity element x of a knot group G , there exists a homomorphism Φ of G onto some finite group such that $\Phi(x) \neq 1$, or equivalently there exists a normal subgroup of finite index in G which does not contain x . This at least gives some heuristic reason for hoping that the method of mapping knot groups onto finite groups will be successful at distinguishing the groups. Thurston proves that the complements of a wide class of prime knots, namely those complements which are atoroidal and which are not Seifert fibre spaces, admit a complete hyperbolic metric of finite volume. It follows that the fundamental group of such a complement is isomorphic to a group of isometries of hyperbolic 3-space, which, being a group of matrices, is residually finite. The fundamental groups of Seifert fibre spaces are known to be residually finite for different reasons ([**Hem**], p.177), whereas knots which are not atoroidal, i.e. whose complements contain non-trivial tori (satellite knots), are dealt with by factorization into companions.

4 Covering spaces

The study of coverings of S^3 branched over knots is of central importance in knot theory, and provides a significant connection with the theory of 3-manifolds. A striking example of this connection is the result that each closed orientable 3-manifold is a 3-sheeted covering space of S^3 branched over some knot [**(Hil)**, **(Mon)**]. More striking still is the very recent result of Hilden, Montesinos and Maria-Teresa Lozano that each such manifold is a finite-sheeted covering of S^3 branched over the knot 9_{46} in **(AB)**'s catalogue!

According to the classical theory of covering spaces [see for instance **(Spa)**], given any connected, locally path-connected, semi-locally simply connected space X, there is a one-to-one correspondence between conjugacy classes of subgroups of $\pi_1(X,x_0)$, and equivalence classes of covering projections with base space X. More explicitly, given any subgroup H of $\pi_1(X,x_0)$, there exists a covering projection $p: (\tilde{X},\tilde{x}_0) \to (X,x_0)$ such that the induced monomorphism p_* of fundamental groups maps $\pi_1(\tilde{X},\tilde{x}_0)$ onto H. Moreover, if $p_1: (\tilde{X},\tilde{x}_0) \to (X,x_0)$, $p_2: (\tilde{Y},\tilde{y}_0) \to (X,x_0)$ are covering projections, the images in $\pi_1(X,x_0)$ of p_{1*}, p_{2*} are conjugate in $\pi_1(X,x_0)$ if and only if there exists a homeomorphism $h: \tilde{X} \to \tilde{Y}$ such that $p_2 \circ h = p_1$. In the special case that $\tilde{X} = \tilde{Y}$, h is called a *covering translation* of $\tilde{X}$. The set of covering translations of $\tilde{X}$ forms a group under composition isomorphic to $N(H)/H$, where $H = p_*(\pi_1(\tilde{X},\tilde{x}_0))$ as above, and $N(H)$ is the normalizer of H in $\pi_1(X,x_0)$. If H is normal in $\pi_1(X,x_0)$, the associated covering is said to be *regular*; otherwise it is *irregular*.

Here, we are principally concerned with the case $X = S^3 - K$ for some knot K. Let $G = \pi_1(S^3 - K)$, as usual. The commutator subgroup G' of G determines a regular covering X_∞ of $S^3 - K$, called the *infinite cyclic cover*, whose group of covering translations is isomorphic to $G/G' \cong Z$. This leads us to what is probably the most celebrated invariant of knots. The *Alexander polynomial* $\Delta(t)$ of a knot K can be described as follows. The group $H_1(X_\infty) \cong G'/[G',G']$ has a neat module structure over the ring of Laurent polynomials $Z[t,t^{-1}]$: the coefficients of the polynomials act as the elements of the coefficient group Z, and t acts as the automorphism of $H_1(X_\infty)$ induced by a generator of the infinite cyclic group of covering translations of X_∞. $H_1(X_\infty)$ is finitely generated as a $Z[t,t^{-1}]$ - module, and furthermore admits a square presentation matrix, whose determinant, suitably normalized, is the Alexander polynomial. Here is a list of some properties of $\Delta(t)$.

1) $\Delta(1) = \pm 1$, because the order ideal of $H_1(X_\infty)$, (Δ), maps to the entire ring of integers under trivialization (see **(CF)**). $\Delta(t)$ is usually normalized to the form $a_0 + a_1 t + \ldots + a_n t^n$, and so that $\Delta(1) = 1$.

2) $\Delta(t) = t^{\deg\Delta}\Delta(t^{-1})$, i.e. the coefficients of $\Delta(t)$ form a palindromic sequence. This is due to the fact that the $Z[t,t^{-1}]$ - module $H_1(X_\infty)$ is isomorphic to its conjugate (see **(Sei$_2$)**, **(Fox$_3$)**, **(Gor)**). For a proof of property 2) using Reidemeister torsion, see **(Mil$_1$)**.

3) $\Delta(t)$ is of even degree. This follows from 1) and 2).

4) $|\Delta(-1)|$, the *determinant* of the knot, is an odd integer (consider

property 1) modulo 2) which is not exceeded by the number of crossings of any alternating projection of the knot [**(Ban)**, **(Cro)**]. The determinant of a knot equals the order of the first homology group of its 2-fold branched cover (see **(Rol)**).

5) The Alexander polynomial of an alternating knot is alternating (!), and its degree is twice the genus of the alternating knot [**(Cro)**, **(Mur$_1$)**]. Murasugi proves further in **(Mur$_2$)** that, for alternating knots, all coefficients of $\Delta(t)$ are non-zero. Properties 4) and 5) give us what are virtually the only known methods of detecting non-alternating knots. It is conjectured that the coefficients of the polynomial of an alternating knot satisfy the further property that $|a_i| \geqslant |a_{i-1}|$ for $1 \leqslant i \leqslant (\deg\Delta)/2$.

6) If $\Delta(t)$ is any polynomial satisfying properties 1) and 2), then there exist infinitely many inequivalent knots, indeed infinitely many inequivalent prime knots, whose Alexander polynomial is $\Delta(t)$. The first proof of the existence of at least one knot with a given $\Delta(t)$ satisfying properties 1) and 2) is attributed to Seifert **(Sei$_2$)**. From this knot, infinitely many inequivalent prime knots can be constructed by expedient use of "tangle algebra". Relevant theorems may be found in **(Con)**, **(Lick)**, **(KT)**. For example, the substitution in a knot diagram of

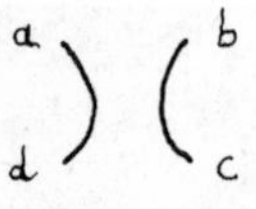

a) (b
d) (c

by the prime tangle

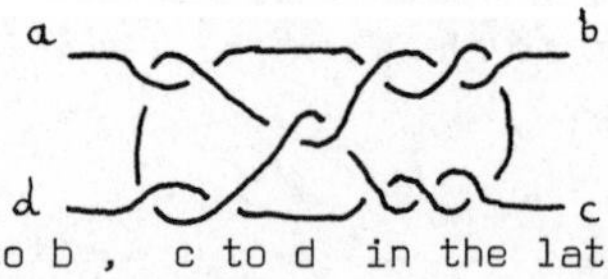

a b
d c

does not alter $\Delta(t)$, since joining a to b, c to d in the latter tangle gives a non-trivial (prime) knot with $\Delta(t) = 1$, whereas joining a to d, b to c gives a trivial link. See §6 for information on prime tangles.

7) If Δ_K denotes the Alexander polynomial of K, then $\Delta_{K \# L} = \Delta_K \Delta_L$ (see for instance **(Rol)**).

Of particular interest are non-trivial knots with $\Delta(t) = 1$ (this implies that $H_1(X_\infty) = 0$ **(Neu)**). These knots are not distinguishable from the unknot by any of the so-called "abelian" invariants (i.e. those knot invariants connected somehow with homomorphisms of the knot group onto cyclic groups); thus the signatures, first homology groups of finite cyclic covers, etc. of such a knot are all the same as those of the unknot. Also, the group of such a knot does not admit a homomorphism onto any non-abelian, finite soluble group **(Ril)**. In §6, a list is given of the 19 knots of up to 13 crossings with $\Delta(t) = 1$.

Other tricky creatures are *mutants*. Distinct knots K , L are mutants of one another if a diagram for L can be obtained from a diagram of K by chopping out a tangle, and then sewing it back after a rotation in the plane through half a turn.

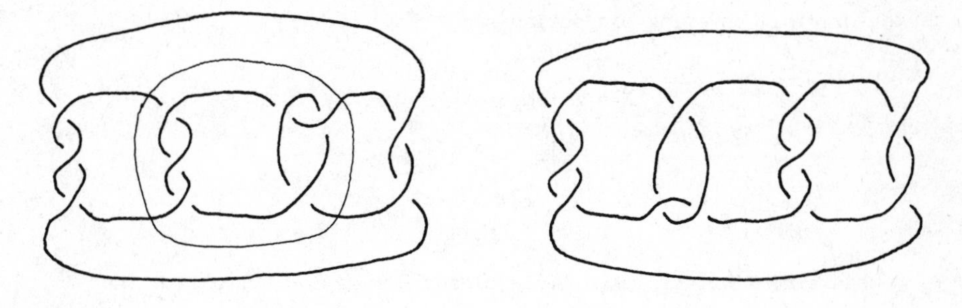

Fig. 4.1. K L

Knots within a mutancy class are not distinguishable from one another by any abelian invariant, and in fact have homeomorphic 2-fold branched covers. This last fact may be explained briefly as follows. Let S be a 2-sphere which intersects K transversely in just four points, and which separates the tangle from the rest of the knot. If S separates K into two non-trivial tangles (otherwise L will equal K), S is known as a *Conway 2-sphere*; the feint closed curve in the diagram for K above represents the intersection of such an S with the plane. In the 2-fold branched cover $\tilde{X}_K$ of K, S is covered by a torus T say (see §5 for details), so we can regard X_K as $\tilde{X}_1 \cup_f \tilde{X}_2$, where $\partial \tilde{X}_1 = \partial \tilde{X}_2 = T$ and $f: T \to T$ is the "glueing" homeomorphism. Then the 2-fold branched cover $\tilde{X}_L$ of L is $\tilde{X}_1 \cup_{f'} \tilde{X}_2$, for some homeomorphism f' isotopic to f.

Fortunately, the non-abelian invariants we shall describe shortly have little respect for such niceties.

Henceforth, we shall assume that H is a subgroup of *finite* index in $G = \pi_1(S^3 - K)$. If $\tilde{X}$ is the covering space of $X = S^3 - K$ corresponding to such a subgroup, then there is a natural way of adding a 1-dimensional *branch set* $\tilde{K}$ to $\tilde{X}$, so as to obtain a closed 3-manifold $\Sigma = \tilde{X} \cup \tilde{K}$, and a map $p: \Sigma \to S^3$ whose restriction to $\tilde{X}$ is the original covering projection $p: \tilde{X} \to X$:-

$$\begin{array}{ccc} \tilde{X} = \Sigma - \tilde{K} & \hookrightarrow & \Sigma \\ \downarrow p & & \downarrow p \\ X = S^3 - K & \hookrightarrow & S^3 \end{array}$$

$\tilde{K}$ consists of finitely many simple closed curves in Σ, called the *branch curves*. For a formal definition, and proof of uniqueness of Σ in a general setting, see **(Fox$_2$)**.

If x lies on K, the preimage in Σ of x will consist of finitely many points $x_1, \ldots, x_k$ (where k does not exceed the index of H in G), such that each x_i lies on some branch curve (the number of branch curves in Σ might be less than k : see **(Per$_1$)** for an example). If U is a suitably small neighbourhood in S^3 of x, $p^{-1}(U)$ will be a union of components $U_1, \ldots, U_k$, such that

$x_i \in U_i$ $(1 \leqslant i \leqslant k)$. If $y \in U - K$, then the cardinality of $p^{-1}(\{y\}) \cap U_i$ depends only on the branch curve which U_i meets, and is called the *branching index* of that branch curve.

Let the right cosets of H in $G = \pi_1(S^3 - K)$ be H_1 , ... , H_d . G acts transitively on this set of cosets by right multiplication; this action determines, for each $g \in G$, a permutation of the subscripts 1 , ... , d . This assignment of permutations to elements of G is in fact a homomorphism Φ of G onto a transitive subgroup Γ of the symmetric group S_d . It is easily checked that if Φ' is a homomorphism determined by the same subgroup H with a different numbering of its cosets, or by a subgroup H' conjugate to H , then Φ , Φ' are *equivalent* in the sense that $\Phi' = \tau \circ \Phi$ for some inner automorphism τ of S_d . Conversely, if Φ , Φ' are equivalent homomorphisms of G into S_d , and H , H' are stabilizers with respect to Φ , Φ' of symbols i , i' , then H is conjugate to H' . Therefore we have a one-to-one correspondence between conjugacy classes of subgroups of G of index d , and equivalence classes of transitive permutational representations of G of degree d .

H is different in general from the *kernel* of the permutational representation Φ : H is an arbitrary subgroup of finite index in G , not necessarily normal. The index of H in G is equal to the degree d of the permutational representation Φ , whereas the index of the kernel of Φ is equal to the order of $\Gamma = \Phi(G)$. The reader can check that the following are equivalent: (i) H is normal in G ; (ii) H is the kernel of the associated representation Φ ; (iii) $\Gamma = \Phi(G)$ is embedded in the symmetric group S_d as a regular representation of itself. In particular, H is normal in G if Γ is cyclic.

Given a presentation of G, and a transitive permutational representation Φ of G, there is a standard technique for obtaining a presentation of the associated subgroup H, known as the *Reidemeister-Schreier rewriting process*. There is an excellent account in **(Fox_3)** of this algorithm, as applied to knot groups, which we shall not duplicate here. A standard reference for matters concerning group presentations is **(MKS)**. A presentation of the fundamental group of the associated branched covering space Σ can then be obtained by adding extra relators, corresponding to insertion of the branch curves. The group $H_1(\Sigma)$ can then be computed by abelianizing.

Consequently, there is an effective procedure for calculating the first homology groups of all n-sheeted covering spaces of S^3 branched over a given knot. Most fortunately for the knot enumerator, these homology groups constitute a powerful invariant: with the exception of one pair of 13-crossing knots, H_1 of branched covering spaces of up to only 6 sheets, together with the Alexander polynomial, suffice to classify all 12965 knots of up to 13 crossings! (These knots have only 5639 different polynomials). The exceptional pair have groups which admit no homomorphism into S_6, but these groups are distinguished by the fact that they admit different numbers of homomorphisms onto A_7 mapping meridians to 7-cycles. This good fortune can be considered as being in two parts: firstly, the knot groups are nearly all very rich in subgroups of small index; secondly, the homology groups are very diverse. For examples, see **(Per_4)**, and also §6 of this article. It is hoped that a complete list of the invariants used to classify these 12965 knots can be published sometime, in microfiche form.

The *cyclic* coverings of $S^3 - K$, corresponding to the kernels of homomorphisms of the knot group onto cyclic groups, and their branched counterparts, admit much algebraic structure, yielding the rich theory of abelian knot invariants, which is described in depth in **(Gor)**.

We shall now describe a useful way of cellulating coverings of S^3 branched over knots. We shall then use this cellulation to illustrate Perko's method of calculating the linking of branch curves. Let us assume that a knot K is presented as a realization of an n-crossing diagram, as in Fig. 3.1. Also, assume that we are given a homomorphism Φ of $G = \pi_1(S^3 - K)$ onto a transitive subgroup of the symmetric group S_d ; thus, to each arc α_i $(1 \leqslant i \leqslant n)$ is associated a permutation σ_i in S_d , namely the image of the meridian μ_i .

Let C be the (contractible) 2-complex formed by taking the join of K with a point v situated a suitably large distance below the diagram of K . $S^3 - C$ is homeomorphic to an open 3-ball. By means of standard cutting and pasting techniques, a pair $(\Sigma, \tilde{C})$ can be constructed from (S^3, C) , together with a covering projection $p\colon \Sigma \to S^3$ branched over K , such that (i) Σ is a branched covering space corresponding to the homomorphism Φ ; (ii) $\Sigma - \tilde{C}$ is the disjoint union of d open 3-cells; (iii) $\tilde{C} = p^{-1}(C)$.

There is an obvious way of cellulating C as the union of n 2-cells $E_1, \ldots, E_n$, so that $K \cap \partial E_i$ contains α_i , part of β_{i-1} and part of β_i (Fig. 4.2a). The interior of each 2-cell E_i contains a number $(\geqslant 0)$ of lines of double points of C , corresponding to intersections with other 2-cells beneath crossing points in α_i . We orient each E_i compatibly with the orientation of K . For simplicity, we now write $\eta_i = K \cap \partial E_i$.

The next stage in the construction process is to "cut open" C to form a new complex C' , homeomorphic to S^2 , together with a natural projection $p\colon C' \to C$. C' can be visualized as an interlocking system

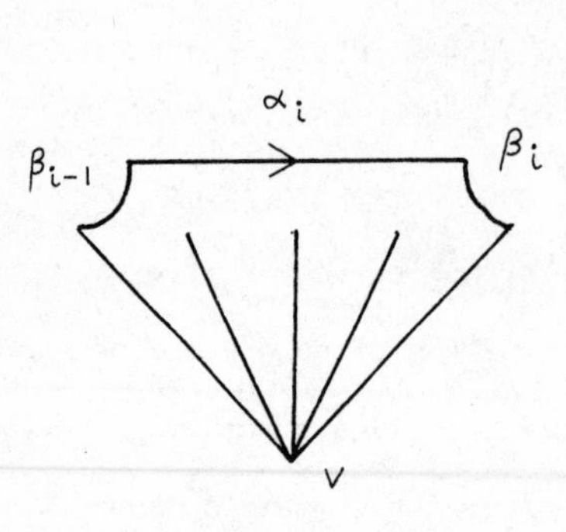

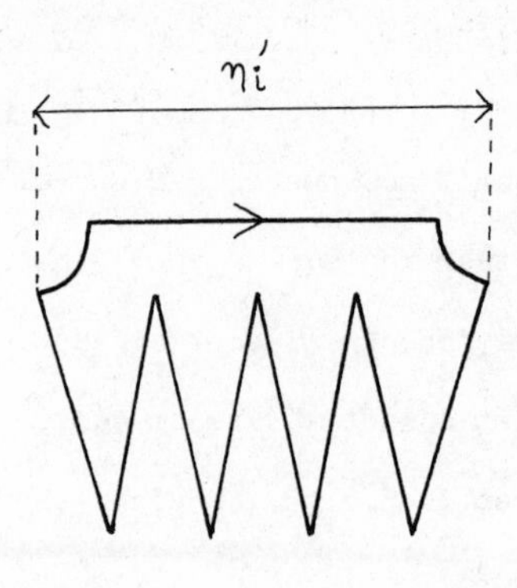

a) A typical 2-cell E_i, showing double lines.

b) $E_i^+ = E_i^- \subset C'$

Fig. 4.2.

of gabled roofs in S^3, constituted from $2n$ 2-cells E_i^+, E_i^- $(1 \leqslant i \leqslant n)$, where E_i^+, E_i^- are identical copies of the 2-cell formed by cutting E_i along its internal double lines (Fig. 4.2b). E_i^+, E_i^- are joined together along their common top edge η_i'. The apexes of the roofs, i.e. $\bigcup_i \eta_i'$, form the original knot except that the halves of each β_i have been pulled apart slightly (Fig. 4.3a).

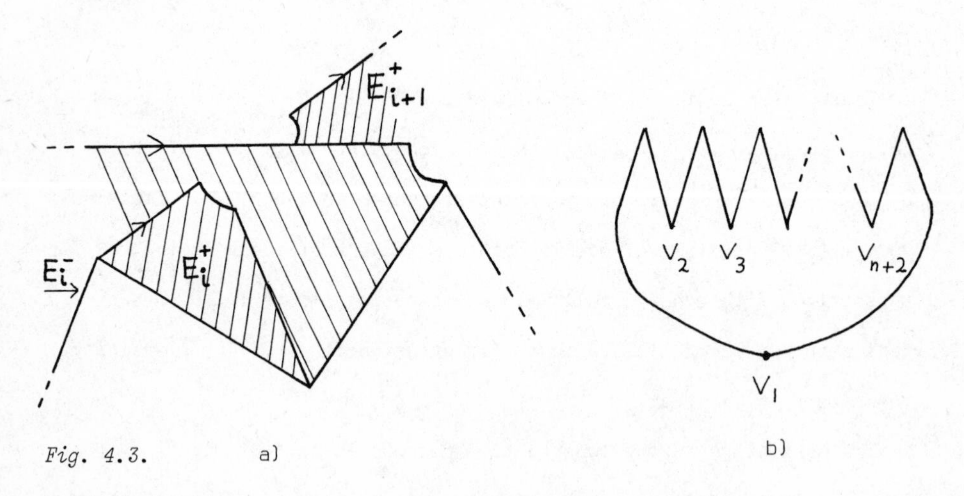

Fig. 4.3. a) b)

As one walks along η_i' in the direction of the knot, E_i^+ slopes down to the right, E_i^- to the left. The "valley rafters" sloping into the space under a particular region of the knot diagram meet at a point v_j say; there are $n+2$ such points v_j, one for each region of the knot diagram (the unbounded region presents no problem). Fig. 4.3b) is a schematic view of C', from a distant point in the plane $z = 0$, which illustrates the fact that C' is homeomorphic to S^2.

There is, of course, a natural projection $p: C' \to C$ which, for each i, maps E_i^+, E_i^- canonically onto E_i. Now let

$X = (S^3 - C) + C'$, endowed with the smallest topology such that

(i) $1 + p : (S^3 - C) + C' \to (S^3 - C) + C = S^3$ is continuous;

(ii) (X, C') is homeomorphic to a ball-sphere pair.

We finally form Σ by pasting together d copies of X along their boundaries, according to the rule: E_i^+ in the jth copy of X is identified canonically with E_i^- in the $(j.\sigma_i)$th copy of X. The image of this E_i^+ in Σ will be called E_{ij}, and the image of the jth copy of X will be denoted X_j. The 2-skeleton of this cellulation of Σ is $\tilde{C} = \bigcup_{i,j} E_{ij}$. Each $X_j - \tilde{C}$ is an open 3-cell, which we denote $\mathring{X}_j$.

The branched covering projection $p: \Sigma \to S^3$ is the obvious one: simply map each $\mathring{X}_j$ canonically to $S^3 - C$, and map each E_{ij} canonically to $E_i \subset C$. The branch set $\tilde{K}$ in Σ is simply the image of all the η_i' under the pasting process. From the definition of Σ, E_{ij} and $E_{i,\, j.\sigma_i}$ share the same piece of branch curve. The branching index of this curve, i.e. the number of distinct $\mathring{X}_j$ to which it is incident, is therefore the number of distinct integers in the set $\{j, j.\sigma_i, j.\sigma_i^2, \ldots\}$; if σ_i is written as a product of disjoint cycles, this number is the length of the cycle containing j.

It is not hard to imagine a journey inside the space Σ : basically, all one has to remember is that if one is in the 3-cell $\mathring{X}_j$ and then happens to pass through the 2-cell E_{ij}, one lands up in the 3-cell $\mathring{X}_{j.\sigma_i}$;

also, $E_{i,\ j.\sigma_i^{-1}}$ is a gateway from X_j to $X_{j.\sigma_i^{-1}}$. We leave it to the reader to check, using the fact that Φ is a homomorphism, that Σ really is a manifold.

We should remark that the *cyclic* covers of S^3 branched over knots can be constructed more simply, by pasting together copies of a space obtained by cutting S^3 along a Seifert surface, i.e. an orientable surface spanning the knot (for details, see **(Rol)**).

As an application of this explicit construction of Σ , we conclude this section with an illustration of Perko's method of calculating the linking of branch curves, thereby proving that the knot 7_3 **(AB)** is not amphicheiral. Another way of showing this is by means of the signature **(Mur$_3$)** (the signature of an amphicheiral knot is zero, as the signature changes sign under orientation-reversal of S^3); see also the discussion of 2-bridged knots in §5.

Let κ, λ be simple oriented closed curves in an oriented 3-manifold M , which carry homology classes of finite order in $H_1(M)$. Then some multiple $m\kappa$ of κ bounds a 2-chain Γ say. Let λ' be a curve isotopic in $M-\kappa$ to λ , which meets Γ transversely. Then the intersection number, n say, of λ' with Γ is defined, and the *linking number* of κ with λ , written $\ell k(\kappa,\lambda)$, is the rational number n/m . This linking invariant differs from the usual homological linking form, which takes values in the quotient ring Q/Z ; here there is no indeterminacy, as the link $\kappa \cup \lambda$ is given up to isotopy. $\ell k(\kappa,\lambda)$ is in fact symmetric in κ , λ ; also, like the signature of a knot in S^3 , $\ell k(\kappa,\lambda)$ changes sign if the orientation of M is reversed.

It so happens that the group G of the 7-crossing knot $K = 7_3$ admits just one homomorphism onto S_5 , mapping meridians to elements of cycle type (abc)(de) . This knot, together with the associated permutations $\sigma_1, \ldots, \sigma_7$, is depicted in Fig. 4.4. From the cycle type of the permutations, Σ contains two branch curves, one of index 2 and the other of index 3. Let $\kappa = \kappa_1 \cup \ldots \cup \kappa_7$ be the index 2 branch curve, and $\lambda = \lambda_1 \cup \ldots \cup \lambda_7$ be the index 3 branch curve. Using the method described earlier, one can calculate that $H_1(\Sigma) \cong Z_{11}$. Therefore $[\kappa]$, $[\lambda]$ are guaranteed to be of finite order in $H_1(\Sigma)$, and so $\ell k(\kappa,\lambda)$ is defined. Since the linking of branch curves is an invariant of the isotopy class of K, and since an orientation-reversal of S^3 induces one of Σ, K will be proved to be non-amphicheiral if it turns out that $\ell k(\kappa,\lambda) \neq 0$.

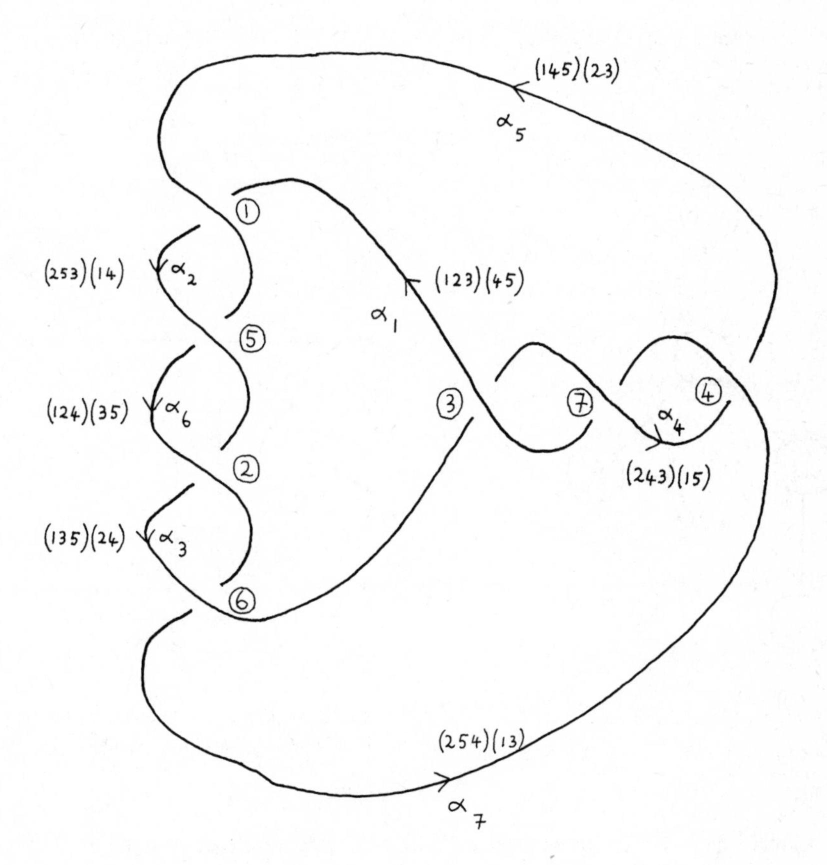

Fig. 4.4. The knot 7_3 together with a homomorphism of its group onto S_5.

In order to find a 2-chain with integer coefficients whose boundary is $m\kappa$ for some m, it is sufficient to find a linear combination $\Sigma\nu_{ij}E_{ij}$ with rational coefficients, whose boundary is κ. m will then be the l.c.m. of the denominators of the rational numbers ν_{ij}, when these are expressed in their lowest terms.

Perko observes that one can remove an open 3-ball from Σ without affecting its 1-dimensional homology. A maximal such subspace consists of all the open 3-cells $X_1, \ldots, X_5$, together with the open 2-cells $E_{13}, E_{11}, E_{51}, E_{54}$. The connections between these cells is indicated in Fig. 4.5a. Therefore we can "throw away" these 2-cells, and assume from the start that $\nu_{13} = \nu_{11} = \nu_{51} = \nu_{54} = 0$. Also, looking at coefficients of branch curves, if $\sigma_i = (abc)(de)$, we have $\nu_{ia} + \nu_{ib} + \nu_{ic} = 0$, $\nu_{id} + \nu_{ie} = 1$. Therefore $\nu_{12} = \nu_{55} = 0$.

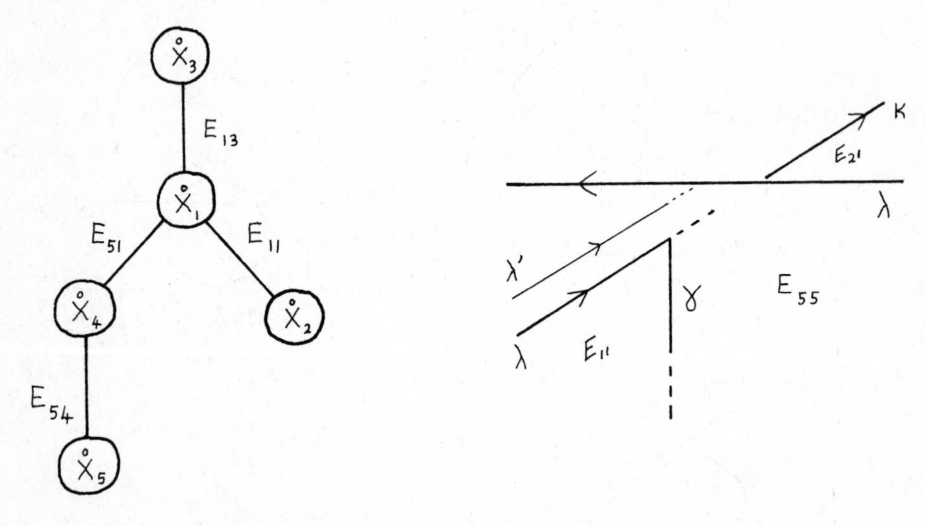

a) A maximal open 3-cell in Σ.

b) View of part of the 2-skeleton of Σ, from inside the 3-cell X_1.

Fig. 4.5.

Note that G is generated by the two meridians associated with the segments α_1, α_5 of the diagram of K. Let $\nu_{14} = p$, $\nu_{52} = q$.

Then, from the "branch" equations, $\nu_{15} = 1-p$, $\nu_{53} = 1-q$. The preimage in Σ of the "vertical" 1-cell under crossing no.1 consists of five 1-cells, each in the boundary of four 2-cells. One of these five 1-cells is shown in Fig. 5.4b, marked γ. The coefficient of each of these 1-cells in our 2-chain must be zero. This gives us the following set of equations:-

$$\begin{aligned}
\nu_{11} + \nu_{55} - \nu_{25} - \nu_{53} &= 0\\
\nu_{12} + \nu_{53} - \nu_{23} - \nu_{52} &= 0\\
\nu_{13} + \nu_{52} - \nu_{22} - \nu_{55} &= 0\\
\nu_{14} + \nu_{51} - \nu_{21} - \nu_{54} &= 0\\
\nu_{15} + \nu_{54} - \nu_{24} - \nu_{51} &= 0
\end{aligned},$$

which allow us to work out each ν_{2j} in terms of p and q.

Continuing in this fashion, using similar equations at crossings nos. 5, 2, 6, 3, we arrive at the following table of values of the ν_{ij} (in terms of p and q) :-

i \ j	1	2	3	4	5
1	0	0	0	p	1-p
2	p	q	1-2q	1-p	-1+q
3	-2+p+2q	-1+2p+q	2-3q	2-2p-q	-p+q
4	-2+3p+q	2-p-3q	-2+p+2q	q	3-3p-q
5	0	q	1-q	0	0
6	p-q	-1+p+q	2-2q	1-2p	-1+2q
7	-2+p+3q	p	3-p-3q	2-3p-q	-2+2p+q

Then the corresponding equations at crossing no.7 (or no.4) give us $p = q$, $6p+5q = 6$, from which $p = q = 6/11$. Therefore our required 2-chain is $\Gamma = \sum_{i,j} (11\nu_{ij})E_{ij}$, and it transpires that the branch curve κ carries a class of order 11 in $H_1(\Sigma)$.

Now to find $\ell k(\kappa,\lambda)$ we need the intersection number of Γ with some curve, isotopic in $\Sigma - \kappa$ to λ, which passes through the interior of each 2-cell E_{ij} which it intersects. Starting in X_1 just above the segment λ_1 of λ , following λ' as in Fig. 4.5b, we pass through the 2-cells E_{55} , E_{63} , E_{12} , E_{74} , E_{21} , E_{35} , E_{41} , all negatively (↻↗) . Since we land up back in X_1 , the journey is complete.

Therefore $\ell k(\kappa,\lambda) = -\nu_{55}-\nu_{63}-\nu_{12}-\nu_{74}-\nu_{21}-\nu_{35}-\nu_{41} = -16/11$.

This linking invariant is a very reliable weapon for establishing non-amphicheirality of knots. It was used by Perko for settling all questions regarding amphicheirality for knots of up to 10 crossings **(Per$_2$)**. Perko computes all his invariants by hand; the present author has written a computer program for calculating linking numbers of branch curves, and another for calculating homology groups (a program for the latter purpose has also been written by Riley **(Ril)**). The linking of branch curves figures also in Cappell and Shaneson's formula for the Rohlin μ-invariant **(CS)**.

5 Torus knots, 2-bridged knots and algebraic knots

Historically, the first infinite class of knots to be classified rigorously was that of torus knots, by Schreier in 1923 **(Schr)**.

Let T be a standardly embedded torus in $\mathbb{R}^3$, and let x, y be meridional and longitudinal generators respectively of $\pi_1(T) \cong Z + Z$, as in Fig. 5.1a. Given relatively prime integers p, q, there exists a simple, oriented, closed curve $T_{p,q}$ in T which is homologous to $qx + py$, and which is called a *torus knot of type p,q*. For convenience, any knot in S^3 equivalent to this curve in the oriented category will also be denoted $T_{p,q}$.

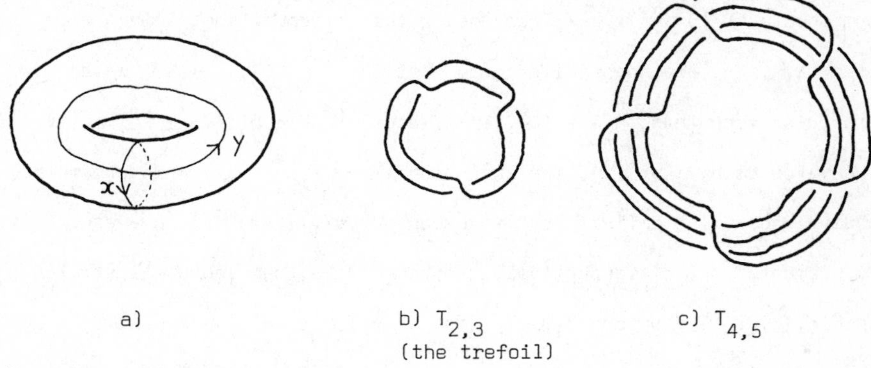

a) b) $T_{2,3}$ (the trefoil) c) $T_{4,5}$

Fig. 5.1.

An alternative definition of $T_{p,q}$ $(p,q > 0)$ is that it is the set of points in $S^3 = \{(z_1, z_2) \in \mathbb{C}^2 : |z_1|^2 + |z_2|^2 = 1\}$ such that $z_1^p + z_2^q = 0$. If p, q are not relatively prime, the set $T_{p,q}$ thus defined is a *torus link* of $d > 1$ components, where d is the greatest common divisor of p and q.

Certain properties of $T_{p,q}$ are immediately apparent:-

(i) $T_{1,q}$, $T_{p,1}$ are unknotted;

(ii) $T_{-p,-q}$ is the reverse of $T_{p,q}$, and since there exists a rotation of R^3 which maps T to itself and which maps x, y to $-x$, $-y$ respectively, it follows that the knot $T_{p,q}$ is invertible;

(iii) a suitable reflection maps $T_{p,q}$ to $T_{p,-q}$, so $T_{p,q}$ is equivalent to $T_{p,-q}$ in the unoriented sense;

(iv) $T_{p,q}$ is isotopic to $T_{q,p}$, as there is an orientation-preserving autohomeomorphism h of S^3 which maps T to itself and which interchanges x and y (h interchanges the two components of $S^3 - T$). Without any loss of generality apart from the exclusion of trivial cases, we shall henceforth assume that $2 \leqslant p < |q|$.

The group $\pi_1(S^3 - T_{p,q})$ has a particularly simple presentation, obtainable by means of van Kampen's theorem. Let X_1, X_2 be the components of $S^3 - T$, X_2 say being the component containing ∞. Then X_1, X_2 are open solid tori. Let $h: T \times (-1,1) \to S^3$ be an embedding such that $h(T \times \{0\}) = T$, and let $N = h((T - T_{p,q}) \times (-1,1))$. Then, for each $i = 1,2$, the inclusion $X_i \hookrightarrow Y_i = X_i \cup N$ is a homotopy equivalence. The infinite cyclic groups $\pi_1(Y_1)$, $\pi_1(Y_2)$ are generated by elements a, b respectively, represented by the loops illustrated in Fig. 5.2. The union $Y_1 \cup Y_2$ is $S^3 - T_{p,q}$. $Y_1 \cap Y_2$ is homeomorphic to an open solid torus; the infinite cyclic fundamental group of $Y_1 \cap Y_2$ is generated by an element c say, which maps (visibly) under the inclusion-induced homomorphisms $\pi_1(Y_1 \cap Y_2) \to \pi_1(Y_1)$, $\pi_1(Y_1 \cap Y_2) \to \pi_1(Y_2)$ to a^p, b^q respectively. The Y_i are open in $Y_1 \cup Y_2 = S^3 - T_{p,q}$, so van Kampen may be applied to give the presentation

$\pi_1(S^3 - T_{p,q}) = \langle a,b \mid a^p = b^q \rangle$. We note that this is not a Wirtinger presentation for $2 \leqslant p < |q|$, as the generators a , b are not meridians.

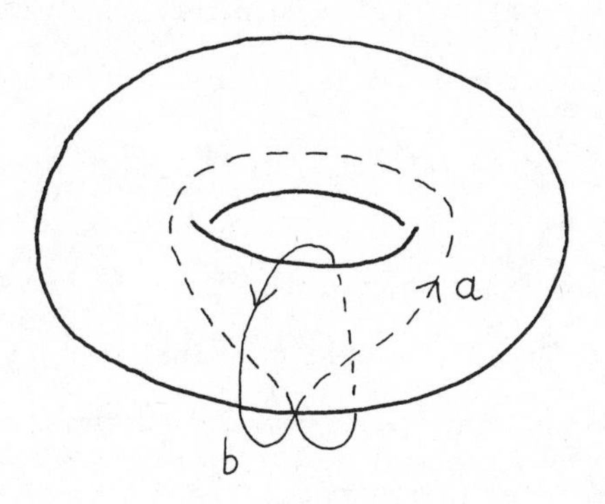

Fig. 5.2.

It is now a simple matter to classify torus knots up to unoriented equivalence. $G_{p,q} = \pi_1(S^3 - T_{p,q})$ has an infinite cyclic centre, generated by $a^p = b^q$ (Burde and Zieschang have proved the striking result that torus knots are the *only* non-trivial knots whose groups have a non-trivial centre **(BZ)**). The group of inner automorphisms of $G_{p,q}$ is therefore isomorphic to $G_{p,q} / \langle a^p \rangle$, which in turn is isomorphic to the free product of two cyclic groups of orders p and q respectively. Hence, under the hypothesis $2 \leqslant p < |q|$, $T_{p,q}$ is non-trivial, and $T_{p,q}$ is equivalent to $T_{p',q'}$ if and only if $p = p'$ and $|q| = |q'|$.

Schreier proved that $T_{p,q}$ is not amphicheiral for $2 \leqslant p < |q|$, by examining automorphisms of $G_{p,q}$ (in essence, Schreier examined the peripheral structure of $G_{p,q}$). Alternatively, one may use the fact that $T_{p,q}$ has non-zero signature (see Litherland's formula for twisted and classical signatures of torus knots in **(Lith)**): recall from §4 that a necessary condition for amphicheirality is that the signature should be zero.

The Alexander polynomial of $T_{p,q}$ is $\frac{(1-t)(1-t^{pq})}{(1-t^p)(1-t^q)}$ (see for instance **(Rol)**). An examination of the zeros of this polynomial reveals that torus knots are classified up to unoriented equivalence by their polynomials. As the polynomial of $T_{p,q}$ has some zero coefficients for $p,q > 2$, it follows from property 5) of Alexander polynomials that $T_{p,q}$ is non-alternating for $p,q > 2$.

The next family of knots to be classified were the 2-bridged knots, by Schubert in 1956 **(Schu$_3$)**. The *bridge number*, or *bridge index* b(D) of a knot diagram D is the number of maximal overpasses it contains, i.e. the number of arcs α_i (in the sense of Fig. 3.1) with at least one overcrossing. Then the *bridge number* of a knot K is

$b(K) = \inf \{b(D): D \text{ is a diagram of } K\}$.

A little experimentation reveals that any 2-bridged knot diagram can be reduced, using Reidemeister moves I and II, either to a diagram with no crossings, or else to a diagram D(p,q) of the type illustrated in Fig. 5.3.

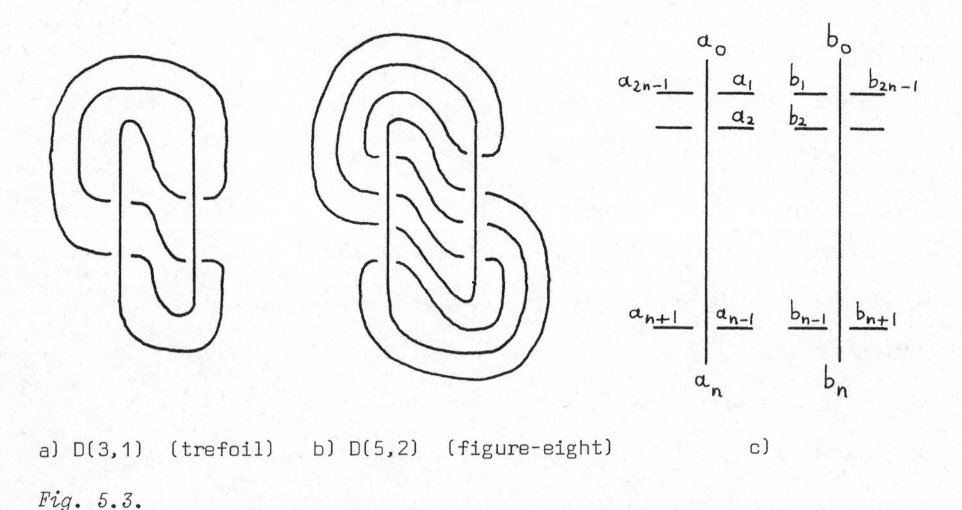

a) D(3,1) (trefoil) b) D(5,2) (figure-eight) c)

Fig. 5.3.

To form D(p,q) in general, join each "end" a_i to b_{i+q} in Fig. 5.3c), where suffixes are taken modulo $2n$, without introducing any new crossings. We assume, of course, that $1 \leq q < p$. Then it is readily established that $D(p,q)$ is a diagram of a knot with one component if and only if $(p,q) = 1$ and p is odd. One can also show combinatorially that the diagrams $D(p,q)$, $D(p,q')$ are isotopic if $q \equiv -q' \pmod p$, or $qq' \equiv \pm 1 \pmod p$. Let $K(p,q)$ denote the knot whose diagram is $D(p,q)$.

Schubert computed that the 2-fold cover of S^3 branched over $K(p,q)$ is the lens space $L(p,q)$. Reidemeister had classified lens spaces (with the obvious piecewise linear structure) up to piecewise linear homeomorphism, by means of Reidemeister torsion, in 1935 **(Reid)**, and it followed from the work of Moise in 1952 that 3-manifolds were topologically homeomorphic if and only if they were piecewise linearly homeomorphic **(Moi)**. Specifically, $L(p,q)$ is homeomorphic to $L(p',q')$ if and only if (i) $p = p'$, and (ii) either $q \equiv \pm q' \pmod p$, or $qq' \equiv \pm 1 \pmod p$. Therefore 2-bridged knots are classified up to unoriented equivalence by their 2-fold branched covers. Hodgson has shown that in fact 2-bridged knots are the only knots whose 2-fold branched covers are lens spaces **(Hodg)**. For an alternative topological classification of lens spaces, see **(Bro)** (1960) : Brody uses the fact that non-homeomorphic lens spaces have different knot theories. We note, incidentally, that the determinant of $K(p,q)$ is p , as $\pi_1(L(p,q)) = Z_p$.

An easy calculation shows that if we draw $D(p,q)$ with an orientation assigned to the knot, then the orientations of the two overpasses (as drawn in Fig. 5.3) are parallel or antiparallel according as q is respectively odd or even. From the previous paragraph, we can take q to be odd without loss of generality, in which case a rotation of $D(p,q)$ through half a turn in the plane reveals that $K(p,q)$ is invertible.

The oriented classification of 2-bridged knots may be completed by using the fact that the lens space $L(p,q)$ admits an orientation-reversing autohomeomorphism if and only if $q^2 \equiv -1 \pmod p$ (see **(Mil$_2$)**). Therefore a necessary condition for amphicheirality of $K(p,q)$ is $q^2 \equiv -1 \pmod p$; sufficiency of this condition is established by direct construction, but this is most easily seen from the alternative presentation of 2-bridged knots described below.

Here, then, is another way of looking at 2-bridged knots. We can represent a 2-bridged knot by a closed curve with just two maxima with respect to the coordinate z . Then, as viewed from the side, the knot will be a closed "4-string plait" or "viergeflecht" (Fig. 5.4a). One of the four plait strings can be "freed" by pulling it tight and pushing the other strings tangled with it out of the way, without introducing new maxima. The result is a picture such as Fig. 5.4b.

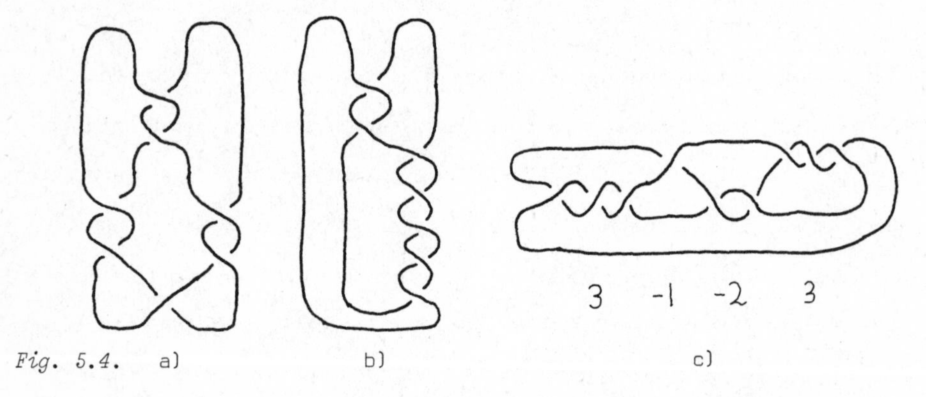

Fig. 5.4. a) b) c)

Any plait manufactured in this way can be denoted by a sequence $c_1\ c_2\ \dots\ c_r$ of signed integers: working from right to left, let c_r be the number of clockwise half-twists of strings 1 and 2, c_{r-1} be the number of anticlockwise half-twists of strings 2 and 3, and so on (see Fig. 5.4c). The convention regarding signs ensures that a sequence of positive integers represents an alternating diagram. The sequence $c_1\ c_2\ \dots\ c_r$ is the *Conway symbol* for the diagram **(Con)**. The symbols

3 , 2 2 represent the trefoil and the figure-eight, respectively. By adjusting the corresponding diagram if necessary, it can always be arranged that a Conway symbol does not begin or end with ±1 .

There is a nice formulation, as a continued fraction in terms of the c_i , for the parameters p , q of the 2-bridged knot $K = K(p,q)$. This formulation is obtained by regarding (S^3, K) as $(B_1, A_1) \cup_f (B_2, A_2)$, where each B_i is a 3-ball, each A_i is a pair of spanning, parallel straight-line segments, and $f: \partial B_1 \to \partial B_2$ is a suitable glueing homeomorphism which maps ∂A_1 onto ∂A_2 .

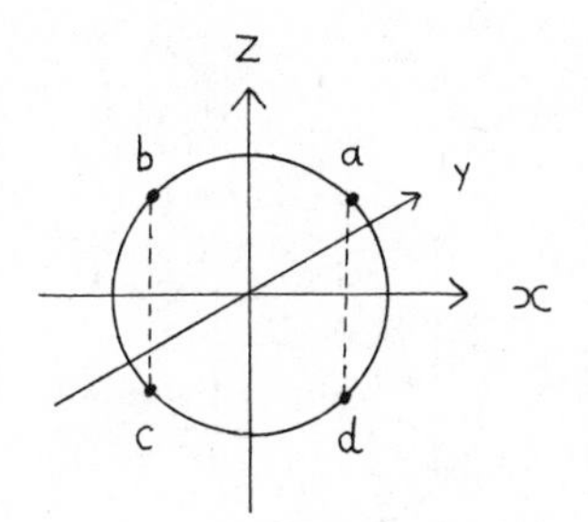

Fig. 5.5.

We proceed with a number of definitions. Let B be the standard 3-ball $\{\mathbf{x} \in \mathbb{R}^3 : |\mathbf{x}| \leqslant 1\}$. Let a, b, c, d be the points on the boundary of B where $y = 0$, and $|x| = |z| = 1/\sqrt{2}$; let $A \subset B$ consist of the straight-line segments ad , bc (Fig. 5.5). Let s be the isometry of $(B, \partial B)$ defined by the formula $s: (x,y,z) \to (-z,-y,x)$; s is the composition of a rotation through $\pi/2$ about the y-axis, transforming a to b , and a reflection in the plane $y = 0$. Let t be an autohomeomorphism of $(B, \partial B)$ which twists the upper half of B , defined in cylindrical coordinates by

$$t: (\rho,\phi,z) \to \begin{cases} (\rho,\phi,z) & (z \leqslant 0) \\ (\rho,\phi+\sqrt{2}\pi z,z) & (z > 0) \end{cases}$$

(ρ , ϕ are defined by $x = \rho\cos\phi$, $y = \rho\sin\phi$). The map t interchanges a and b , and fixes c and d . For any sequence σ of integers $c_1, c_2, \ldots, c_r$, let f_σ be the composite $t^{c_r}s \ldots st^{c_2}st^{c_1}$.

Then $(B , f(A)) \cup_{1_{\partial B}} (B , A) \cong (B , A) \cup_{f_\sigma|\partial B} (B , A) \cong (S^3 , K)$, where K is the 2-bridged knot with Conway symbol $c_1\ c_2\ \ldots\ c_r$. The geometrical motivation for this construction should be clear from the sequence of pictures in Fig. 5.6, which illustrates the case $r = 2$, $c_1 = 3$, $c_2 = 2$.

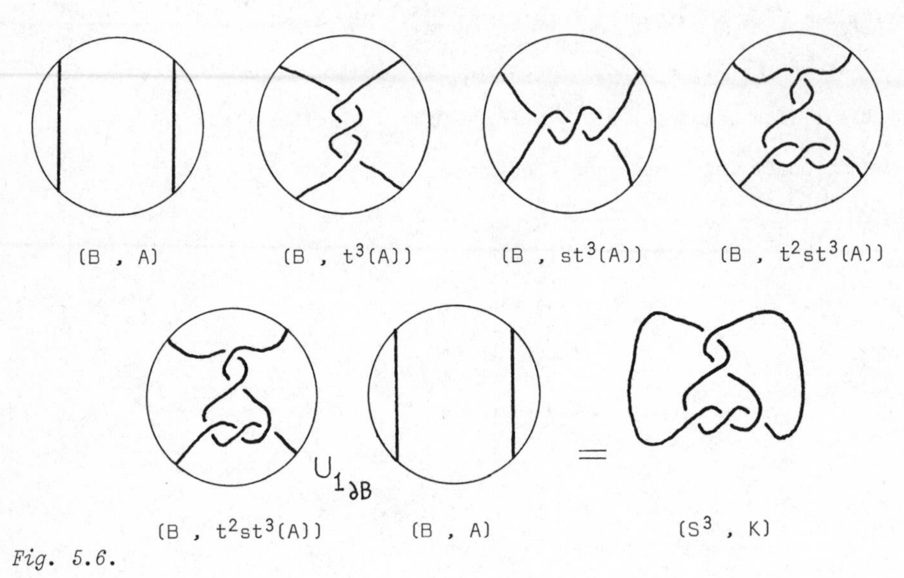

Fig. 5.6.

A 2-fold cover of B branched over A may conveniently be formed by pasting together two copies of a space obtained by suitably cutting B (Fig. 5.7). The result is a manifold V homeomorphic to a solid torus, whose boundary ∂V is the 2-fold cover of the 2-sphere ∂B , branched over the four points $A \cap \partial B = \{a, b, c, d\}$.

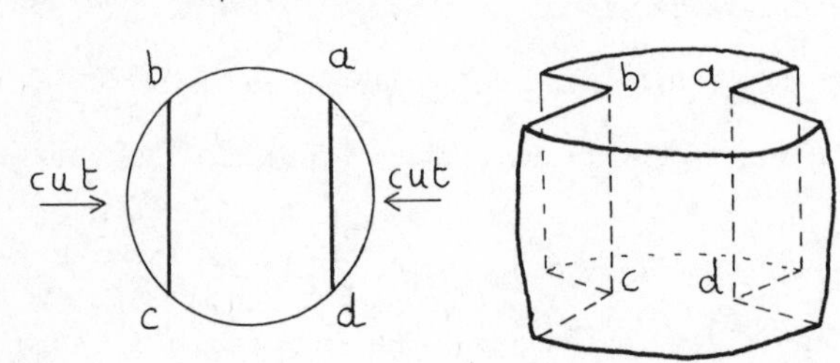

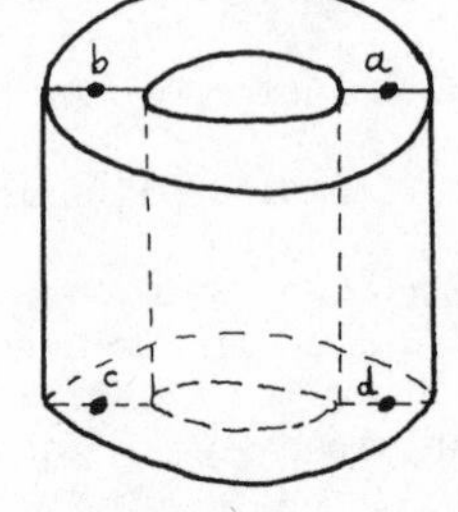

Fig. 5.7. (B , A) A space X , obtained by cutting open B . Two copies of X glued together to form V .

The homeomorphisms s , t map the branch set $\{a, b, c, d\}$ to itself, hence are covered by maps $\tilde{s}$, $\tilde{t}$: $\partial V \to \partial V$. Therefore the 2-fold cover of S^3 branched over K is the lens space formed by glueing together two copies of V by the homeomorphism

$$\tilde{f}_\sigma = \tilde{t}^{c_r}\tilde{s} \ldots \tilde{s}\tilde{t}^{c_2}\tilde{s}\tilde{t}^{c_1} : \quad \partial V \to \partial V .$$

Let $\{\mu,\lambda\}$ be a meridian-longitude basis for $\pi_1(\partial V) \cong Z + Z$. Then the matrices of the induced homomorphisms of fundamental groups $\tilde{s}_*$, $\tilde{t}_*$, with respect to the ordered basis $\{\mu,\lambda\}$, are $\begin{pmatrix}0 & 1\\ 1 & 0\end{pmatrix}$, $\begin{pmatrix}1 & 0\\ 1 & 1\end{pmatrix}$ respectively ($\tilde{t}_*$ is a so-called *longitudinal twist*).

Therefore the matrix of $\tilde{f}_{\sigma *}$ is

$$\begin{pmatrix}1 & 0\\ 1 & 1\end{pmatrix}^{c_r}\begin{pmatrix}0 & 1\\ 1 & 0\end{pmatrix} \ldots\ldots \begin{pmatrix}0 & 1\\ 1 & 0\end{pmatrix}\begin{pmatrix}1 & 0\\ 1 & 1\end{pmatrix}^{c_2}\begin{pmatrix}0 & 1\\ 1 & 0\end{pmatrix}\begin{pmatrix}1 & 0\\ 1 & 1\end{pmatrix}^{c_1} = \begin{pmatrix}q & s\\ p & r\end{pmatrix} \quad \text{say,}$$

where p/q is the value of the continued fraction

$$c_r + \frac{1}{c_{r-1}+}\;\frac{1}{c_{r-2}+}\;\cdots\; + \frac{1}{c_1} .$$

But the statement $\tilde{f}_{\sigma *}(\mu) = q\mu + p\lambda$ is equivalent to the statement $V \cup_{\tilde{f}_\sigma} V = L(p,q)$. Therefore the lens space parameters p , q are the numerator and denominator respectively of the continued fraction

$$c_r + \frac{1}{c_{r-1}+}\;\frac{1}{c_{r-2}+}\;\cdots\; + \frac{1}{c_1} .$$

For readers who are familiar with the language of surgery diagrams, the 2-fold cover of S^3 branched over a knot with Conway symbol $c_1\ c_2 \ldots c_r$ has surgery presentation

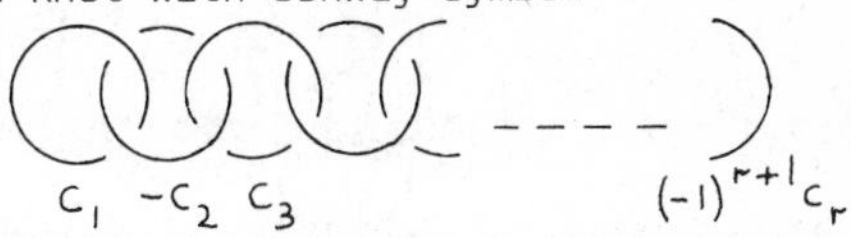

c_1 $-c_2$ c_3 ---- $(-1)^{r+1}c_r$

which reduces under surgery moves to p/q in Rolfsen's fractional notation, where p , q are as before (see **(Rol)** for a similar example).

Let us say that a Conway symbol for $K(p,q)$ is *standard* if (i) each c_i is positive; (ii) c_1 , $c_r > 1$. From the above classification, and from uniqueness properties of continued fractions, each 2-bridged (unoriented) knot type has a standard Conway symbol which is unique up to reversal of the sequence $c_1 \ldots c_r$. Also, for a given p/q the sum $\sum_{i=1}^{r} |c_i|$ is minimal when the c_i are all positive, so a "4-string plait" diagram of a 2-bridged knot has minimal crossing-number when it is alternating. It is conjectured that this minimal sum is in fact the crossing-number of the knot.

There is a particularly pleasing characterization of the standard Conway symbols of amphicheiral 2-bridged knots. It is a fact that if $p/q = c_1 + \frac{1}{c_2 +} \ldots \frac{1}{c_r}$, $p'/q' = c_r + \frac{1}{c_{r-1} +} \ldots \frac{1}{c_1}$, then $p = p'$ and $qq' \equiv (-1)^{r+1} \pmod p$ (see for instance **(Olds)** ex.7 p.26, together with theorem 1.4). Recall that $K(p,q)$ can only be amphicheiral if $q^2 \equiv -1 \pmod p$. Using the uniqueness property of standard Conway symbols, it follows that a necessary condition for amphicheirality of $K(p,q)$ is that its standard Conway symbol contain an even number of terms, and be palindromic. That this condition is also sufficient in general is indicated by the sequence of diagrams of Fig. 5.8.

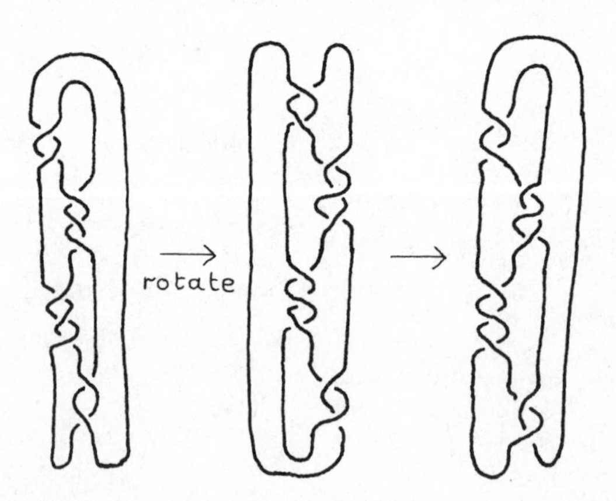

Fig. 5.8.

Algebraic knots, introduced by Conway in **(Con)**, are the knots representable by 4-string plaits which extend over a *planar tree*, as in Fig. 5.9a (for the special case of 2-bridged knots, the tree consists merely of a line segment). Alternatively, an algebraic knot may be presented as the boundary of a surface F in S^3 constructed by starting from an embedded band (I-bundle over S^1), and then successively attaching bands by *plumbing*, as in Fig. 5.9b. The surface F must be embedded in such a way that it is possible to replace each "plumbing patch" 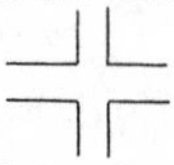 in F by 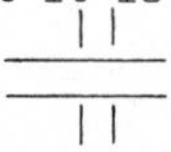 or 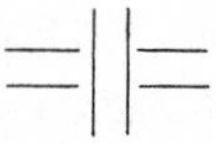 , so as to get a surface consisting of disjoint bands which are allowed to be twisted, but not knotted or linked. Note that attaching a band with n clockwise twists amounts to replacing  somewhere in the knot by .

n

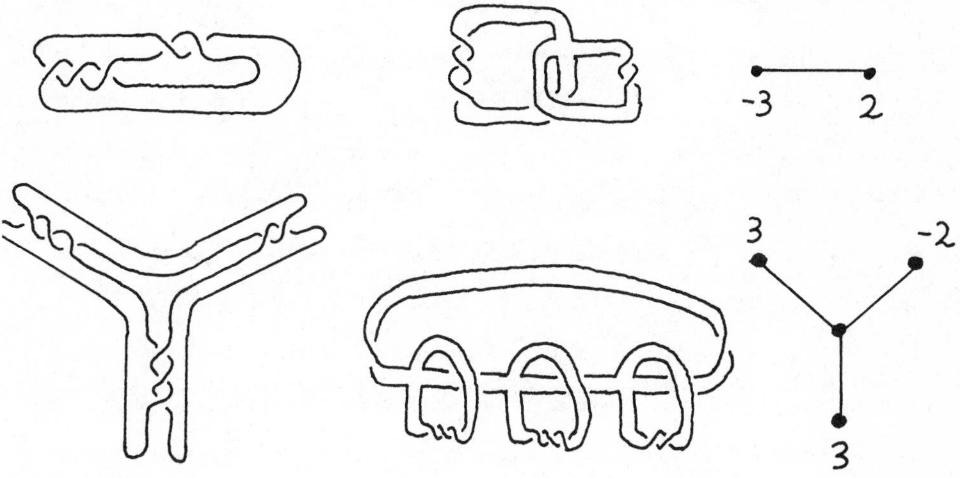

-3 2

3 -2

3

Fig. 5.9. a) b) c)

Algebraic knots have been studied exhaustively by F. Bonahon and L. Siebenmann. Their work is due to appear as a volume of the London Mathematical Society Lecture Note Series **(BS)**. Their analysis deals with

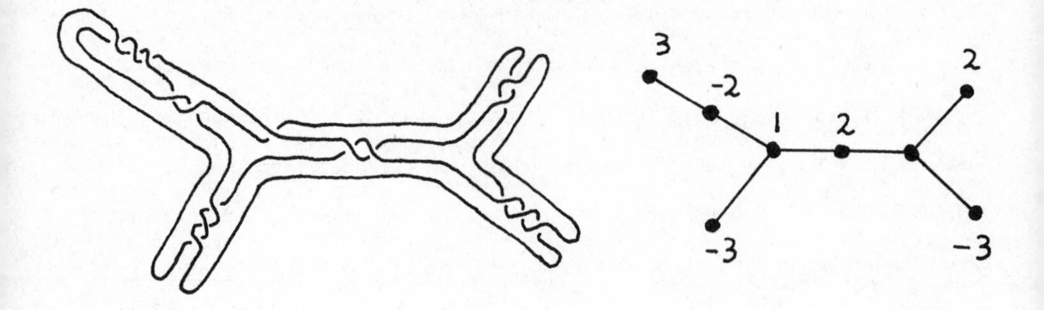

Fig. 5.10. A more complicated example of an algebraic knot.

knots and links simultaneously; they denote algebraic knot diagrams by *weighted planar trees* as in Fig. 5.9c. Note that their sign convention is different from Conway's: a positive coefficient (weight) always corresponds to a sequence of positive twists. Bonahon and Siebenmann have developed an effective procedure, using Tait's flyping operation and a few other types of move, for reducing any weighted planar tree to one from a small, finite canonical set of trees for the knot. As their procedure works in the oriented category, they can determine the symmetry type of an algebraically presented knot. Motivation for the reduction procedure stems from an examination of the 2-fold cover of S^3 branched over the knot. This branched covering space possesses a *characteristic variety* (see **(JS)**, **(Joh)**), consisting of a finite disjoint union (possibly empty) of incompressible tori, unique up to isotopy; the variety may be taken to be equivariant with respect to the covering involution, so that the incompressible tori project to *Conway 2-spheres*, enclosing prime tangles in the knot. Each component of the complement of the characteristic variety is a Seifert fibre space, and the number of tori in the variety is minimal with respect to this property. The problem of mutancy (recall from §4 that mutants have homeomorphic 2-fold branched covers) was dealt with by insisting that homeomorphisms of 2-fold branched covers be equivariant: it is clear that *any* two knots are equivalent if and only if their 2-fold branched covers are equivariantly homeomorphic.

The significance of algebraic knots is twofold. Firstly, the great majority of knots and links of up to 11 crossings are algebraic (the hordes of non-algebraic knots appear to take over thereafter; one would reasonably conjecture that any ratio of form $\frac{\text{no. of nice prime knots with n crossings}}{\text{no. of prime knots with n crossings}}$ was asymptotically zero). Secondly, there is the following geometrical interpretation. Bonahon and Siebenmann, again using the characteristic variety theorem, define up to isotopy, for each pair (S^3, K), whether or not algebraic, an *algebraic part* $(A^3, A^3 \cap K)$ and a *non-algebraic part* $(N^3, N^3 \cap K)$ of (S^3, K). K is algebraic if and only if N^3 is empty. According to W. Thurston's hyperbolization theorem for 3-manifolds, the non-algebraic part of (S^3, K) is in some sense, at least when A^3 is non-empty, a maximal sub-pair admitting a complete π-*hyperbolic* metric of finite volume: $(N^3, N^3 \cap K)$ is locally isometric to $(H^3/\tau$, axis of $\tau)$, where τ is a rotation through π of hyperbolic 3-space, and H^3/τ is the orbit space. Certain knots K are known to be π-*hyperbolic*, in the sense that A^3 is empty, and (S^3, K) has a π-hyperbolic structure as above; an example of such a knot is the "Turk's head" knot 8^* (see below for a picture); indeed, Thurston reports that he has proved that any knot, whose algebraic part is empty, is π-hyperbolic.

Fig. 5.11. The π-hyperbolic knot 8^*.

6 Twentieth century tabulations

Apart from M. Haseman's enumeration of 12-crossing amphicheirals in 1917, no attempt was made to extend the nineteenth century knot tables during the first half of this century. Alexander and Briggs, however, did initiate a rigorous classification **(AB)**; using the first homology groups of branched cyclic coverings, they distinguished all tabled knots of up to 9 crossings, except for three pairs. Reidemeister then completed the classification of knots of up to 9 crossings, using the linking of branch curves in irregular coverings associated with homomorphisms onto dihedral groups.

Surprisingly, nothing further was done until the 1960's, when Conway instilled fresh vigour into the subject by inventing a brand new notation for knots (and links), very faintly reminiscent of Kirkman's notation. Conway used his notation to enumerate, by hand, knots of up to 11 crossings and links of up to 10 crossings **(Con)**. There are a few errors in Conway's tables, but his notation has had significant influence on both theoretical and practical aspects of the subject. Meanwhile, Dowker was working on a completely different notation, based on Tait's "alphabetical" notation, with a view to computerizing the enumeration of knots. This computerization was carried out by the author in 1980. So the competition between the "geometric" types of notation (Kirkman's, Conway's) and the "cyclic" types (Gauss's, Tait's alphabetical, Dowker's) continues to the present day.

The striking features of Conway's notation are that it is extremely compact for knots of up to 11 crossings, and that it often expresses important structural properties of the knot. Also, it treats links with equal ease. On the debit side, the components of each knot symbol come from a fairly large list, and are assembled according to a fairly large number of conventions. Consequently, the notation takes a little time to master, and does not lend itself very readily to computer programming. Dowker's notation, on the other hand, is extremely simple and uniform: the symbol for an n-crossing knot is a sequence of n even integers. It is probably non-structural, but is ideal for computer use, both in the enumeration of knots and the calculation of invariants. Also, it is easily mastered in a few minutes by anyone, even a small child.

As Conway's notation is very much connected with the subject of tangles, we shall briefly describe Lickorish's presentation of "tangle calculus" **(Lick)**. As explained in §4, a *tangle* is a pair (B, A), where B is a 3-ball and A is a pair of disjoint arcs properly embedded in B (i.e. $A \quad B = \quad A$).

Equivalence of tangles is defined as for knots: (B_1, A_1) is *equivalent* to (B_2, A_2) if there is a homeomorphism of pairs from (B_1, A_1) to (B_2, A_2). A tangle is *trivial*, or *untangled* if it is equivalent to the tangle (B, A) pictured in Fig. 5.6; note that all the tangles pictured in Fig. 5.6 are trivial, according to this definition. A tangle (B, A) is *prime* if (i) it is non-trivial, and (ii) any 2-sphere in B, which meets A transversely in two points, bounds in B a ball meeting A in an unknotted spanning arc.

Lickorish defines two ways of combining tangles, the first of which we have already seen in §5. If two tangles (B_1, A_1), (B_2, A_2) are glued together by a homeomorphism $h: (B_1, A_1) \to (B_2, A_2)$, the result is a pair (S^3, L), where L is a knot, or a link with two components, called a *sum* of (B_1, A_1), (B_2, A_2). Essentially, h ranges over the classical braid group of four strings in $S^2 \times I$. Fig. 5.6 illustrates the summing together of two trivial tangles, to produce the 2-bridged knot 3 2. An elementary, but significant and very useful theorem in **(Lick)** states that any sum of two prime tangles is a prime knot, or link. Also, it is proved in **(KL)** that *any* knot, whether or not prime, can be expressed as a sum of a prime tangle and a trivial tangle.

The second method of combining two tangles (B_1, A_1), (B_2, A_2) identifies a (disc, pair of points) in (B_1, A_1) with a (disc, pair of points) in (B_2, A_2) by means of some homeomorphism, so that the result is still a tangle. Lickorish proves that any partial sum of two prime tangles is a prime tangle.

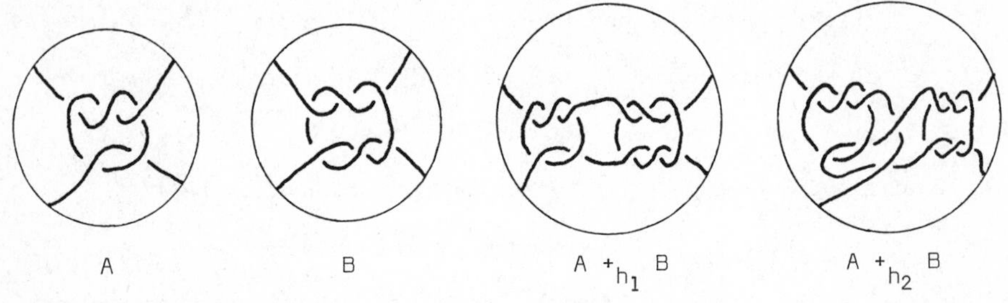

Fig. 6.1. Two different partial sums of a given pair of (prime) tangles.

We can now define an *algebraic tangle* as a tangle obtainable by partially summing together a collection of trivial tangles according to some bracket arrangement. Thus, if τ denotes a trivial tangle and $+_h$ denotes partial summation using some homeomorphism h of (disc, pair of points) pairs, then examples of algebraic tangles would be $((\tau +_{h_1} \tau) +_{h_2} \tau) +_{h_3} \tau$, $((\tau +_{h_1} \tau) +_{h_2} (\tau +_{h_3} \tau))$ (see Fig. 6.2.)

There is an intimate connection between algebraic tangles and algebraic knots: a prime knot is algebraic if and only if it is the sum of an algebraic tangle with a trivial tangle (see also **(Lick)**, p. 331). With minor adaptations, Bonahon and Siebenmann's method of classifying algebraic knots applies equally well to the classification of algebraic tangles. Of course, if one wishes to consider links of more than one component, a more general sort of tangle is required, where the subspace A of (B, A) consists of two arcs as above, together with a number $\geqslant 0$ of simple closed curves in B.

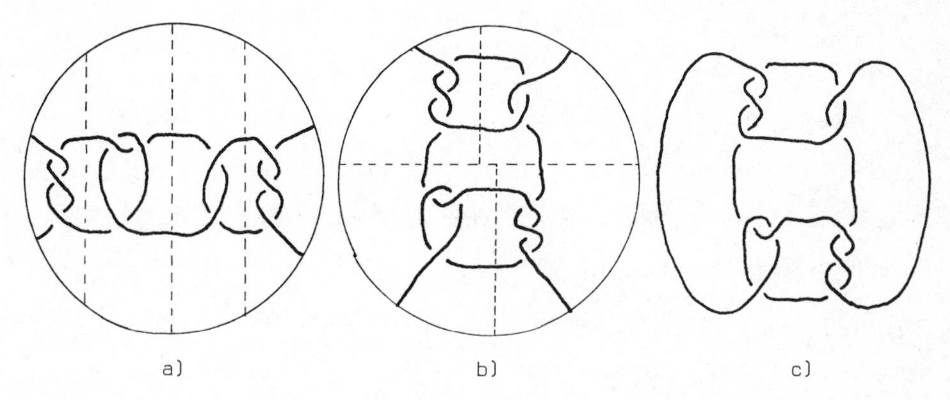

Fig. 6.2. a) An algebraic tangle of form $((\tau +_{h_1} \tau) +_{h_2} \tau) +_{h_3} \tau$

b) An algebraic tangle of form $((\tau +_{h_1} \tau) +_{h_2} (\tau +_{h_3} \tau))$

c) An algebraic knot obtained by summing tangle b) with a trivial tangle

The essential idea behind Conway's notational system is that any knot diagram may be constructed, either by summing an algebraic tangle with a trivial tangle to form an algebraic knot as in Fig. 6.2c, or by "substituting" diagrams of algebraic tangles into a *basic polyhedron*, i.e. a 4-valent graph in the plane with no region bounded by less than 3 edges. In Kirkman's language, a basic polyhedron is a "solid knot". The substitution process, which is in essence a generalization of the process of summing tangles, is illustrated in Fig. 6.3a. Justification of this method relies on the observation that a tangle is algebraic if and only if it admits a projection such that successive replacements of maximal strings of flaps by single crossings eventually reduce the entire projection to one with a single crossing (Fig. 6.3b). Therefore, if we successively remove flaps in this way from a regular knot projection which is not cut non-trivially in just two points by any circle in the plane, we eventually arrive either at , or at a basic polyhedron; the portions of the projection which have been "collapsed" are projections of algebraic tangles.

Conway's notation, then, consists of (i) symbols for algebraic tangles which express the way in which they are constituted from trivial tangles; (ii) symbols for basic polyhedra; (iii) a way of combining these symbols, using various punctuation marks and the like, for expressing the way in which the algebraic tangles are substituted into the basic polyhedra. Space forbids further details, but in any event we would only be duplicating Conway's own description in **(Con)**.

One reason for the enormous success of Conway's notation, quite apart from its structural nature, is that up to 11 crossings there are only seven basic polyhedra (the numbers begin to increase markedly thereafter).

By way of contrast, Dowker's notation is the ultimate in simplicity. Suppose we are given a regular projection p of a knot K into the plane. Having chosen a starting-point and a direction, we can travel round K, labelling $1,2, \ldots ,2n$ the points of K which project to the n crossing points of $p(K)$. Clearly, we lose nothing by assuming that $n \geqslant 3$. The projection p defines an obvious parity-reversing involution τ on $\{1,2, \ldots ,2n\}$: $\tau(i) = j$ if and only if $p(i) = p(j)$ and $i \neq j$.

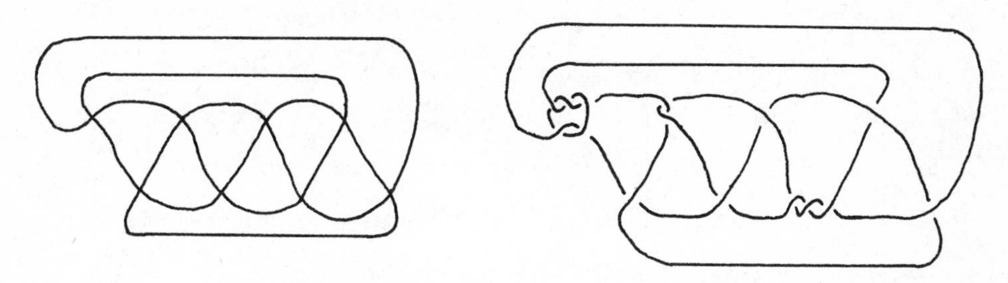

a) Substitution of algebraic tangles into a basic polyhedron

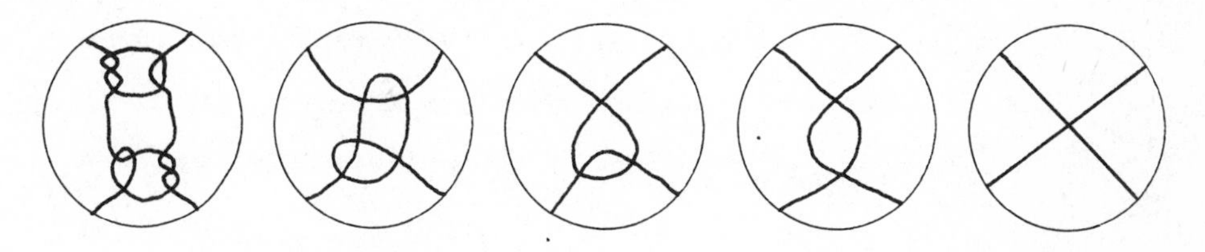

b) Reduction of an algebraic tangle projection by removing flaps

Fig. 6.3.

Let us write τ_i for $\tau(i)$. This involution τ is completely specified by the sequence of even integers $\tau_1, \tau_3, \dots, \tau_{2n-1}$. In order to denote over- and undercrossings, we associate to each point on K labelled i a sign σ_i which is positive if and only if i is an overcrossing. Then the knot diagram associated with the projection p may be denoted by the sequence of signed integers $\sigma_1\tau_1$, $\sigma_3\tau_3$, ... , $\sigma_{2n-1}\tau_{2n-1}$. Note that the terms of the sequence all have the same sign if and only if the diagram is alternating. This sequence depends, of course, on a choice of initial point, and on direction. Here are some examples:-

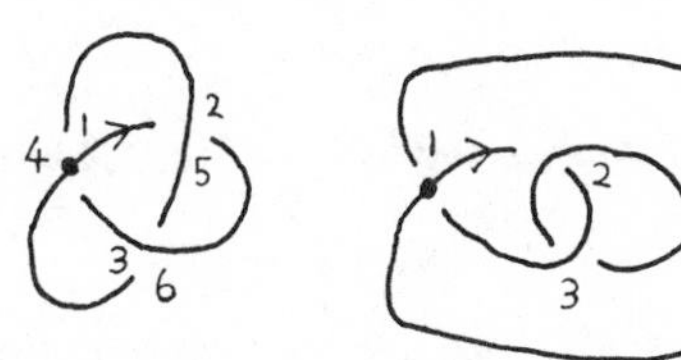

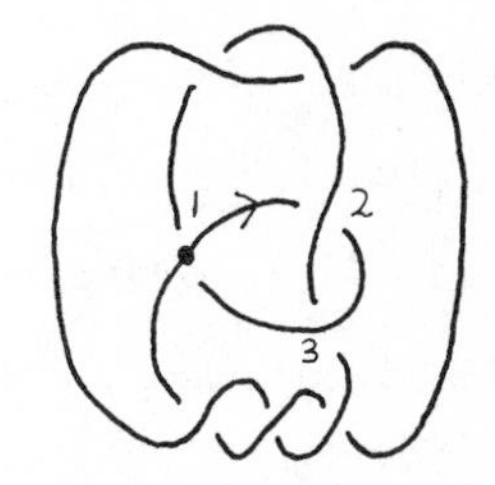

Fig. 6.4. 4 6 2 4 6 8 2 4 8 -12 2 14 16 -6 10

It is convenient to order sequences according to the following rule. If S is a sequence $\sigma_1\tau_1 , \sigma_3\tau_3 , \ldots , \sigma_{2n-1}\tau_{2n-1}$ as above, let S' be the sequence $n , \tau_1 , \tau_3 , \ldots , \tau_{2n-1} , \sigma_1 , \sigma_3 , \ldots , \sigma_{2n-1}$. Then we say that $S_1 < S_2$ if S_1' precedes S_2' lexicographically. In its basic form, this notation is only intended to denote diagrams up to homeomorphism of the extended plane, so we may assume that the sign σ_1 is $+1$. The *standard sequence* for a knot diagram D is $\inf\{S : S$ is a sequence denoting $D\}$, and the standard sequence for a knot K is $\inf\{S : S$ is a sequence denoting a diagram of $K\}$.

Many parity-reversing involutions τ do not arise in this way from knot diagrams (for example the involution giving rise to 4 8 2 10 6), and some may be rejected on the grounds that they are obviously non-standard, or else do not represent prime knots. Thus we exclude any involution which maps a proper subinterval (mod 2n) of $\{1,2, \ldots ,2n\}$ to itself, as any diagram associated with such a τ would admit a circle in the plane cutting it non-trivially in just two points; the special case $\tau(i) = i\pm1 \pmod{2n}$ implies the existence of a "nugatory" loop in the diagram.

Let us say that a parity-reversing involution on $\{1,2, \ldots ,2n\}$ is *admissible* if it has none of these bad properties. Then it is shown in **(DT)** that admissible involutions determine images of regular knot projections up to homeomorphism of the extended plane. Also, an efficient (though mildly intricate) algorithm is given which filters out sequences of even integers which do not arise from knot projections. This algorithm also determines for admissible sequences the signs (in the sense of §1) of the crossings, relative of course to the sign of the first crossing. It is these results which make the notation usable for enumerating knots.

It is then relatively straightforward to write a computer program which generates sequences in ascending order, and which filters out all inadmissible sequences. Procedures can be written which perform transformations on the sequences corresponding to flypes and passes, thus detecting the vast majority of sequences which are not minimal for a given knot. These "non-standard" sequences are rejected forthwith. It is convenient in practice, when enumerating knots of a given crossing-number, to enumerate first the alternating knots (whose sequences consist of positive terms), and then to generate non-alternating knots by prefixing minus signs.

The resulting crude list of knots of up to some previously agreed crossing-number is then subjected to a large-scale calculation of invariants, with the hope of distinguishing all distinct pairs of knots. When classifying knots of up to 13 crossings, the author first calculated the Alexander polynomial of each knot, together with a 12-component vector $v(K)$ whose entries were the numbers (up to inner automorphism of S_5) of homomorphisms of various distinguishable sorts of the knot group into the symmetric group S_5 :-

	Image of homomorphism	Cycle type of image of meridians
1.	S_3	(ab)
2.	S_4	(ab)
3.	S_5	(ab)
4.	A_4	(abc)
5.	A_5	(abc)
6.	M	(abcd)
7.	S_4	(abcd)
8.	S_5	(abcd)
9.	A_5	(abcde)
10.	D_5	(ab)(cd)
11.	A_5	(ab)(cd)
12.	S_5	(abc)(de)

M denotes the "metacyclic" group of order 20, i.e. the 1-dimensional affine group over the field with 5 elements. It is representable as the subgroup of S_5 generated by the permutations (1234) , (1425) . Note the use of the peripheral structure: one can often distinguish knots by looking at the respective cycle types of images of meridians, even when the image subgroups are the same.

This "ragbag" invariant v(K) might seem crude, but whereas for the 12965 knots of up to 13 crossings there are only 5639 different polynomials, the application of v(K) leaves just over a thousand unresolved cases. The computation of $\Delta(t)$ and v(K) each take less than one second for a typical 13-crossing knot. There are in fact some interesting relationships between the entries in v(K) (for instance $v_1 + v_2 = v_7$!) which the author hopes to discuss elsewhere.

The next stage in the battle was to calculate the first homology groups of the branched covering spaces associated with the homomorphisms counted in v(K) . This reduced the number of unresolved cases to 176 pairs, 4 triples, 2 quadruples and one sextuple. The calculation of numbers of homomorphisms into S_6 , and associated homology groups, reduced the number of cases to 37 pairs and 3 triples. One further pair was resolved by counting the numbers of homomorphisms onto A_7 , mapping meridians to 7-cycles. Finally, the residual list of unresolved cases was shown by explicit construction to consist of 36 duplicates and 3 triplicates respectively. Thus the classification of knots of up to 13 crossings was complete, at least up to clerical or programming error.

Some of the final duplications were remarkably hard to demonstrate. A particularly awkward case is illustrated in Fig. 6.5. This method of classifying knots is really an elaboration of the time-honoured method used by Alexander, and Reidemeister. Perhaps the first knot theorist to appreciate the power of the technique of mapping the knot group onto symmetric and alternating groups was Perko (see **(Per$_4$)**). Virtually nothing is known in general about these homomorphisms.

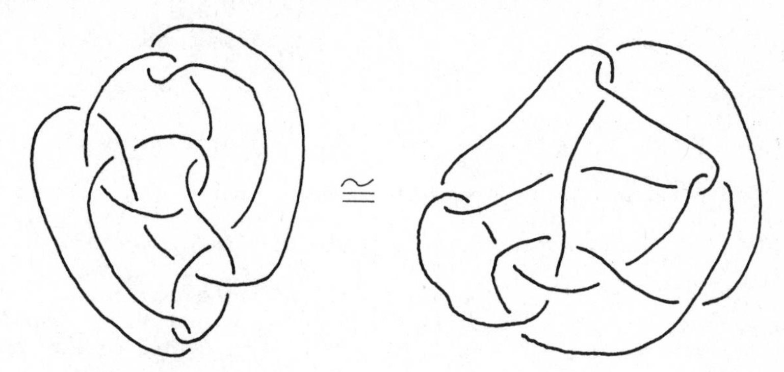

Fig. 6.5. Exercise.

As mentioned earlier, there is still the problem of proving that each of these 12965 knots is prime. There are several useful theorems here. Perhaps the most spectacular is W. Menasco's result that if K has an alternating diagram which is not cut non-trivially in two points by any circle in the plane, then K is prime **(Men)**. Therefore knots with alternating presentations are prime if and only if "they look as if they are". One can say the same for algebraically-presented knots, as these succumb to the Bonahon-Siebenmann machinery. Therefore one only has to look carefully at knots with no known alternating or algebraic presentation.

All but very few of these remaining knots are 3-bridged. By Schubert's (non-trivial) theorem on the additivity of bridge-number, namely $b(K\#L) = b(K) + b(L) - 1$, a 3-bridged composite knot must be the connected sum of two 2-bridged knots. There are only finitely many 2-bridged knots of given polynomial (or determinant), so the possibilities for factors K , L are severely limited by property 7) of Alexander polynomials, §4. One would then hope to settle problems of primality for these 3-bridged knots by means of the usual invariants. Even 4-bridged knots are reasonably tractable, as at least one factor of a composite 4-bridged knot is 2-bridged. One can then use the amalgamated-free-product structure to compare homomorphisms of the group of the knot under investigation with those of any knot of form $K \# L$, where K is from a finite list of possible 2-bridged knots. A preliminary survey of (apparently) 4-bridged, non-alternating, non-algebraic knots of up to 13 crossings suggests that the great majority will in fact be proved to be prime by means of Lickorish's theorem on sums of prime tangles.

The reader might feel uneasy about the partly empirical nature of these computations. One redeeming feature is undoubtedly the vast quantity of knot invariants which are recorded. Astonishingly, there *is* a uniform method of classifying knots, and also knot groups, arising from the Haken theory of irreducible, sufficiently large manifolds **(Hak)**, and a certain missing step supplied by Hemion **(He)**. A very readable account of the underlying theory is given in Waldhausen's expository article **(Wal)**. *Very* briefly, Haken proves that the exterior of the knot (i.e. the complement in S^3 of an open tubular neighbourhood of the knot) has only finitely many hierarchies of a certain sort, and he then proceeds to find them all. This involves triangulating the knot exterior, and necessitates the solution of a sizeable system of linear equations. Haken has demonstrated his method in certain simple cases, but to the author's knowledge no-one has yet written a computer program to implement the algorithm.

We conclude this section with a complete list of the knots of up to 13 crossings with $\Delta(t) = 1$ (modulo, as always, clerical error!), together with sufficient invariants to distinguish them from one another. Here, first, is a list of the 19 knots, together with the invariant $v(K)$.

The reader can easily construct diagrams of the knots from the sequences.

	K	v(K)
1.	4 8 12 2 -16 -18 6 -20 -22 -14 -10	0 0 0 0 1 0 0 0 0 0 0 0
2.	4 8 12 2 -18 -20 6 -10 -22 -14 -16	0 0 0 0 1 0 0 0 0 0 0 0
3.	4 10 12 -20 -16 2 -22 -8 -24 -14 -6 -18	0 0 0 0 0 0 0 0 2 0 0 0
4.	4 10 -14 -22 -16 2 20 -24 -8 -6 12 -18	0 0 0 0 1 0 0 0 0 0 1 0
5.	4 8 10 14 2 -20 -18 6 -22 -26 -24 -16 -12	0 0 0 0 0 0 0 0 0 0 1 1
6.	4 8 10 14 2 -22 -20 6 -12 -24 -16 -26 -18	0 0 0 0 0 0 0 0 0 0 1 1
7.	4 8 14 2 -18 -24 -20 6 -22 -26 -16 -12 -10	0 0 0 0 1 0 0 1 3 0 1 0
8.	4 8 14 2 -22 -18 -24 6 -12 -10 -26 -16 -20	0 0 0 0 1 0 0 0 3 0 1 0
9.	4 10 12 -16 2 8 -22 -6 -26 -24 -14 -20 -18	0 0 0 0 0 0 0 0 0 0 0 0
10.	4 10 14 -16 2 -20 -24 22 -12 -26 -8 6 -18	0 0 0 0 0 0 0 0 4 0 0 0
11.	4 10 14 -16 2 -20 -24 22 -12 -26 8 -6 -18	0 0 0 0 2 0 0 0 0 0 0 0
12.	6 8 14 16 4 -18 -24 2 -22 -12 -26 -10 -20	0 0 0 0 2 0 0 1 0 0 1 0
13.	6 8 14 16 4 -24 -20 2 -22 -10 -26 -18 -12	0 0 0 0 2 0 0 1 0 0 1 0
14.	6 8 16 14 4 -18 -22 2 -24 -12 -26 -10 -20	0 0 0 0 2 0 0 1 0 0 1 0
15.	6 8 20 22 4 -16 -26 -10 -24 -14 2 -12 -18	0 0 0 0 2 0 0 1 0 0 1 0
16.	6 8 20 22 4 -24 -18 -10 -26 -14 2 -16 -12	0 0 0 0 2 0 0 1 0 0 1 0
17.	6 8 22 20 4 -16 -26 -10 -24 -12 2 -14 -18	0 0 0 0 2 0 0 1 0 0 1 0
18.	6 10 -22 12 -16 2 26 20 -8 24 -4 14 18	0 0 0 0 0 0 0 1 1 0 0 1
19.	6 -10 -22 14 -2 18 8 26 24 12 16 -4 20	0 0 0 0 2 0 0 0 1 0 1 3

The unresolved cases are: (1,2) ; (5,6) ; (12,13,14,15,16,17) .

Let us denote the group Z_n by n , and the trivial group by 0 .

The homology groups corresponding to the homomorphisms counted in v(K)

are as follows:-

1.	7			
2.	7			
5.	21 ,	2+924		
6.	21 ,	2+924		
12.	0 ,	0 ,	5+20 ,	3
13.	0 ,	0 ,	5 ,	3
14.	0 ,	0 ,	35 ,	3
15.	0 ,	0 ,	5+20 ,	3
16.	0 ,	0 ,	5 ,	3
17.	0 ,	0 ,	10 ,	3

The unresolved cases are now: (1,2) ; (5,6) ; (12,15) ; (13,16) .

Knots numbers 1 , 2 are in fact distinguished by the fact that the group of knot number 1 admits a homomorphism onto S_6 mapping meridians to 6-cycles, whereas the group of knot number 2 doesn't.

However, all these knots are easily dealt with by homology groups associated with homomorphisms onto A_6 mapping meridians to 5-cycles. For amusement, here are the groups:-

1.	87 , 15 , 3 , 3
2.	87 , 51 , 147 , 75
5.	3+51 , 2+42 , 6+42 , 939 , 2+66 , 681 , 2+294 , 303
6.	3+45 , 6+60 , 4+12 , 2+384 , 2+192 , 87 , 147 , 3
12.	2+54 , 2+2586 , 3+54 , 339 , 2+528 , 195 , 105 , 3+60 , 147 , 63
15.	387 , 2+384 , 3+54 , 93 , 2+336 , 375 , 75 , 3+42 , 63 , 147
13.	3+15 , 2+12 , 2+216 , 72 , 147 , 63 , 6 , 6+12 , 2+270 , 2+2046
16.	3+129 , 2+156 , 207 , 72 , 147 , 63 , 2+2+24 , 6+66 , 2+108 , 4+348

Knots numbers 1 and 2 are the famous 11-crossing mutants, one of which was discovered by Kinoshita and Terasaka **(KT)**, and the other by Conway. The problem of distinguishing them is discussed in **(Ril)**. As they are algebraic, they fall under the Bonahon-Siebenmann classification scheme. It is perhaps not altogether obvious that knots numbers 12 to 17 inclusive are also algebraic, and are all in the same mutancy class. This is best seen at the 15-crossing level. For instance, knot number 13 has the following 15-crossing diagram:-

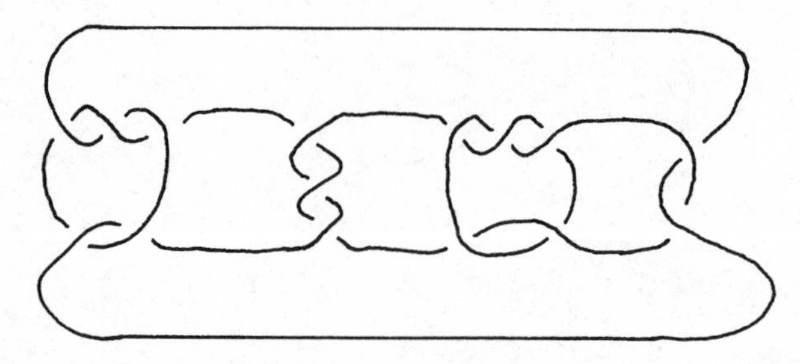

Fig. 6.6.

References

AB J.W. Alexander and G.B. Briggs, *On types of knotted curve,* Annals of Math. 28 (1927), 562-586.

Ban C. Bankwitz, *Uber der Torsionzahlen der zyklischen Uberlagerungsräume des Knotenaussenraumes,* Annals of Math. 31 (1930), 131-133.

Bro E.J. Brody, *The topological classification of the lens spaces,* Annals of Math. 71 (1960), 163-185.

BS F. Bonahon and L.C. Siebenmann, *New geometric splittings of classical knots (algebraic knots),* to appear.

BZ G. Burde and H. Zieschang, *Eine Kennzeichnung der Torusknoten,* Math. Ann. 167 (1966), 169-176.

Cau A. Caudron, *Classification des noeuds et des enlacements,* Publications math. d'Orsay, 1980.

CF R.H. Crowell and R.H. Fox, *Introduction to knot theory,* Springer 1963.

Con J.H. Conway, *An enumeration of knots and links,* Computational problems in abstract algebra (ed. Leech), Pergamon Press, 1969, 329-358.

Cro R.H. Crowell, *Genus of alternating link types,* Annals of Math. 69 (1959), 258-275.

CS S.E. Cappell and J.L. Shaneson, *Invariants of 3-manifolds,* Bull. Amer. Math. Soc. 81 (1975), 559-562.

De_1 M. Dehn, *Die beiden Kleeblattschlingen,* Math. Ann. 75 (1914), 402-413.

De_2 ______, *Uber Kombinatorische Topologie,* Acta Math. 67 (1936),123-168.

DT C.H. Dowker and M.B. Thistlethwaite, *Classification of knot projections,* Topology and its Applications 16 (1983), 19-31.

Fox_1 R.H. Fox, *On the complementary domains of a certain pair of inequivalent knots,* Indag. Math. 14 (1952), 37-40.

Fox_2 ______, *Covering spaces with singularities,* Proc. Symposium in honor of S. Lefschetz, Princeton Math. Series 12 (1957), 243-257.

Fox_3 ______, *A quick trip through knot theory,* Topology of 3-manifolds and related topics, Prentice-Hall 1962, 120-167.

Gor C. McA. Gordon, *Some aspects of classical knot theory,* Springer lecture notes in math. 685 (1978), 1-60.

Hak W. Haken, *Theorie der Normalflächen,* Acta Math. 105 (1961), 245-375.

Has M.G. Haseman, *On knots, with a census of the amphicheirals with twelve crossings,* Trans. Roy. Soc. Ediburgh 52 (1918), 235-255.

He G. Hemion, *On the classification of homeomorphisms of 2-manifolds and the classification of 3-manifolds*, Acta Math. 142 (1979), 123-155.

Hem J. Hempel, *3-manifolds*, Annals of Math. Studies 86, Princeton 1976.

Hil H. Hilden, *3-fold branched coverings of S^3*, Amer. J. Math. 98 (1976), 989-997.

Hodg C. Hodgson, *Involutions and isometries of lens spaces*, MSc dissertation, Melbourne 1981.

Joh K. Johannson, *On exotic homotopy equivalences of 3-manifolds*, Proc. Georgia Topology Conf. (ed. Cantrell), Academic Press 1979, 101-112.

JS W. Jaco and P.B. Shalen, *Seifert fibered spaces in 3-manifolds*, Proc. Georgia Topology Conf. (ed. Cantrell), Academic Press 1979, 91-100.

Kirk T.P. Kirkman, *The enumeration, description, and construction of knots with fewer than 10 crossings*, Trans.Roy.Soc.Edin.32 (1885), 281-309.

KL R.C. Kirby and W.B.R. Lickorish, *Prime knots and concordance*, Math. Proc. Cambridge Phil. Soc. 86 (1979), 437-441.

KT S. Kinoshita and H. Terasaka, *On unions of knots*, Osaka Math. Journal 9 (1957), 131-153.

Lick W.B.R. Lickorish, *Prime knots and tangles*, Trans. Amer. Math. Soc. 271(1) (1981), 321-332.

Lis J.B. Listing, *Vorstudien zur Topologie*, Göttingen Studien, 1847.

Lit_1 C.N. Little, *On knots, with a census for order 10*, Trans. Connecticut Acad. Sci. 18 (1885), 374-378.

Lit_2 __________, *Non-alternate ± knots of orders eight and nine*, Trans. Roy. Soc. Edinburgh 35 (1889), 663-664.

Lit_3 __________, *Non-alternate ± knots*, Trans. Roy. Soc. Edinburgh 39 (1900), 771-778.

Lith R.A. Litherland, *Signatures of iterated torus knots*, 1979 Sussex Conf. (ed. Fenn), Springer Lecture Notes in Math. 722, 71-84.

Men W. Menasco, *Closed incompressible surfaces in alternating knot and link complements*, Topology 23 no.1 (1984), 37-44.

Mil_1 J. Milnor, *A duality theorem for Reidemeister torsion*, Annals of Math. 76 (1962), 137-147.

Mil_2 __________, *Whitehead torsion*, Bull. Amer. Math. Soc. 72 (1966), 358-426.

MKS W. Magnus, A. Karrass and D. Solitar, *Combinatorial group theory*, Interscience 1966.

Moi E.E. Moise, *Affine structures in 3-manifolds V: the triangulation theorem and Hauptvermutung,* Annals of Math. 55 (1952), 96-114.

Mon J. Montesinos, *3-manifolds as 3-fold branched coverings of* S^3, Quart. Journal Math. Oxford 27 (1976), 85-94.

Mur_1 K. Murasugi, *On the genus of the alternating knot I, II,* Journal Math. Soc. Japan 10 (1958), 94-105 and 235-248.

Mur_2 __________, *On the Alexander polynomial of the alternating knot,* Osaka Math. Journal 10 (1958), 181-189.

Mur_3 __________, *On a certain numerical invariant of link types,* Trans. Amer. Math. Soc. 114 (1965), 377-383.

Neu L.P. Neuwirth, *Knot groups,* Annals of Math. Studies 56, Princeton 1965.

Olds C.D. Olds, *Continued fractions,* Random House, Yale 1963.

Papa C.D. Papakyriakopoulos, *On Dehn's lemma and the asphericity of knots,* Annals of Math. 66 (1957), 1-26.

Per_1 K.A. Perko, *An invariant of certain knots,* Princeton senior thesis, Princeton University, Princeton N.J., 1964.

Per_2 __________, *On 10-crossing knots,* unpublished.

Per_3 __________, *On dihedral covering spaces of knots,* Inventiones Math. 34 (1976), 77-82.

Per_4 __________, *Invariants of 11-crossing knots,* Publications Math. d'Orsay, 1980.

Reid K. Reidemeister, *Homotopieringe und Linsenräume,* Abh. Math. Sem. Hamburg 11 (1935), 102-109.

Ril R. Riley, *Homomorphisms of knot groups on finite groups,* Math. Comp. 25 (1971), 603-619.

Rol D. Rolfsen, *Knots and links,* Publish or Perish Inc., Berkeley Ca., 1976.

RR R.C. Read and P. Rosenstiehl, *Principal edge tripartition of a graph,* Annals of Discrete Math. 3 (1978), 195-226.

Schr O. Schreier, *Uber die gruppen* $A^aB^b=1$, Abh. Math. Sem. Hamburg 3 (1923), 167-169.

$Schu_1$ H. Schubert, *Die eindeutige Zerlegbarkeit eines Knotens in Primknoten,* S.-B. Heidelberger Akad. Wiss. Math. Natur K1, 3 (1949), 57-104.

$Schu_2$ __________, *Knoten und Vollringe,* Acta Math. 90 (1953), 131-286.

Schu_3 __________, *Knoten mit zwei Brücken*, Math. Zeit. 65 (1956), 133-170.

Spa E.H. Spanier, *Algebraic Topology*, McGraw-Hill 1966.

Seif_1 H. Seifert, *Verschlingungsinvarianten*, S.-B. Preuss. Akad. Wiss. 26 (1933), 811-828.

Seif_2 __________, *Uber das Geschlecht von Knoten*, Math. Ann. 110 (1934), 571-592.

Sha P.B. Shalen, *Infinitely divisible elements in 3-manifold groups*, Knots, groups and 3-manifolds, Annals of Math. Studies 84 (1975), 293-335.

Tai P.G. Tait, *On knots I, II, III*, Scientific Papers Vol. I, Cambridge University Press, London, 1898, 273-347.

Tro H.F. Trotter, *Noninvertible knots exist*, Topology 2 (1964), 275-280.

Wal F. Waldhausen, *Recent results on sufficiently large 3-manifolds*, Proc. Symp. Pure Math. Vol.32 (2), Amer. Math. Soc. 1978, 21-38.

Wen H. Wendt, *Die Gordische Auflösung von Knoten*, Math. Zeit. 42 (1937), 680-696.

HOW GENERAL IS A GENERALIZED SPACE?

Peter T. Johnstone

Introduction

The notion of topos was introduced by Grothendieck around 1960 as a natural generalization of that of a topological space, in the context of the "non-topological" cohomology theories needed in his work in algebraic geometry. However, Grothendieck and his followers saw toposes as a potentially useful generalization of spaces in a context much wider than this, as is demonstrated by the following passage from the introduction to [11], exposé 4:

> "On peut donc dire que la notion de topos, dérivé naturel du point de vue faisceautique en Topologie, constitue à son tour un élargissement substantiel de la notion d'espace topologique, englobant un grand nombre de situations qui autrefois n'étaient pas considérées comme relevant de l'intuition topologique."

If this sentence is viewed as a prophecy, then it has surely been amply fulfilled by the work of Lawvere, Tierney and those who have followed them, in demonstrating how sheaf-theoretic and topological ideas can indeed be applied through the medium of topos theory to contexts where an earlier generation of mathematicians would never have suspected their utility.

However, the dictum "A topos is a generalized space" is not entirely free from oversimplification. Lawvere [24] sought to "turn it backward" by arguing that a topos is the "algebra of continuous (set-valued) functions" on a generalized space, rather than the generalized space itself. More important, for our present purposes, is the fact that a topos is not so much a generalized space as a generalized locale (or frame). Although the theory of locales (or "pointless topology" [17]) is often viewed as a generalization of classical point-set topology, it does not contain a fully faithful representation of the latter: in effect, one sacrifices a small amount of pathology (non-sober spaces) in order to achieve a category that is more smoothly and purely "topological" than the category of spaces itself. And it is this category of locales (whose importance Hugh Dowker was one of the first topologists to recognize [8]), and not the classical category of spaces, which embeds naturally and pleasingly in the category of toposes.

All this is by way of preamble to the question which stands as the title of this article: how big a leap do we in fact make when we pass from spaces (or locales) to toposes? In the early days of topos theory, and still more after its "elementary" rebirth in 1969-70, the emphasis was all on its generality, and the (sometimes bewildering) multiplicity of examples from diverse sources that could be accommodated within this simple axiomatic framework. However, recent work (a good deal of it unpublished) has produced a number of theorems which tend to suggest that the gap is not as wide as it may have appeared: that, in fact, the notion of topos is faithfully representable in terms of locales and a small number of auxiliary concepts (notably that of a localic group or groupoid).

These theorems have "topological" statements which assert the existence of open covering maps with particular properties; but (in a fashion which is typical of the fusion of geometric and logical ideas that constantly occurs in topos theory) their simplest proofs are model-theoretic in character -- albeit invoking a "categorical model theory" which is as yet unfamiliar to a great many logicians. Nevertheless, I think it is not inappropriate that this survey should appear in a volume dedicated to the memory of Hugh Dowker, who would (I believe) be delighted to see how traditional topological ideas (provided they are framed in the "pointless" way that he did much to popularize) can turn out to have an applicability that reaches far beyond the traditional demesne of point-set topology.

In concluding this introduction, I must declare my indebtedness to the topos-theorists whose work I am here recording; only a small part of it is my own, and the rest is taken from correspondence and conversations with Peter Freyd, Martin Hyland, André Joyal, Michael Makkai, Andrej Ščedrov and Myles Tierney, among others. I have done my best to give original references for all the main ideas, but since many of them have not been published by their original authors I have found it an unusually difficult task; I trust I shall be forgiven for any accidental oversights. Finally, I must thank Ioan James and Erwin Kronheimer for the opportunity to contribute this article to Hugh Dowker's memory; I hope it will not be thought altogether unworthy.

1. Toposes, Theories and Sets

One reason why topos theory, viewed from the outside, appears a difficult subject is that the notion of topos has many different aspects, and that most of its spectacular applications can only be understood in terms of the interaction of two or more of these aspects. Thus a topos theorist has to acquire the skill of looking at the same idea simultaneously from several different angles, which is perhaps contrary to the mental habits of most mathematicians today. In this article we shall be concerned exclusively with Grothendieck toposes (i.e. with toposes defined and bounded over a fixed topos $\mathcal{S}$ of constant sets), and there are three particular aspects of the topos notion which will predominate; we devote this section to introducing and briefly describing each of them.

The first aspect, and the one that requires least comment, is the one which has already been mentioned in the introduction: the idea that a topos is a generalized topological space, or better a generalized locale. Thus we have the possibility of taking concepts from topology and applying them to toposes; the most important topological properties of spaces (global and local compactness, connectedness, etc.) all have their topos-theoretic analogues, and similarly the important properties of continuous maps have their analogues for geometric morphisms of toposes. The particular instances of this analogy which we shall need will be discussed in detail in the next section.

The second aspect is that a topos is the "embodiment" of a geometric theory. Recall that a formula (in an infinitary first-order language) is said to be <u>geometric</u> if it is built up from atomic formulae by the use of finite conjunction, arbitrary disjunction (but with the proviso that the number of free variables in any subformula remains finite), and existential quantification. A theory is said to be <u>geometric</u> if it can be axiomatized by sentences of the form $\forall \vec{x}\ (\varphi \Rightarrow \psi)$, where φ and ψ are geometric formulae and $\vec{x}$ is a finite string of variables including all those free in φ and/or ψ. (If only finite disjunctions are required, the theory is said to be <u>coherent</u>; if no disjunctions occur, it is called <u>regular</u>.) Given such a theory, we may study its models in any Grothendieck topos, and they are preserved by inverse image functors; we write $\mathbb{T}(\mathcal{E})$ for the category of $\mathbb{T}$-models in a particular topos $\mathcal{E}$ and homomorphisms between them.

The link between geometric theories and Grothendieck toposes is provided by

<u>Theorem 1.1</u>. (i) For any geometric theory $\mathbb{T}$, the pseudofunctor

$$\mathbb{T}(-): (\underline{\underline{\mathrm{BTop}}}/\mathcal{S})^{op} \to \underline{\underline{\mathrm{Cat}}}$$

is representable; i.e. there exists a topos $\mathcal{S}[\mathbb{T}]$, called the <u>classifying topos</u> of $\mathbb{T}$, and a particular (<u>generic</u>) $\mathbb{T}$-model $M_{\mathbb{T}}$ in $\mathcal{S}[\mathbb{T}]$, such that for each Grothendieck topos $\mathcal{E}$ the assignment $f \mapsto f^*(M_{\mathbb{T}})$ defines an equivalence of categories

$$\underline{\mathrm{BTop}}/\mathcal{S}\ (\mathcal{E}, \mathcal{S}[\mathbb{T}]) \to \mathbb{T}(\mathcal{E}) .$$

(ii) For any Grothendieck topos $\mathcal{E}$, there exists a geometric theory $\mathbb{T}$ such that $\mathcal{E} \simeq \mathcal{S}[\mathbb{T}]$.

<u>Proof</u>. See [12], 7.45 and 7.47, or [27], passim. □

Theorem 1.1 tells us that geometric theories are "essentially the same thing" as Grothendieck toposes; but one has to exercise a little care in interpreting this statement. The point is that a topos is a "presentation-independent" way of describing a theory, i.e. it does not have a distinguished set of "primitive" types, function-symbols and predicates from which others may be constructed by the usual rules of logic, but it treats all possible predicates on the same footing. (One could thus say that topos theory stands in the same relation to (the geometric fragment of) traditional model theory as Lawvere's algebraic theories [23] do to traditional universal algebra.) Now many of the important concepts of model theory are strongly presentation-dependent, and so one cannot expect them to correspond exactly to their topos-theoretic counterparts; this seems to be the principal reason why topos-theoretic model theory is as yet so little developed.

We shall return to this aspect of topos theory in section 3, where we shall study some particular examples of geometric theories which arise in the proofs of the open covering theorems of section 4.

The third aspect of toposes which we wish to mention, and perhaps the most confusing, arises from the idea that <u>every</u> topos is the category of sets. The use of the word "the" rather than "a" is not a misprint: of course, we do assume that there is a particular topos $\mathcal{S}$ of constant sets, which can be distinguished by its external properties (for example, the fact that its terminal object is a generator), but the point is that from the point of view of its own internal logic any topos $\mathcal{E}$ provides as good a set-theoretic foundation for mathematics as any other -- thus it is the topos of sets for that particular logic.

The particular consequence of this viewpoint which concerns us here is that there is a certain confusion between objects and morphisms in the category of Grothendieck toposes. On the one hand, a (bounded) geometric morphism $f: \mathcal{F} \to \mathcal{E}$ is nothing more than a Grothendieck topos when seen from the viewpoint that $\mathcal{E}$ is the topos of sets; thus any attribute which can be applied to Grothendieck toposes can also be applied to morphisms. On the other hand, a Grothendieck topos $\mathcal{E}$ is distinguished as such amongst elementary toposes by the existence of a (particular) bounded geometric morphism $\gamma: \mathcal{E} \to \mathcal{S}$ [12, 4.41]; thus any attribute of bounded morphisms also applies to Grothendieck toposes. This confusion will be evident in the next section, where we study various "topological" properties of geometric morphisms; our terminology for these properties is rather unsystematically borrowed sometimes from properties of spaces, and sometimes from properties of continuous maps.

The confusion also affects our view of toposes in relation to geometric theories. For example, if $\mathbb{T}'$ is an extension of a theory $\mathbb{T}$ (i.e. is obtained from $\mathbb{T}$ by adding further types, primitive symbols and axioms), then the natural forgetful functor $\mathbb{T}'(\mathcal{E}) \to \mathbb{T}(\mathcal{E})$ corresponds to a morphism of representing objects $\mathcal{S}[\mathbb{T}'] \to \mathcal{S}[\mathbb{T}]$. As explained above, we may view this morphism as a Grothendieck topos relative to $\mathcal{S}[\mathbb{T}]$; hence by Theorem 1.1 we may view the extension of theories as a single theory expressed in a logic where the terms, types, etc. definable in $\mathbb{T}$ have all become actual "constant" entities, constructed from the generic model $M_{\mathbb{T}}$. This point of view will be very important in sections 3 and 4.

2. Some classes of morphisms

In this section, our aim is to gather together a number of results about the "topological" aspect of geometric morphisms, all of which may be found in the existing literature. As explained in the last section, the terminology which we adopt for particular classes of morphisms is derived sometimes from existing terminology for spaces and sometimes from terminology for continuous maps. It will be convenient (though admittedly rather inelegant) to introduce the word "if¾f" meaning "if and almost only if"; when talking about continuous maps of spaces, we shall use "P if¾f Q" to mean "Q always implies P, and P implies Q provided the spaces involved satisfy certain separation conditions, which we do not bother to state precisely". (Details of the appropriate separation conditions in each case will be found in the references cited.)

We begin with the two most familiar classes of geometric morphisms:

(a) $f: \mathcal{F} \to \mathcal{E}$ is an <u>inclusion</u> if f_* is full and faithful.

(b) $f: \mathcal{F} \to \mathcal{E}$ is a <u>surjection</u> if f^* is faithful.

It is well known that every geometric morphism admits an essentially-unique factorization as a surjection followed by an inclusion [12, 4.14]; since a continuous map $f: X \to Y$ of spaces induces an inclusion (resp. a surjection) $Sh(X) \to Sh(Y)$ if¾f f is the inclusion of a subspace (resp. surjective) [12, 4.12(iv)], this factorization generalizes a familiar one for topological spaces. However, the factorization for geometric morphisms is less well-behaved than its topological counterpart: in particular, surjections are far from being epimorphic in BTop/$\mathcal{S}$ [12, Ex. 4.3], and the class of surjections is not stable under pullback [12, Ex. 7.1]. (The second of these problems, though not the first, is already present when we pass from spaces to locales.) It is therefore appropriate to consider various possible strengthenings of condition (b).

One obvious possibility is

(c) $f: \mathcal{F} \to \mathcal{E}$ is connected if f^* is full and faithful.

The reason for this name is that the unique morphism $Sh(X) \to \mathcal{S}$ is connected iff X is a connected space [12, Ex. 4.8]. More generally, $f: X \to Y$ induces a connected morphism $Sh(X) \to Sh(Y)$ iff f is a monotone quotient map [19, 1.12], i.e. a quotient map with connected fibres. However, the class of connected maps is still not stable under pullback in BTop/$\mathcal{S}$, and a further strengthening was introduced in [15]:

(d) $f: \mathcal{F} \to \mathcal{E}$ is hyperconnected if f^* is full and faithful, and its image is closed under subobjects (equivalently, under quotients) in $\mathcal{F}$.

The notion of hyperconnected map is linked with that of a localic map:

(e) $f: \mathcal{F} \to \mathcal{E}$ is localic if every object of $\mathcal{F}$ is a subquotient (= a subobject of a quotient, or equivalently a quotient of a subobject) of one in the image of f^*.

In [15] it was shown that every geometric morphism has an essentially-unique factorization as a hyperconnected map followed by a localic map; in particular, a map which is both hyperconnected and localic must be an equivalence. A morphism $\mathcal{E} \to \mathcal{S}$ is localic iff $\mathcal{E}$ is equivalent to the topos of sheaves on a locale [12, 5.37]; from this it follows easily that any morphism $Sh(X) \to Sh(Y)$ is localic.

The remaining conditions we wish to consider in this section are all "smoothness" conditions: the most restrictive of these is the notion of a local homeomorphism.

(f) $f\colon \mathcal{F} \to \mathcal{E}$ is a <u>local homeomorphism</u> if there exists an object X of $\mathcal{E}$ such that $\mathcal{F}$ is equivalent (as an $\mathcal{E}$-topos) to $\mathcal{E}/X$.

It follows easily from the well-known connection between sheaves and local homeomorphisms that a continuous map $f\colon X \to Y$ induces a local homeomorphism $Sh(X) \to Sh(Y)$ iff it is a local homeomorphism in the usual sense. An intrinsic characterization of local homeomorphisms is given in [12, 1.47], from which in particular we may deduce that they are atomic:

(g) $f\colon \mathcal{F} \to \mathcal{E}$ is <u>atomic</u> if f^* is a logical functor (i.e. preserves Ω and exponentials).

(The name "atomic" is due to M. Barr [2]; from our present point of view it is not particularly well chosen, since it fails to suggest smoothness, but it is now well established.)

It is clear that a logical inverse image functor $f^*\colon \mathcal{E} \to \mathcal{F}$ commutes up to isomorphism (in an obvious sense) with the functor $\prod_\alpha$ (the right adjoint of the pullback functor α^*) for any morphism α of $\mathcal{E}$. By a theorem essentially due to Mikkelsen (cf. [31, p. 123]), this is equivalent to saying that f^* has an $\mathcal{E}$-indexed left adjoint $f_!$, as well as its usual right adjoint f_*.

(h) $f\colon \mathcal{F} \to \mathcal{E}$ is <u>locally connected</u> if f^* has an $\mathcal{E}$-indexed left adjoint.

Once again, the name "locally connected" derives from the fact that $Sh(X) \to \mathcal{S}$ has this property iff X is a locally connected space [12, Ex. 0.7]. For a characterization of locally connected morphisms $Sh(X) \to Sh(Y)$, see [16, 2.6]. Locally connected morphisms were also studied by Barr and Paré [3], who called them "$\mathcal{E}$-essential" or "molecular".

As a further weakening of this condition, we may demand that

f^* commute not with the functors $\prod_\alpha$ but with their restrictions to subobjects, the universal quantification functors $\forall_\alpha$ [12, 5.29]. In [14], this was shown to be equivalent to demanding that the comparison map $f^*(\Omega_{\mathcal{E}}{}^X) \to \Omega_{\mathcal{F}}{}^{f^*X}$ should be mono for all objects X of $\mathcal{E}$, and that this condition holds for a morphism Sh(X) → Sh(Y) iff the corresponding continuous map X → Y is open. Accordingly,

(j) $f: \mathcal{F} \to \mathcal{E}$ is <u>open</u> if the comparison maps $f^*(\Omega_{\mathcal{E}}{}^X) \to \Omega_{\mathcal{F}}{}^{f^*X}$ are all monomorphisms.

Open maps were also studied extensively by Joyal and Tierney [22].

<u>Lemma 2.1</u>. The following implications hold between the concepts introduced above:

(i) Connected atomic => hyperconnected => connected => surjective.

(ii) Local homeomorphism <=> atomic and localic.

(iii) Atomic => locally connected => open.

(iv) Hyperconnected => open.

<u>Proof</u>. Most of these implications have been discussed above; we comment only on those which have not. For the first implication of (i), note that if f is connected and atomic then we have natural bijections

$$\hom(X, \Omega_{\mathcal{E}}) \cong \hom(f^*X, f^*\Omega_{\mathcal{E}}) \cong \hom(f^*X, \Omega_{\mathcal{F}})$$

which imply that every subobject of f^*X in $\mathcal{F}$ is in the image of f^*. For the right-to-left implication of (ii), note that every locally connected morphism f factors as one which is connected and locally connected followed by a local homeomorphism [16, 4.6]; if f is atomic and localic, then the first half of this factorization must be hyperconnected and localic, and hence an equivalence. (iv) is proved in [15, 1.6]. □

<u>Lemma 2.2</u>. (i) All nine classes of morphisms introduced in this section are stable under composition.

(ii) All the classes except (b) and (c) are stable under pullback.

(iii) The classes (b) and (c) are not stable under pullback, even along inclusions.

(iv) The class of open surjections, and the class of connected and locally connected maps, are stable under pullback.

Proof. (i) is trivial in every case from the definition given.

(ii): We cite appropriate references in each case: (a) [12, 4.47(b)]; (d) [15, 2.3]; (e) [15, 2.1]; (f) [12, 4.37(i)]; (g) [22, VII 4.1]; (h) [30, Thm. 26]; (j) [14, 4.7] or [22, VII 1.3].

(iii): As already mentioned, a counterexample for (b) is given in [12, Ex. 7.1]; we give a counterexample for (c). Let X be the closed unit interval $[0,1]$ with its usual topology, Y the same set with the topology whose nontrivial open sets are the intervals $(t,1]$ $(0 \leq t < 1)$, $f\colon X \to Y$ the identity map. f is continuous, and it is not hard to verify that the inverse image functor $f^*\colon \mathrm{Sh}(Y) \to \mathrm{Sh}(X)$ is full as well as faithful, i.e. $f\colon \mathrm{Sh}(X) \to \mathrm{Sh}(Y)$ is connected. Now let $S = \{0,1\}$ be the Sierpiński space (in which $\{1\}$ is open but $\{0\}$ is not), and $i\colon S \to Y$ the inclusion map. If we form the pullback

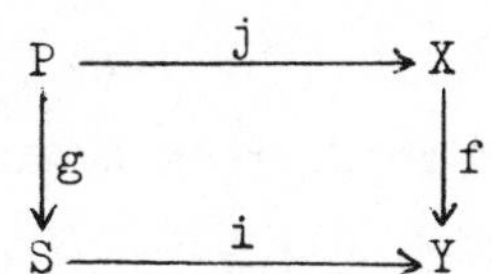

in the category of locales (equivalently, in $\mathbf{BTop}/\mathcal{S}$), it is not immediately apparent (though it is in fact true) that P is the discrete space $\{0,1\}$ (i.e. the pullback in the category of spaces); but we do not need to prove this in order to show that P is disconnected. For since the open sets $(\frac{1}{3},1]$ and $(\frac{1}{2},1]$ in Y are identified by the functor i^*, and since in X we have $[0,\frac{1}{2}) \cup (\frac{1}{3},1] = X$, $[0,\frac{1}{2}) \cap (\frac{1}{2},1] = \emptyset$, it is clear that $[0,\frac{1}{2})$ and $(\frac{1}{2},1]$ are mapped by j^* to complementary open subsets of P. Since S is connected and P is not, it follows from (i)(c) that $g\colon \mathrm{Sh}(P) \to \mathrm{Sh}(S)$ cannot be connected.

(iv): For open surjections, see [14, 4.7] or [22, VII 1.3]. For connected and locally connected maps, note that a locally connected $f: \mathcal{F} \to \mathcal{E}$ is connected iff $f_!(1) = 1$ [16, 4.5]; the fact that this condition is preserved under pullback follows easily from the fact that pullbacks of locally connected morphisms satisfy the Beck condition (cf. [16, 2.4]). □

If an $\mathcal{E}$-topos $f: \mathcal{F} \to \mathcal{E}$ is defined by an internal site (C,J) in $\mathcal{E}$ (i.e. an internal category with a Grothendieck topology), then it is useful to have conditions on (C,J) which ensure that f satisfies one or more of the conditions we have considered in this section. We conclude the section with three examples of such conditions.

<u>Lemma 2.3</u>. Let (C,J) be an internal site in a topos $\mathcal{E}$, and let $f: \mathcal{F} \to \mathcal{E}$ be the corresponding $\mathcal{E}$-topos. If every J-covering sieve (on each object of C) is inhabited, then f is open. If further C has a terminal object, then f is surjective.

<u>Proof</u>. This is proved in the case when C is a partial order in [22, V 3.2]; the general case does not present any extra difficulties. Note that objects of C all of whose covers are inhabited correspond to <u>$\mathcal{E}$-positive</u> objects of $\mathcal{F}$, i.e. objects Y such that, if Y is covered by an f^*X-indexed family of subobjects, then X must be inhabited. □

A very similar criterion applies for local connectedness. We shall say that a sieve R on an object c of C is <u>connected</u> if it is connected when regarded as a full subcategory of C/c, i.e. if any pair of arrows in R can be linked by a zigzag of the form

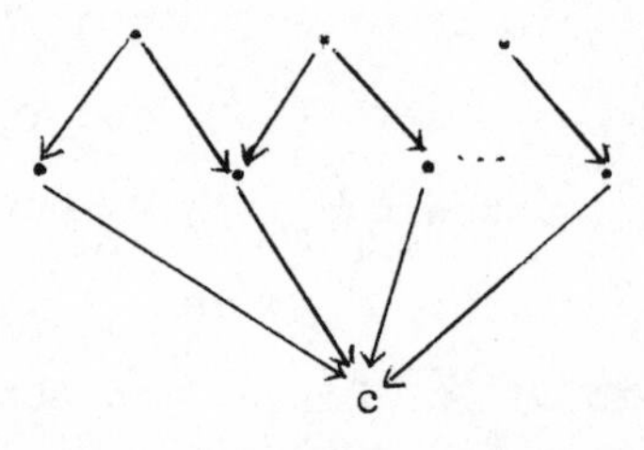

in which every arrow with codomain c is in R.

Lemma 2.4. With the same notation as Lemma 2.3, if every J-covering sieve is connected, then f is locally connected. If further C has a terminal object, then f is connected.

Proof. Essentially similar to 2.3; the key step is to observe that an object of C whose coverings are all connected gives rise to a connected object (as seen from $\mathcal{E}$) of $\mathcal{F}$. Hence the given condition implies that every object of $\mathcal{F}$ can be covered by connected objects, from which it follows easily that every object decomposes as an $\mathcal{E}$-indexed sum of "connected components". □

Finally, we give the Barr-Diaconescu characterization of "atomic sites":

Lemma 2.5. With the same notation as Lemma 2.3, suppose that

(i) every morphism of C is a regular epimorphism;

(ii) every pair of morphisms of C with common codomain can be completed to a commutative square;

(iii) every inhabited sieve is J-covering, and conversely.

Then f is atomic. If further C has a terminal object, then f is (hyper)connected.

Proof. See [2, Theorem A]. □

It should be mentioned that each of (2.3), (2.4) and (2.5) has a partial converse, asserting that every bounded $\mathcal{E}$-topos satisfying the given condition on f is definable by a site satisfying the given condition on (C,J) (although, of course, not every site of definition for it will satisfy this condition). However, we shall not need these converse results, since we shall be applying (2.3) - (2.5) to particular sites arising from particular presentations of geometric theories.

3. Examples of Geometric Theories

In this section we shall describe some particular geometric

theories which will play a role in the open covering theorems of the next section, and their classifying toposes. Once again, most of the results we shall need can be found in the existing literature. In presenting particular theories, we shall sometimes make use of primitive function-symbols as well as primitive predicates; but it is useful to bear in mind that the former can always be eliminated, by the device of replacing an n-ary function-symbol by an (n+1)-ary predicate together with (geometric) axioms which say that (the extension of) the predicate is the graph of a function of its first n variables. Thus in our discussion of general constructions which we can perform on theories, we shall feel free to assume that they are presented using predicates alone.

We begin with a very simple example, the theory $\mathbb{O}$ of objects. This has a single primitive type (or sort), no predicates (other than equality), and no axioms; so for any topos $\mathcal{E}$ we have $\mathbb{O}(\mathcal{E}) \cong \mathcal{E}$. Its classifying topos $\mathcal{S}[\mathbb{O}]$ may be taken to be the functor category $[S_f, \mathcal{S}]$ where S_f is (a skeleton of) the category of finite sets and functions; the generic object is the inclusion functor $S_f \to \mathcal{S}$ (cf. [12, 6.33] or [21, 4.1]).

Slightly less trivially, the theory $\mathbb{D}$ of decidable objects has a binary predicate # and axioms

$$(\forall x,y)(\top \Rightarrow x = y \vee x \# y)$$

$$(\forall x)(x \# x \Rightarrow \bot)$$

where $\top$ and $\bot$ denote respectively the empty conjunction and the empty disjunction. For a topos $\mathcal{E}$, $\mathbb{D}(\mathcal{E})$ is the category of <u>decidable</u> objects of $\mathcal{E}$ (i.e. objects X such that the diagonal $X \to X \times X$ is a complemented subobject of $X \times X$) and monomorphisms between them. Once again, $\mathcal{S}[\mathbb{D}]$ may be taken to be a functor category, namely $[S_{fm}, \mathcal{S}]$ where S_{fm} is the category of finite sets and injections between them. (See [21, 4.12]; the reason why $\mathcal{S}[\mathbb{D}]$ is a presheaf topos is essentially that $\mathbb{D}$ is a <u>disjunctive theory</u> in the sense of [13].)

The theory $\mathbb{D}_\infty$ of infinite decidable objects is obtained by adding to $\mathbb{D}$ the axioms

$$\top \Rightarrow (\exists x)(x = x)$$

and for each $n > 0$

$$(\forall x_1,\ldots,x_n)(\top \Rightarrow (\exists x_{n+1})(\bigwedge_{i=1}^{n} x_{n+1} \# x_i)) .$$

Its classifying topos may be described as $Sh(S_{fm}{}^{op},J)$, where the topology J consists of all inhabited sieves on objects of $S_{fm}{}^{op}$; note that this site satisfies the hypotheses of Lemma 2.5, so that $\mathcal{S}[\mathbb{D}_\infty]$ is connected and atomic. As has been observed by Blass and Ščedrov [5], the reason for this is essentially that $\mathbb{D}_\infty$ is complete and $\aleph_0$-categorical. It follows from the work of Makkai [26, 3.3] and of Joyal and Tierney [22, VIII 3.1] that $\mathcal{S}[\mathbb{D}_\infty]$ has an alternative description as the topos $\mathcal{C}(G_0)$ of (discrete) sets equipped with a continuous action of the topological group G_0 of all automorphisms of the countable model of $\mathbb{D}_\infty$ (i.e., in this case, the group of all permutations of the natural numbers, topologized as a subspace of Baire space). Direct proofs of the equivalence of these two descriptions have been given by Ščedrov [34, 6.1.3] and Johnstone [18, 1.8].

We shall also be concerned with the theory $\mathbb{L}$ of (trichotomous) linearly ordered objects and $\mathbb{L}_\infty$ of dense linear orders without endpoints: for $\mathbb{L}$ we take a binary predicate $<$ and the axioms

$$(\forall x,y)(\top \Rightarrow x = y \vee x < y \vee y < x)$$

and

$$(\forall x,y)(x < y \wedge y < x \Rightarrow \bot),$$

and for $\mathbb{L}_\infty$ we add the axioms

$$(\forall x)(\top \Rightarrow (\exists y)(x < y) \wedge (\exists z)(z < x))$$

and

$$(\forall x,y)(x < y \Rightarrow (\exists z)(x < z \wedge z < y)) .$$

The classifying toposes of these theories are respectively

$$\mathcal{S}[\mathbb{L}] \simeq [\Delta_m,\mathcal{S}] \quad \text{and} \quad \mathcal{S}[\mathbb{L}_\infty] \simeq Sh(\Delta_m{}^{op},J) \simeq \mathcal{C}(G_1) ,$$

where Δ_m is the category of finite ordinals and order-preserving injections between them, J is the topology on Δ_m^{op} in which every inhabited sieve covers, and G_1 is the topological group of order-preserving permutations of the rationals. The proofs are similar to those for $\mathbb{D}$ and $\mathbb{D}_\infty$; some details may be found in [34, 6.2].

So far, we have considered only single-sorted theories; but Theorem 1.1 allows us to consider also theories having several primitive sorts. For example, the theory $(\mathbb{D} \twoheadrightarrow \mathbb{O})$ of objects with a decidable cover has two sorts X and Y, a function-symbol f: $X \to Y$ and a binary predicate # on X, such that # satisfies the axioms for $\mathbb{D}$ and f satisfies

$$(\forall y \in Y)(\top \Rightarrow (\exists x \in X)(f(x) = y)) .$$

Note, however, that $(\mathbb{D} \twoheadrightarrow \mathbb{O})$ is equivalent to a single-sorted theory $(\mathbb{D}/\equiv)$ (in the sense that it has an equivalent category of models in any topos, and hence an equivalent classifying topos), where $(\mathbb{D}/\equiv)$ is obtained from $\mathbb{D}$ by adjoining a new binary predicate $\equiv$ and axioms which say that $\equiv$ is an equivalence relation.

We shall say that two theories $\mathbb{T}$ and $\mathbb{T}'$ are Morita-equivalent if they have equivalent classifying toposes. The phenomenon which we have just observed for $(\mathbb{D} \twoheadrightarrow \mathbb{O})$ is part of a wider pattern:

Proposition 3.1. Any geometric theory is Morita-equivalent to a single-sorted one.

Proof. Let $\mathbb{T}$ be a geometric theory, and suppose its primitive sorts are $(X_i \mid i \in I)$. Our new theory $\mathbb{T}'$ will have a single sort X and a family of unary predicates $(\varphi_i \mid i \in I)$, together with the axioms

$$(\forall x)(\top \Rightarrow \bigvee_{i \in I} \varphi_i(x))$$

and $$(\forall x)(\varphi_i(x) \wedge \varphi_j(x) \Rightarrow \bot)$$

whenever $i \neq j$, so that X is the coproduct of the interpretations of the φ_i. Each primitive predicate of $\mathbb{T}$ is replaced by a new predicate with appropriate axioms; for example, a binary $\psi \subseteq X_i \times X_j$ is replaced by a binary predicate $\bar{\psi}$ with the axiom

$$(\forall x,y)(\bar{\psi}(x,y) \Rightarrow \varphi_i(x) \wedge \varphi_j(y)) .$$

The remaining details are straightforward. □

What does this result mean in topos-theoretic terms? Let us call a theory $\mathbb{T}'$ a <u>propositional</u> <u>extension</u> of $\mathbb{T}$ if it is obtained by adding new predicates and axioms, but no new types. The reason for this terminology is that a propositional theory in the classical sense is just a propositional extension of the empty theory (if we have no types, then the only predicates we can add are nullary ones, i.e. propositions). And a propositional extension $\mathbb{T}'$ of $\mathbb{T}$ <u>is</u> just a propositional theory as seen from the classifying topos $\mathcal{S}[\mathbb{T}]$, in the manner explained at the end of section 1; it is the theory of "$\mathbb{T}'$-enrichments" of a particular $\mathbb{T}$-model, namely $M_{\mathbb{T}}$.

Now it is well known that the classifying topos of a propositional geometric theory is localic over $\mathcal{S}$ (it is the topos of sheaves on the "Lindenbaum algebra" of equivalence classes of geometric propositions), and similarly a propositional extension $\mathbb{T}'$ of $\mathbb{T}$ induces a localic morphism $\mathcal{S}[\mathbb{T}'] \to \mathcal{S}[\mathbb{T}]$. Thus Proposition 3.1, which says that any geometric theory is equivalent to a propositional extension of $\mathbb{O}$, immediately implies

<u>Corollary 3.2</u>. For any Grothendieck topos $\mathcal{E}$, there exists a localic geometric morphism $\mathcal{E} \to \mathcal{S}[\mathbb{O}]$. □

Corollary 3.2 was first proved by Joyal and Tierney [22, VII 3.1]; a slightly different proof was given by Johnstone [18, 1.7].

<u>Remark 3.3</u>. In the next section we shall in fact require a slight strengthening of Corollary 3.2, namely that every Grothendieck topos $\mathcal{E}$ admits a localic morphism to $\mathcal{S}[\mathbb{O}_1]$, where $\mathbb{O}_1$ is the theory of inhabited objects. We can achieve this by a slight modification of the construction in the proof of (3.1), to ensure that the base sort

$\mathcal{K}$ of $\mathbb{T}'$ is inhabited. To do this, we add a further unary predicate φ_0 (where $0 \notin I$) to our family $(\varphi_i \mid i \in I)$, and the axioms

$$\top \Rightarrow (\exists x)(\varphi_0(x))$$

and
$$(\forall x,y)(\varphi_0(x) \wedge \varphi_0(y) \Rightarrow x = y)$$

which say that the interpretation of φ_0 is a singleton (which plays no other part in the theory $\mathbb{T}'$).

4. The Open Covering Theorems

In this section we shall prove our two main theorems, both of which have the following general form:

"Let $\mathcal{E}$ be an arbitrary Grothendieck topos. Then there exists a geometric morphism $f: \mathcal{F} \to \mathcal{E}$ having property P, such that the topos $\mathcal{F}$ (or equivalently, the morphism $\mathcal{F} \to \mathcal{S}$) has property Q."

Before we go on to discuss and justify our choice of the particular properties P and Q which will appear in the theorems, let us pause to observe that, for suitable P and Q, Corollary 3.2 enables us to reduce the problem of proving such a theorem for an arbitrary $\mathcal{E}$ to that of proving it for the particular topos $\mathcal{S}[\mathbb{O}]$. For if the property P is stable under pullback, and Q is stable under composition with localic morphisms, then given a solution $f_0: \mathcal{F}_0 \to \mathcal{S}[\mathbb{O}]$ of the problem for $\mathcal{S}[\mathbb{O}]$ we may form the pullback

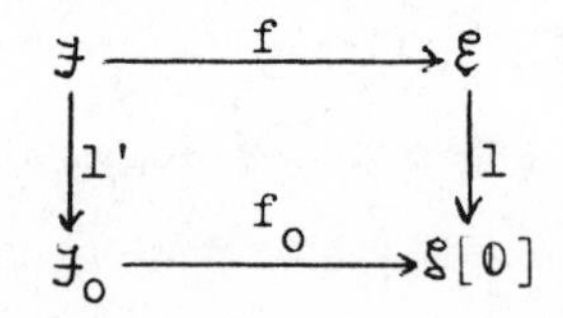

where l is a localic morphism as in Corollary 3.2; then l' is localic and f has property P, so we have a solution of the problem for $\mathcal{E}$.

For the first of our theorems, we take property Q to be "localic" and ask how strong a restriction we can impose on P. The first result of this kind was proved by Barr [1], who showed that one can always find a surjection $f: \mathcal{F} \to \mathcal{E}$ with $\mathcal{F}$ localic over $\mathcal{S}$; later Diaconescu [7] gave a simpler proof of Barr's result. In an unpublished analysis of Diaconescu's proof, M. Coste [6] observed that the reason why it works is that the surjection Diaconescu constructs is in fact open, although the concept of "open map" did not appear explicitly in Coste's work. The first _explicit_ proof that f may be taken to be an open surjection was given by Joyal and Tierney [22, VII 3.1]; subsequently Joyal improved this by showing that one can actually take f to be connected and locally connected. This is the result which we shall prove below.

Before doing so, however, let us note that "connected and locally connected" is actually the best possible choice for P (at least among the properties considered in section 2), given that Q is "localic". In the first place, we cannot improve "connected" to "hyperconnected"; for since $\mathcal{F} \to \mathcal{S}$ is localic its first factor $\mathcal{F} \to \mathcal{E}$ must also be localic, and hence if it were hyperconnected it would be an equivalence, i.e. $\mathcal{E}$ itself would be localic over $\mathcal{S}$. Similarly, we cannot improve "locally connected" to "atomic", even if we weaken "connected" to "surjective"; for if f were atomic and localic it would be a local homeomorphism by Lemma 2.1(ii), and hence $\mathcal{E}$ would be an étendue in the sense of [25] (which differs from [11] and [12] only in not requiring that an étendue should have enough points). Since there are plenty of toposes which are not étendues (and _a fortiori_ not localic over $\mathcal{S}$), this is clearly impossible.

With these preliminaries out of the way, we are ready to prove

Theorem 4.1. For any Grothendieck topos $\mathcal{E}$, there exists a connected and locally connected map $f: \mathcal{F} \to \mathcal{E}$ such that $\mathcal{F}$ is localic over $\mathcal{S}$.

Proof. As observed at the beginning of this section, it is sufficient to prove the result in the case $\mathcal{E} = \mathcal{S}[\mathbb{O}]$. To do this, we introduce a geometric theory ($\mathbb{N} \twoheadrightarrow_{\infty} \mathbb{O}$) of "objects with an infinite-to-one partial enumeration", as follows: ($\mathbb{N} \twoheadrightarrow_{\infty} \mathbb{O}$) has one sort X, and a unary predicate "$q(n) = x$" for each natural number n. The axioms are

$$(\forall x,y)((q(n) = x) \wedge (q(n) = y) \Rightarrow x = y)$$

for each n, and

$$(\forall x)(\top \Rightarrow \bigvee_{n \notin F} (q(n) = x))$$

for each finite subset $F \subseteq \mathbb{N}$.

We note first that, in the same way that the theory ($\mathbb{D} \twoheadrightarrow \mathbb{O}$) is equivalent to a single-sorted theory, ($\mathbb{N} \twoheadrightarrow_{\infty} \mathbb{O}$) is equivalent to a propositional theory generated by propositions of the form "$q(n) = q(m)$" ($m,n \in \mathbb{N}$), with axioms which say that this relation on $\mathbb{N}$ is symmetric and transitive (though not necessarily reflexive) and

$$\bigvee_{m \in \mathbb{N}} (q(n) = q(m)) \Rightarrow \bigvee_{m \notin F} (q(n) = q(m))$$

for each n and each finite $F \subseteq \mathbb{N}$. Thus its classifying topos $\mathcal{F}_0 = \mathcal{S}[\mathbb{N} \twoheadrightarrow_{\infty} \mathbb{O}]$ is localic over $\mathcal{S}$.

However, if viewed as a topos over $\mathcal{S}[\mathbb{O}]$, $\mathcal{F}_0$ becomes the classifying topos for the theory of infinite-to-one partial enumerations of a <u>particular</u> object, namely the generic object (X, say). It is straightforward to verify that a site of definition for this classifying topos may be constructed as follows: take the poset M whose objects are diagrams of the form ($q: F \to X$) (F a finite subset of $\mathbb{N}$), ordered by $(q_1: F_1 \to X) \leqslant (q_2: F_2 \to X)$ iff $F_1 \supseteq F_2$ and $q_2 = q_1|_{F_2}$, and impose the topology whose covering sieves on $q: F \to X$ are the sets

$$R(q,K) = \{q': F' \to X \mid K \subseteq \mathrm{im}(q'|_{F'-F})\}$$

where K runs over all finite subsets of X. It is clear that M has a terminal object, namely the unique map $0 \to X$; so by Lemma 2.4 we have to verify that each $R(q,K)$ is connected as a sub-poset of M/q.

Let $(q_1: F_1 \to X)$ and $(q_2: F_2 \to X)$ be two elements of $R(q,K)$.

We may choose an epimorphism $r: G \twoheadrightarrow K$ where G is a finite cardinal (cf. [12, 9.20]), and then choose an embedding $G \rightarrowtail N$ whose image is disjoint from $F_1 \cup F_2$. We may then extend q_1 and q_2 to maps defined on $F_1 \cup G$ and $F_2 \cup G$ respectively, and so construct a diagram

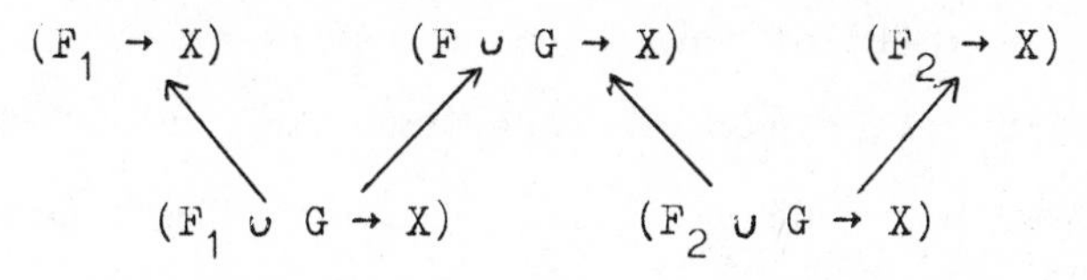

connecting q_1 and q_2 in $R(q,K)$. So the result is proved. □

The reader may have wondered why we used partial rather than global enumerations in defining $\mathcal{F}_0$, i.e. why we did not add the axioms

$$\top \Rightarrow (\exists x)(q(n) = x)$$

for each $n \in N$ to our theory. Apart from the trivial objection that the classifying topos of this theory does not even map surjectively to $\mathcal{S}[\mathbb{O}]$ (since the empty set **has** no global enumeration), which could have been avoided by working with $\mathcal{S}[\mathbb{O}_1]$ rather than $\mathcal{S}[\mathbb{O}]$, the reason is that the additional coverings in our site M which arise from these axioms are not connected -- and in fact the locale which they define is zero-dimensional. (However, it is still true -- provided we replace $\mathbb{O}$ by $\mathbb{O}_1$ -- that every covering sieve is inhabited, so we do get an open surjection; and it was this surjection which was used by Joyal and Tierney in their proof of the weaker form of Theorem 4.1 to which we alluded earlier.)

The second open covering theorem was first proved by P. Freyd [10], and imposes a stronger condition (atomicity) on the map $\mathcal{F} \to \mathcal{E}$ at the expense of a weaker condition on $\mathcal{F} \to \mathcal{S}$. Freyd began from a point of view rather different from our present one; we must now digress to explain it.

In set theory, there are two principal methods employed for constructing new models of set theory so as to obtain independence results: the method of permutation models (Fraenkel-Mostowski models) and that of forcing (or equivalently of Boolean-valued extensions). In topos-theoretic terms, these correspond (roughly) to the passage from $\mathcal{S}$ to the topos $\mathcal{C}(G)$ of continuous actions of a topological group G, and to the topos Sh(B) of sheaves on a complete Boolean algebra. To a topos-theorist, it is natural to extend the latter construction to that of sheaves on a locale (i.e. to consider "Heyting-valued extensions" rather than Boolean-valued ones). Freyd then posed the question: can every Grothendieck topos be obtained from $\mathcal{S}$ by a combination of these two constructions?

The answer is "no", but Freyd showed that the difference is not very great: in fact any Grothendieck topos $\mathcal{E}$ is equivalent to an exponential variety (i.e. a full subcategory closed under arbitrary limits, colimits and power-objects) in a topos $\mathcal{F}$ which is localic over some $\mathcal{C}(G)$. Moreover, it is not necessary to consider arbitrary topological groups G; for any $\mathcal{E}$, we may take G to be either the group G_0 of all permutations of N, or the group G_1 of order-preserving permutations of Q, which were mentioned in the last section in connection with the theories $\mathbb{D}_\infty$ and $\mathbb{L}_\infty$. (It is doubtless significant that G_0 is (essentially) the group used by Fraenkel [9] in his original proof of the independence of the axiom of choice, and G_1 was used by Mostowski [29] to show the independence of AC from the ordering principle.)

Now if $\mathcal{E}$ is an exponential variety in a topos $\mathcal{F}$, then the inclusion $\mathcal{E} \to \mathcal{F}$ is logical and has adjoints on both sides; thus it is the inverse image of a connected atomic morphism $\mathcal{F} \to \mathcal{E}$. The converse implication is also clearly true; thus Freyd's result may be stated in the following form, which conforms to that given at the beginning of this section:

<u>Theorem 4.2</u>. (i) For any Grothendieck topos $\mathcal{E}$, there exists a connected atomic map $f\colon \mathcal{F} \to \mathcal{E}$ where $\mathcal{F}$ is localic over $\mathcal{G}(G_0)$.

(ii) For any $\mathcal{E}$, there exists a connected atomic map $f'\colon \mathcal{F}' \to \mathcal{E}$ where $\mathcal{F}'$ is localic over $\mathcal{G}(G_1)$.

Freyd's proof of this result [10] was based on an explicit calculation of sites of definition for $\mathcal{E}$ and $\mathcal{F}$; but shortly afterwards Joyal provided a proof using categorical model theory, which at first sight looks very different from Freyd's but is in fact little more than a translation of it into more conceptual terms. It is this proof which we now give, decorated with a few additional observations of our own.

<u>Proof of Theorem 4.2</u>. (i) By Remark 3.3 and the argument at the beginning of this section, it suffices to prove the result for the particular topos $\mathcal{S}[\mathbb{O}_1]$. To do this, we construct a theory $(\mathbb{D} \twoheadrightarrow_\infty \mathbb{O}_1)$ as follows: the theory has two sorts X and Y, a binary predicate # on X and a function-symbol $q\colon X \to Y$. The axioms are as follows:

(a) # satisfies the axioms for $\mathbb{D}$;

(b) $$\top \Rightarrow (\exists y \in Y)(y = y) \ ;$$

(c) for each natural number n, the axiom
$$(\forall x_1, \ldots, x_n, y)(\top \Rightarrow (\exists x \in X)(q(x) = y \wedge \bigwedge_{i=1}^{n} x \# x_i)).$$

A model of $(\mathbb{D} \twoheadrightarrow_\infty \mathbb{O}_1)$ is thus an epimorphism $X \to Y$ whose domain is decidable, whose codomain is inhabited, and whose fibres are all infinite; in particular the domain itself must be infinite, so we can regard $(\mathbb{D} \twoheadrightarrow_\infty \mathbb{O}_1)$ as an extension of $\mathbb{D}_\infty$ as well as of $\mathbb{O}_1$. Moreover, it is (equivalent to) a propositional extension of $\mathbb{D}_\infty$, since as in our previous examples we may replace the type Y and the function-symbol q by a binary predicate $\equiv$ on X satisfying suitable axioms. It thus remains to prove that $\mathcal{S}[\mathbb{D} \twoheadrightarrow_\infty \mathbb{O}_1] \to \mathcal{S}[\mathbb{O}_1]$ is connected and atomic, which we do by constructing a site of definition for it and appealing to Lemma 2.5.

From the point of view of $\mathcal{S}[\mathbb{O}_1]$, $(\mathbb{D} \twoheadrightarrow_{\infty} \mathbb{O}_1)$ becomes the theory of infinite-to-one decidable coverings of a particular (inhabited) object Y. By analogy with the theory $\mathbb{D}_{\infty}$ (which may be regarded as the special case Y = 1), its classifying topos may be obtained as follows: let C be the category whose objects are maps f: $[m] \to Y$ from finite cardinals to Y, and whose morphisms from f: $[m] \to Y$ to g: $[n] \to Y$ are commutative triangles

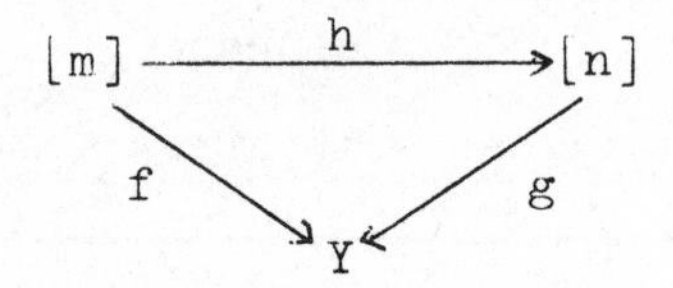

with h injective. Then impose on C^{op} the topology in which every inhabited sieve covers; and just as in the case Y = 1 it is straightforward to verify that the conditions of Lemma 2.5 are satisfied. So the result is established.

To prove part (ii), we proceed in exactly the same way, except that we use the theory $(\mathbb{L} \twoheadrightarrow_{\infty} \mathbb{O}_1)$ whose models consist of epimorphisms q: $X \to Y$ such that X is linearly ordered, Y is inhabited, and each fibre of q is dense in the linear order on X. □

Before leaving this section, we make a few remarks about the classes of toposes for which the coverings f and f' of Theorem 4.2 may be reduced to equivalences. For the situation of Theorem 4.2(i), this was investigated in [18], in which we showed

<u>Proposition 4.3</u>. The following conditions on a topos $\mathcal{E}$ are equivalent:

(i) $\mathcal{E}$ is localic over $\mathcal{B}(G_0)$.

(ii) $\mathcal{E}$ is localic over some topos of the form $\mathcal{B}(G)$.

(iii) $\mathcal{E}$ is localic over some Boolean topos.

(iv) Every object of $\mathcal{E}$ is a quotient of a decidable object.

Sketch of proof. The implications (i) => (ii) => (iii) are trivial. (iii) => (iv) follows from the definition of a localic morphism, since decidability is preserved by inverse image functors and inherited by subobjects. The key to proving (iv) => (i) is to show that (iv) allows us to express $\mathcal{E}$ as the classifying topos of a propositional extension of $\mathbb{D}_{\alpha}$. The argument is not dissimilar to that used in proving Proposition 3.1; we refer the reader to [18] for the details. □

Note that Proposition 4.3 (specifically, the implication (ii) => (iii) therein) does require the assumption that our base topos $\mathcal{S}$ is Boolean (an assumption which we have not used hitherto). To obtain a similar characterization of the toposes which are localic over $\mathcal{B}(G_1)$, we require a further assumption on $\mathcal{S}$:

Proposition 4.4. Assume that the Ordering Principle holds in $\mathcal{S}$, i.e. that every set can be linearly ordered. Then the following conditions on a Grothendieck topos $\mathcal{E}$ are equivalent:

(i) $\mathcal{E}$ is localic over $\mathcal{B}(G_1)$.

(ii) $\mathcal{E}$ is localic over a topos in which every object can be linearly ordered.

(iii) Every object of $\mathcal{E}$ is a quotient of a linearly orderable object.

Sketch of proof. The proofs of (ii) => (iii) => (i) are similar to those of (iii) => (iv) => (i) in the last Proposition. For (i) => (ii) we have to show that every object of $\mathcal{B}(G_1)$ can be linearly ordered. Since $\mathcal{B}(G_1)$ is an atomic topos, each object of it admits a canonical decomposition

$$X \cong \coprod_{i \in I} X_i$$

as a (set-indexed) coproduct of atoms; since by assumption we can linearly order the indexing set I, it suffices to show that we can choose linear orderings for each of the atoms X_i. But every atom in

$\mathcal{E}(G_1)$ is embeddable as a subobject of some finite power M^n, where M is (the underlying object of) the generic model of $\mathbb{L}_\infty$, and so inherits a (lexicographic) linear ordering from M^n. Moreover, each atom has only finitely many distinct linear orderings (in fact the nth atom has $2^n.n!$ such orderings), and so to choose orderings for each of the X_i we need only the axiom of choice for families of finite sets -- which is well known to follow from the Ordering Principle. (For more information about linearly ordered objects in $\mathcal{E}(G_1)$, see [20].) □

Note that every Boolean topos satisfies the conditions of Proposition 4.3, but not every Boolean topos satisfies those of Proposition 4.4; for example, in $\mathcal{E}(G_0)$ itself only the constant objects $\gamma^*(S)$ (where γ: $\mathcal{E}(G_0) \to \mathcal{S}$) can be linearly ordered, and any quotient of a constant object is again constant. Moreover, the conjunction of (4.4) and Booleanness does not imply that every object can be linearly ordered; the following counterexample is closely related to recent work of G.P. Monro [28] and J.L. Bell [4].

Example 4.5. Let ($\mathbb{L}_\infty \twoheadrightarrow \mathbb{D}_\infty$) be the theory whose models are pairs (q: $X \twoheadrightarrow Y$, $<$) such that Y is an infinite decidable object, $<$ is a linear order on X and each fibre of q is dense in this linear order. It is easy to see that ($\mathbb{L}_\infty \twoheadrightarrow \mathbb{D}_\infty$) is complete and satisfies the other hypotheses of [5, Corollary 1]; hence its classifying topos is of the form $\mathcal{E}(G_2)$, where G_2 is the topological group of automorphisms of the countable model ($X \twoheadrightarrow Y$) of ($\mathbb{L}_\infty \twoheadrightarrow \mathbb{D}_\infty$). Also, $\mathcal{E}(G_2)$ is localic over $\mathcal{E}(G_1)$, since ($\mathbb{L}_\infty \twoheadrightarrow \mathbb{D}_\infty$) may be presented as a propositional extension of $\mathbb{L}_\infty$. However, it is easy to see that (each open subgroup of) G_2 acts at least 2-transitively on (a cofinite subset of) Y, and hence that Y cannot (even internally) be linearly ordered when we regard it as an object of $\mathcal{E}(G_2)$ in the obvious way.

Finally, let us comment on a theorem of Rosenthal [33, 3.1] which is in a sense intermediate between our Theorem 4.2 and (a weak version of) Theorem 4.1. Recall that a topos $\mathcal{F}$ is called an <u>étendue</u> [32] if there is a surjective local homeomorphism $\mathcal{F}/X \to \mathcal{F}$ such that $\mathcal{F}/X$ is localic (over $\mathcal{S}$). Rosenthal's result is:

<u>Theorem 4.6</u>. For any Grothendieck topos $\mathcal{E}$, there exists a hyperconnected morphism $f: \mathcal{F} \to \mathcal{E}$ where $\mathcal{F}$ is an étendue.

It is easy to see that any topos which is localic over an étendue is itself an étendue, and so this theorem falls within the general scheme which we outlined at the beginning of this section. Rosenthal's proof involves analysing the particular open surjection $f: \mathcal{F} \to \mathcal{E}$ with $\mathcal{F}$ localic which was constructed by Diaconescu [7], and showing that it can be factored as a surjective local homeomorphism followed by a hyperconnected map. As stated above, (4.6) does not lie strictly between (4.1) and (4.2), since an étendue need not satisfy the conditions of (4.3). However, the particular étendue constructed by Rosenthal does have this property, and even satisfies the conditions of (4.4); in the case $\mathcal{E} = \mathcal{S}[\mathbb{O}]$, it is the functor category $[C, \mathcal{S}]$ where C is the free category on the underlying directed graph of S_f.

It appears that the connected and locally connected $f: \mathcal{F} \to \mathcal{E}$ which we constructed in proving Theorem 4.1 does not admit a factorization similar to that of Diaconescu's surjection. Nevertheless, it would be interesting to know whether the word "hyperconnected" in the statement of (4.6) can be strengthened to "hyperconnected and locally connected" (Rosenthal's proof does not yield this, since Diaconescu's surjection fails to be locally connected).

5. The Representation Theorems

As we indicated in the Introduction, one of the key applications of the open covering theorems (4.1) and (4.2) is to prove representation theorems which say that every Grothendieck topos can be built up in some way from the concepts of "locale" and "group": that, in effect, topos cohomology is a minimal common generalization of the (sheaf) cohomology of spaces and of the (Galois) cohomology of groups. Although the detailed proofs of these theorems would take too long, and involve too many new concepts, to be included here, it seems appropriate to conclude the paper with a section in which we sketch the main ideas of the proofs.

The precursor of the first theorem is a result of Grothendieck and Verdier [11, IV 9.8.2(e)] covering the special case of étendues. They observed that if $\mathcal{E}$ is a topos and X an object of $\mathcal{E}$ such that $\mathcal{E}/X$ is localic (over $\mathcal{S}$), then $\mathcal{E}/X \times X$ is also localic, and the two projections $X \times X \rightrightarrows X$ in $\mathcal{E}$ induce local homeomorphisms $\mathcal{E}/X \times X \rightrightarrows \mathcal{E}/X$ which give $\mathcal{E}/X$ the structure of a groupoid (called a "pre-equivalence relation" in [11]) in the category of locales. Conversely, given any groupoid $\underset{\sim}{G} = (G_1 \rightrightarrows G_0)$ in **Loc** for which the domain and codomain maps $G_1 \to G_0$ are local homeomorphisms, we may construct a topos $\mathcal{G}(\underset{\sim}{G})$ of sheaves on G_0 equipped with an action of G_1, in much the same way that toposes of internal presheaves are constructed in [12, Chapter 2]. And if X has global support in $\mathcal{E}$ (so that $\mathcal{E}/X \to \mathcal{E}$ is surjective) then the topos which arises in this way from $(\mathcal{E}/X \times X \rightrightarrows \mathcal{E}/X)$ is canonically equivalent to $\mathcal{E}$.

This result was extended by Joyal and Tierney in [22, VIII 3.2]. Let $f\colon \mathcal{F} \to \mathcal{E}$ be a geometric morphism, and form the pullback

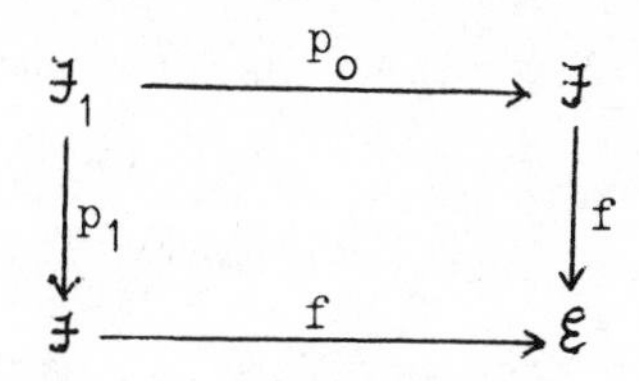

The diagonal map $\mathcal{F} \to \mathcal{F}_1$, the "twist" map $\mathcal{F}_1 \to \mathcal{F}_1$ and the three maps from the triple pullback $\mathcal{F}_2 = \mathcal{F} \times_{\mathcal{E}} \mathcal{F} \times_{\mathcal{E}} \mathcal{F}$ to $\mathcal{F}_1$ give $(\mathcal{F}_1 \rightrightarrows \mathcal{F})$ the structure of a groupoid (in the 2-categorical sense, i.e. "up to canonical coherent isomorphism") in $\underline{\text{BTop}}/\mathcal{S}$. Also, if $\mathcal{F}$ is localic over $\mathcal{S}$ then f is necessarily localic; hence so are p_0 and p_1 and $\mathcal{F}_1$ is also localic over $\mathcal{S}$, i.e. the groupoid may be considered as lying in $\underset{\sim}{\text{Loc}}$.

In the converse direction, let $\underset{\sim}{G} = (G_1 \rightrightarrows G_0)$ be any groupoid in $\underset{\sim}{\text{Loc}}$. If $(E \to G_0)$ is any locale over G_0, an action of G on E is a map (necessarily an isomorphism) $p_0^*(E) \to p_1^*(E)$ over G_1 which satisfies the appropriate conditions of compatibility with the structure of $\underset{\sim}{G}$. We write $\mathcal{B}(\underset{\sim}{G})$ for the category of local homeomorphisms over G_0 (i.e. sheaves on G_0) equipped with such action maps, and morphisms which respect the action. For a general $\underset{\sim}{G}$, it is not clear whether $\mathcal{B}(\underset{\sim}{G})$ is a topos; but if the "Beck condition" holds for the pullback square

$$\begin{array}{ccc} G_2 & \xrightarrow{d_2} & G_1 \\ {\scriptstyle d_0}\downarrow & & \downarrow{\scriptstyle p_0} \\ G_1 & \xrightarrow{p_1} & G_0 \end{array}$$

(i.e. the canonical natural transformation $p_0^* p_{1*} \to d_{2*} d_0^*$ of functors $\text{Sh}(G_1) \to \text{Sh}(G_1)$ is an isomorphism), then it may be shown that $\mathcal{B}(\underset{\sim}{G})$ is comonadic over $\text{Sh}(G_0)$, and hence a topos by [12, 2.32]. In particular, this condition is satisfied when p_0 and p_1 are locally connected. (In another instance, when G_0 is a one-point space and G_1 is a topological group, the definition of $\mathcal{B}(\underset{\sim}{G})$ reduces to that of the topos of continuous G_1-sets which we considered earlier -- hence our re-use of the same notation.)

Now if the pair (p_0, p_1) arises as the kernel-pair of a geometric morphism $f\colon Sh(G_0) \to \mathcal{E}$, then the canonical natural isomorphism $fp_0 \to fp_1$ enables us to lift the functor $f^*\colon \mathcal{E} \to Sh(G_0)$ canonically to a functor $\mathcal{E} \to \mathcal{B}(\underset{\sim}{G})$. We say that f is an effective descent morphism if this functor is an equivalence. The key result of Joyal and Tierney [22, VIII 2.1] is that every open surjection is an effective descent morphism; the proof of this result is extremely involved, but a much simpler proof may be given for locally connected surjections by using [12, 4.15(ii)] and the Beck condition for the pullback square

$$\begin{array}{ccc} Sh(G_1) & \xrightarrow{p_0} & Sh(G_0) \\ \big\downarrow{\scriptstyle p_1} & & \big\downarrow{\scriptstyle f} \\ Sh(G_0) & \xrightarrow{f} & \mathcal{E} \end{array}$$

we thus obtain

Theorem 5.1. For any Grothendieck topos $\mathcal{E}$, there exists a groupoid $\underset{\sim}{G} = (G_1 \rightrightarrows G_0)$ in the category of locales, such that the domain and codomain maps $G_1 \to G_0$ are (connected and) locally connected and $\mathcal{E}$ is equivalent to $\mathcal{B}(G_0)$.

Proof. Combine the ideas above with Theorem 4.1. □

The second representation theorem, arising from Freyd's Theorem 4.2, is rather different in character. It begins with an analysis of the notion of internal locale in the topos $\mathcal{B}(G)$, where G is a topological group. Observe first that a locale in the topos $[G, \mathcal{S}]$ of all G-sets is nothing more than a locale A (in $\mathcal{S}$) equipped with an action of G by automorphisms. Moreover, the coreflection functor $[G, \mathcal{S}] \to \mathcal{B}(G)$ (which sends an arbitrary G-set to the union of its continuous orbits) is the direct image of a geometric morphism, and so preserves internal locales. Thus we have a supply of internal locales in $\mathcal{E}(G)$.

In fact every internal locale in $\mathcal{E}(G)$ arises in this way. If B is such a locale, then B is at least a Heyting algebra in $\mathcal{S}$ (though not necessarily a complete one) on which G acts continuously; and this action extends canonically (though not continuously) to its MacNeille completion, which is a complete Heyting algebra and hence a locale in $[G,\mathcal{S}]$. Moreover, the internal completeness of B in $\mathcal{E}(G)$ says precisely that it is (isomorphic to) the continuous coreflection of its MacNeille completion.

Given an internal locale B in $\mathcal{E}(G)$, Freyd [10] has shown how to construct an explicit site of definition for the corresponding localic extension of $\mathcal{E}(G)$. An object of the site is a pair (b,H) where $b \in B$ and H is an open subgroup of G contained in the stabilizer of b; a morphism $(b,H) \to (b',H')$ is induced by an element γ of G such that $\gamma(b) \leq b'$ and $\gamma H \gamma^{-1} \subseteq H'$, two such elements inducing the same morphism if they lie in the same right coset of H'. Composition of morphisms is induced by multiplication in G; the Grothendieck topology on the site is generated by families $(\gamma_i : (b_i,H_i) \to (b,H) \mid i \in I)$ such that $b = \bigvee\{\gamma_i(b_i) \mid i \in I\}$.

Our second representation theorem thus asserts:

<u>Theorem 5.2</u>. Let $\mathcal{E}$ be a Grothendieck topos. Then there exists a locale A and a (not necessarily continuous) action of a topological group G (which may be taken to be either of the groups G_0 and G_1 considered earlier) on A, such that $\mathcal{E}$ is embeddable as an exponential variety in the topos of sheaves on the site constructed as above from the continuous coreflection B of A. If every object of $\mathcal{E}$ is a quotient of a decidable object (in particular if $\mathcal{E}$ is Boolean), then the embedding may be taken to be an equivalence.

<u>Proof</u>. Combine the ideas above with Theorem 4.2 and Proposition 4.3. □

If a topological (or localic) group G acts on a locale A in such a way that the action is a (continuous) locale map $G \times A \to A$ (which is weaker than demanding that it be continuous in the sense in which we have used the term hitherto, but is apparently not implied by the assertion that A is the MacNeille completion of an internal locale in $\mathcal{C}(G)$), then there is a groupoid $(G \times A \rightrightarrows A)$ in **Loc** whose domain and codomain maps are the action of G and the projection respectively. In this case the representation provided by Theorem 5.2 coincides with that provided by Theorem 5.1; but I do not at present know a characterization of those toposes for which a representation of this type is possible.

In conclusion, it is surely too early to say which of the two representation theorems of this section will prove more important in the future development of topos theory, or even whether either of them will actually have any useful applications (though the latter does seem highly likely, particularly in areas concerned with cohomology). But it is clear already that the two theorems have significantly deepened our understanding of the scope of Grothendieck topos theory, and in particular of where it stands in relation to "pointless topology" and -- through the latter -- to classical topology. And it seems inevitable that this deepened understanding will lead to the transfer of further concepts and theorems from classical topology to topos theory, and their consequent application in non-topological contexts -- which is surely the major justification for developing a theory of "generalized spaces".

References

[1] M. Barr, Toposes without points. J. Pure Appl. Alg. 5 (1974), 265-280.

[2] M. Barr and R. Diaconescu, Atomic toposes. J. Pure Appl. Alg. 17 (1980), 1-24.

[3] M. Barr and R. Paré, Molecular toposes. J. Pure Appl. Alg. 17 (1980), 127-152.

[4] J.L. Bell, Some aspects of the category of subobjects of constant objects in a topos. J. Pure Appl. Alg. 24 (1982), 245-259.

[5] A.R. Blass and A. Ščedrov, Boolean classifying topoi. J. Pure Appl. Alg. 28 (1983), 15-30.

[6] M. Coste, La démonstration de Diaconescu du théorème de Barr. Unpublished manuscript, Séminaire J. Bénabou, 1975.

[7] R. Diaconescu, Grothendieck toposes have Boolean points -- a new proof. Commun. Alg. 4 (1976), 723-729.

[8] C.H. Dowker and D. Papert, Quotient frames and subspaces. Proc London Math. Soc. 16 (1966), 275-296.

[9] A. Fraenkel, Zu den Grundlagen der Cantor-Zermelo'schen Mengenlehre. Math. Ann. 86 (1922), 230-237.

[10] P. Freyd, All topoi are localic, or Why permutation models prevail. Unpublished typescript, University of Pennsylvania, 1979.

[11] A. Grothendieck and J.L. Verdier, Théorie des Topos (SGA 4, tom I). Lecture Notes in Mathematics vol. 269 (Springer-Verlag, 1972).

[12] P.T. Johnstone, Topos Theory. L.M.S. Mathematical Monographs no. 10 (Academic Press, 1977).

[13] P.T. Johnstone, A syntactic approach to Diers' localizable categories. In Applications of Sheaves, Lecture Notes in Mathematics vol. 753 (Springer-Verlag, 1979), 466-478.

[14] P.T. Johnstone, Open maps of toposes. Manuscripta Math. 31 (1980), 217-247.

[15] P.T. Johnstone, Factorization theorems for geometric morphisms, I. Cahiers top. et géom. diff. 22 (1981), 3-17.

[16] P.T. Johnstone, Factorization theorems for geometric morphisms, II. In Categorical Aspects of Topology and Analysis, Lecture Notes in Mathematics vol. 915 (Springer-Verlag, 1982), 216-233.

[17] P.T. Johnstone, The point of pointless topology. Bull. Amer. Math. Soc. (N.S.) 8 (1983), 41-53.

[18] P.T. Johnstone, Quotients of decidable objects in a topos. Math. Proc. Cambridge Philos. Soc. 93 (1983), 409-419.

[19] P.T. Johnstone, Open locales and exponentiation. In Applied Category Theory (Proceedings of A.M.S. Special Session, Denver, Jan. 1983), to appear.

[20] P.T. Johnstone, A topos-theorist looks at dilators. In preparation.

[21] P.T. Johnstone and G.C. Wraith, Algebraic theories in toposes. In Indexed Categories and their Applications, Lecture Notes in Mathematics vol. 661 (Springer-Verlag, 1978), 141-242.

[22] A. Joyal and M. Tierney, An extension of the Galois theory of Grothendieck. Memoirs Amer. Math. Soc., to appear.

[23] F.W. Lawvere, Functorial semantics of algebraic theories. Ph.D. thesis, Columbia University 1963; summarized in Proc. Nat. Acad. Sci. U.S.A. 50 (1963), 869-872.

[24] F.W. Lawvere, Continuously variable sets: Algebraic Geometry = Geometric Logic. In Logic Colloquium 1973, Studies in Logic and the Foundations of Mathematics (North-Holland, 1975), 135-156

[25] F.W. Lawvere, Variable quantities and variable structures in topoi. In Algebra, Topology and Category Theory: a Collection of Papers in Honor of Samuel Eilenberg (Academic Press, 1976), 101-131.

[26] M. Makkai, Full continuous embeddings of toposes. Trans. Amer. Math. Soc. 560 (1982), 167-196.

[27] M. Makkai and G.E. Reyes, First Order Categorical Logic. Lecture Notes in Mathematics vol. 611 (Springer-Verlag, 1977).

[28] G.P. Monro, On generic extensions without the axiom of choice. *J. Symbolic Logic* **48** (1983), 39-52.

[29] A. Mostowski, Über die Unabhängigkeit des Wohlordnungssätzes vom Ordnungsprinzip. *Fund. Math.* **32** (1939), 201-252.

[30] R. Paré, Indexed categories and generated topologies. *J. Pure Appl. Alg.* **19** (1980), 385-400.

[31] R. Paré and D. Schumacher, Abstract families and the adjoint functor theorems. In *Indexed Categories and their Applications*, Lecture Notes in Mathematics vol. 661 (Springer-Verlag, 1978), 1-125.

[32] K.I. Rosenthal, Etendues and categories with monic maps. *J. Pure Appl. Alg.* **22** (1981), 193-212.

[33] K.I. Rosenthal, Quotient systems in Grothendieck topoi. *Cahiers top. et géom. diff.* **23** (1982), 425-438.

[34] A. Ščedrov, Sheaves and forcing and their metamathematical applications. Ph.D. thesis, S.U.N.Y. at Buffalo 1981; to appear in *Memoirs Amer. Math. Soc.*

A SURVEY OF METRIZATION THEORY

Jun-iti Nagata
Osaka Kyoiku University, Osaka, Japan

The purpose of this note is to give a rather brief survey of recent metrization theory while putting emphasis on developments during the last decade. R.E. Hodel's paper [23] is recommended for a quick survey of the early history of metrization theory as well as its development until 1972.

As is well known, a topological space is called **metrizable** iff it is homeomorphic to a metric space. **Metrization theory** is a field of topology where the main objective is to study conditions for a given topological space to be metrizable. In this note we assume that all spaces are at least T_2 (Hausdorff), though some theorems may be true for more general spaces. Generally X denotes a T_2-space. The reader is referred to [10], [40] for standard terminologies in general topology, while less popular concepts will be defined in the text.

1 CONDITIONS IN TERMS OF BASES, SEQUENCES OF COVERS AND SEQUENCES OF NEIGHBOURHOODS

In many classical theorems metrizability conditions were described in terms of bases (Urysohn's theorem, [40] p.196, Nagata-Smirnov theorem, p.194, Bing's theorem, p.198, Arhangelskii's theorem, p.203, etc.), sequences of open covers (Alexandroff-Urysohn theorem, p.184, Bing's theorem, p.199, etc.), sequences of nbds (= neighbourhoods) (Frink's theorem, p.192, Nagata's theorem, p.189, etc.) or similar entities. (Page numbers are, throughout, those of [40] where one can find the respective theorems.) The situation has changed little since then. As for metrizability conditions in terms of bases, the following result due to G. Gruenhage and P.J. Nyikos is especially interesting.

DEFINITION 1. A collection $\mathcal{U}$ of subsets of a space X is said to have **$\mathrm{rank}_x \le m$**, where $x \in X$, iff $|\mathcal{U}'| \le m$ holds for each $\mathcal{U}' \subset \mathcal{U}$ such that $x \in U$ for all $U \in \mathcal{U}'$ and such that $U \not\subset U'$, $U' \not\subset U$ whenever $U, U' \in \mathcal{U}'$ and $U \ne U'$. Then **$\mathrm{rank}\,\mathcal{U} \le m$** means that $\mathrm{rank}_x \mathcal{U} \le m$ for all $x \in X$. $\mathcal{U}$ is said to have a **point-finite rank** iff $\mathrm{rank}_x \mathcal{U} \le m < \infty$ for each $x \in X$, where m may depend on x.

The terminology 'rank' is a modification of 'order', and its original motivation was to give a simple n-dimensionality theorem: **A metric space has dim $\le n$ iff it has a base of rank $\le n+1$** (see [41]). Later this concept was studied by A.V. Arhangelskii, G. Gruenhage, P.J. Nyikos and others in relation to metrizability of spaces.

THEOREM 1 (Gruenhage & Nyikos [17]). **Every compact space X with a base of point-finite rank is metrizable.**

A merit of this theorem is implied by the following corollary, where metrizability and n-dimensionality are characterized by a single condition.

COROLLARY. **A compact space X is metrizable and has dim $\le n$ iff it has a base of rank $\le n+1$.**

(In fact the condition 'compact' can be relaxed to 'locally compact' in this corollary.)

G. Gruenhage and P. Zenor [18], too, obtained interesting metrization theorems for non-compact spaces by the use of the concept of rank.

The following theorem, which is a straighforward extension of the Nagata-Smirnov theorem, gives another metrizability condition in terms of a base.

THEOREM 2 (Burke, Engelking & Lutzer [6]). **A regular space X is metrizable iff X has a σ-hereditarily closure-preserving base $\mathcal{U}$: that is, a base $\mathcal{U} = \cup\{\mathcal{U}_n \mid n = 1,2,\dots\}$, where each $\mathcal{U}_n$ satifies the condition that, whenever a subset $W(U) \subset U$ is chosen for each $U \in \mathcal{U}_n$, the collection $\{W(U) \mid U \in \mathcal{U}_n\}$ is closure-preserving.**

Remarkable theorems were obtained on the metrizability of an M-space, which was defined by K. Morita as follows.

DEFINITION 2. A space X is called an **M-space** iff it has a sequence $\{\mathcal{U}_i \mid i=1,2,\dots\}$ of open covers satisfying i) $\mathcal{U}^*_{i+1} < \mathcal{U}_i$, $i=1,2,\dots$ and ii) whenever $x_i \in St(x, \mathcal{U}_i)$ for $i=1,2,\dots$, then $\{x_i\}$ has a cluster point. If condition i) is omitted, X is called a **wΔ-space**. X is said to have a **G_δ-diagonal** iff $\Delta = \{(x,x) \mid x \in X\}$ is G_δ in $X \times X$. An open collection $\mathcal{U}$ in X is called a **p-base** iff $\cap\{U \mid x \in U \in \mathcal{U}\} = \{x\}$ for each $x \in X$.

THEOREM 3. **A space X is metrizable iff it is an M-space with a G_δ-diagonal.**

THEOREM 4. **A space X is metrizable iff it is an M-space with a p-base $\mathcal{U}$ which is point-countable: that is, each point of X is contained in at most countably many members of $\mathcal{U}$.**

Theorem 3 is essentially due to J. Chaber [7], and Theorem 4 is the result of various contributions, including V.V. Filippov [12], J. Nagata [39] and T. Shiraki [52]. We can see a feature of recent metrization theory in these theorems, where metrizability is factored into several subconditions of metrizability that are themselves good subjects of investigation. J. Chaber [8] also obtained interesting results in this respect.

Meanwhile R.E. Hodel [24] and H.W. Martin [35] studied metrizability in terms of 'weak bases', a generalization of neighbourhood bases introduced by A.V. Arhangelskii.

DEFINITION 3. Suppose a collection $\mathcal{U}(x)$ of subsets of X is assigned to each $x \in X$ satisfying i) $x \in U$ for each $U \in \mathcal{U}(x)$ and ii) $U, U' \in \mathcal{U}(x)$ implies $U'' \subset U \cap U'$ for some $U'' \in \mathcal{U}(x)$. Then $\{\mathcal{U}(x) \mid x \in X\}$ is called a **weak base** of X, provided that $V \subset X$ is open in X iff for each $x \in V$ there is $U \in \mathcal{U}(x)$ such that $U \subset V$.

As shown by Hodel, the Alexandroff-Urysohn theorem (p.184) and Nagata's theorem (p.189), for example, can be generalized by the use of weak bases.

A new phase of metrization theory was opened by A.V. Arhangelskii [2], who proved the following theorem, which contains a hereditary condition, unlike classical theorems.

THEOREM 5. **A regular space which is hereditarily a Lindelöf M-space is (separable) metrizable.**

Later R.E. Hodel & A. Okuyama [26] proved:

THEOREM 6. **A regular wΔ-space with a point-countable p-base which hereditarily satisfies CCC (= countable chain condition) is (separable) metrizable.**

It is known that 'with a point-countable p-base' cannot be dropped from the above condition, meaning that a generalization of Theorem 5 is impossible in this direction.

On the other hand Z. Balogh [4] made a deep study of spaces which are hereditarily paracompact M-spaces (called F_{pp}**-spaces**) and especially of the metrizability of such spaces. As a result he obtained the following generalization of Arhangelskii's theorem.

THEOREM 7. **A space X is metrizable iff it is a perfectly normal F_{pp}-space.**

H.H. Hung [27] proved the following theorem generalizing Nagata's theorem (p.189), whose object was to unify classical metrization theorems. Thus he could derive various metrization theorems more easily from his own.

THEOREM 8. **A space X is metrizable iff there is a collection $\{(B_\xi, \tilde{B}_\xi) \mid \xi < \alpha\}$ of pairs of disjoint subsets of X, where α is a fixed ordinal number with countable cofinality, satisfying: i) for each $x \in X$ and each $\beta < \alpha$, the set $\cap\{X - \tilde{B}_\xi \mid x \in B_\xi, \xi < \beta\}$ is a nbd of x; ii) for each $x \in X$ and each open nbd U of x there are $\beta, \gamma < \alpha$ such that $x \in B_\beta \subset X - \tilde{B}_\beta \subset B_\gamma \subset X - \tilde{B}_\gamma \subset U$.**

(In fact he gave two other similar conditions which can be substituted for ii).)

There are several other attempts to unify basic metrization theorems; but it seems to the author that their ranges of unification do not go much beyond the classical theorems. It could be a worthwhile problem to establish a theory which unifies classical and modern metrization theorems as corollaries of a single main theorem.

Further, H.H. Hung [28] proved the following theorem by the use of Theorem 8 and embedding of a metric space into the countable product of hedgehogs.

THEOREM 9. **X is metrizable iff it has a sequence $\{\mathcal{U}_i \mid i \in N\}$ ($N = \{1,2,\dots\}$) of disjoint open collections $\mathcal{U}_i$ satisfying: given any closed set C of X and any $x \in X - C$, there is a finite subset N' of N such that, if we write**

$$V_i = \cup\{U \mid x \in U \in \mathcal{U}_i\}, \quad W_i = \cup\{U \mid x \notin U \in \mathcal{U}_i\},$$

for $i \in N'$, and

$$V = \cap\{V_i \mid i \in N'\}, \quad W = \cup\{W_i \mid i \in N'\},$$

then we have $x \in V$, $C \subset W$.

H.W. Martin [36], too, generalized Nagata's theorem cited above.

It should be also noted that R.E. Hodel [25] and P.J. Nyikos & H.C. Reichel [44] extended some results of metrization theory to higher cardinality.

As for the famous **normal Moore space problem** ('Is every normal Moore space metrizable?'), there are at least two well-known results (besides Bing's theorem mentioned before): namely F.B. Jones's theorem that **every separable normal Moore space is metrizable under CH** (= continuum hypothesis), and the result that **there is a normal Moore space which is not metrizable under Martin's axiom and the negation of CH** (see Tall [55]). Furthermore, T. Przymusiński & F.D. Tall [48] proved that **the statement 'A normal Moore space satisfying CCC is metrizable' is independent of the axioms of set theory.**

While we refer the reader to M.E. Rudin [50] and F.D. Tall [56] for developments of this aspect, we note that there has been remarkable progress even in the last few years. P.J. Nyikos [42] showed that **the so-called PMEA** (= product measure extension axiom) **implies the metrizability of normal Moore spaces, while the consistency of PMEA implies the consistency of there being a measurable cardinal.** W.G. Fleissner [13, 14] constructed **a normal non-metrizable Moore space assuming CH** and proved that **the consistency of the statement 'All normal Moore spaces are metrizable' implies the consistency of the statement 'There is a measurable cardinal'.**

Let us conclude this section with M.L. Wage's theorem [59] indicating a relation between the normal Moore space problem and the concept of countable paracompactness which was originated by C.H. Dowker [9] and M. Katětov [30]: **If there is a model of set theory in which every countably paracompact Moore space is normal, then the normal Moore space conjecture is true in that model.**

2 OTHER SPECIAL CONDITIONS

In this section we discuss two areas of metrization theory: **a)** the metrizability of spaces endowed with special properties concerning connectedness, dimension, order, etc., and **b)** metrizability conditions expressed in terms of functions.

a) Some of the interesting results in this area are as follows.

THEOREM 10 (Reed & Zenor [49]). **Every locally compact, locally connected, normal Moore space is metrizable.**

THEOREM 11 (Zenor & Chaber [64]). **Every rim-compact, locally connected, perfectly normal Moore space is metrizable.**

The latter theorem is obviously a generalization of the former and among a few good results obtained on the normal Moore space problem making no set-theoretic assumptions.

The following is a generalization of S. Kakutani's famous theorem that **a topological group is metrizable iff it is first-countable.**

THEOREM 12 (Nyikos [43]). **A topological group X is metrizable iff it is weakly first-countable: that is, iff X has a weak base $\{\mathcal{U}(x) \mid x \in X\}$ such that each $\mathcal{U}(x)$ is countable.**

D.J. Lutzer [31], M.J. Faber [11], J.M. van Wouwe [60] and H.R. Bennett & D.J. Lutzer [5] obtained remarkable results on the metrizability of linearly ordered topological spaces and their subspaces (called **GO-spaces**). The following are two examples of these results, which we owe to [31] and [5] respectively.

THEOREM 13. **A GO-space X is metrizable iff X is semi-stratifiable: that is, iff to each open set U of X and $n \in N$ one**

can assign a closed set $F(U,n)$ in such a way that $U = \cup\{F(U,n) \mid n \in N\}$ and $F(U,n) \subset F(U',n)$ whenever $U \subset U'$.

THEOREM 14. **A GO-space is metrizable iff it is hereditarily an M-space.**

The following theorem contains a dimensional condition as part of a metrizability condition, which does not happen often.

THEOREM 15 (Ščepin [51]). **Every finite-dimensional compact absolute neighbourhood retract is metrizable.**

DEFINITION 4. Denote by $C(X)$ ($C^*(X)$) the set of all (bounded) real-valued continuous functions on X and by $\tilde{C}(X)$ the set $\cup\{C(H) \mid H$ is a closed subset of $X\}$. We introduce a topology into $\tilde{C}(X)$ by use of the finite (Vietoris) topology on the graphs of $f \in \tilde{C}(X)$. A map $\Phi : \tilde{C}(X) \to C(X)$ is called an **extender** for X iff $\Phi(f)$ is an extension of each $f \in \tilde{C}(X)$.

According to Tietze's theorem, X is normal iff there is an extender for X. P. Zenor [62] characterized metrizability in terms of continuous extenders as follows.

THEOREM 16. **A compact space X is metrizable iff $X \times [0,1]$ admits a continuous extender, where $[0,1]$ denotes the closed segment.**

It is interesting to compare this theorem with the famous theorem of C.H. Dowker: **X is countably paracompact and normal iff $X \times [0,1]$ is normal.**

In the following theorem due to V.V. Popov [46] metrizability of a space X is related to properties of the space $\mathcal{F}(X)$ of the closed sets of X.

THEOREM 17. **A regular space X is metrizable if the space $\mathcal{F}(X)$ with the Vietoris topology is symmetrizable** (see Definition 5) **or has a G_δ-diagonal, or if every closed set of $\mathcal{F}(X)$ is G_δ.**

b) Some metrizability conditions are described in terms of functions on X, $X \times X$, etc. Classical examples of this kind are Chittenden's theorem (p.270) and J. Nagata's theorem: **A Tychonoff space X is metrizable iff there exist sets $L_n \subset C^*(X)$, $n = 1,2,\ldots$ — where $C^*(X)$ is regarded as a special lattice endowed with**

the infinite operations $\bigvee_\alpha f_\alpha$ **and** $\bigwedge_\alpha f_\alpha$ **defined by** $(\bigvee_\alpha f_\alpha)(x) = \sup_\alpha f_\alpha(x)$ **and** $(\bigwedge_\alpha f_\alpha)(x) = \inf_\alpha f_\alpha(x)$ **— such that:**
i) for each n **and for every family** $\{f_\alpha \mid \alpha \in A\} \subset L_n$ **, both** $\bigvee_\alpha f_\alpha$ **and** $\bigwedge_\alpha f_\alpha$ **exist and belong to** L_n **; ii) for each** $f \in C^*(X)$ **there is a family** $\{f_\beta \mid \beta \in B\} \subset \cup\{L_n \mid n = 1, 2, \ldots\}$ **such that** $f = \bigwedge_\beta f_\beta$ **.** (This theorem was first proved in [38] in a slightly different form; and J.A. Guthrie & M. Henry [19] also obtained a theorem based on a similar idea. To improve this theorem could be a worthwhile problem; for example, it would be nice if one could characterize the metrizability of X in terms of ordinary lattice properties of $C(X)$.)

A.V. Arhangelskii's [1] metrizability condition, using a symmetric, also belongs to the category **b)**:

DEFINITION 5. A function $d: X \times X \to [0, \infty)$ is called a **symmetric** iff it satisfies i) $d(x,y) = 0$ iff $x = y$, ii) $d(x,y) = d(y,x)$, iii) $H \subset X$ is a closed set iff $d(x,H) > 0$ for all $x \notin H$.

Then Arhangelskii's theorem says: **A space is metrizable iff it has a symmetric** d **such that** $d(K,H) > 0$ **whenever** K **is compact,** H **is closed and** $H \cap K = \emptyset$ **.** P.W. Harley & G.D. Faulkner [21] also characterized metrizability by use of a symmetric satisfying a certain condition which is somewhat different from Arhangelskii's.

Various modifications of complete regularity, normality and perfect normality are studied in relation to other topological properties. We shall show in the following some results in this area which are related to metrizability.

DEFINITION 6. Let $\mathfrak{F}(X)$ be the collection of all closed sets of X with the Vietoris topology. A map $\varphi : X \times \mathfrak{F}(X) \to [0,1]$ is called a **PN-operator** iff, for each $G \in \mathfrak{F}(X)$, $G = \{x \in X \mid \varphi(x,G) = 0\}$. Further, define $\mathfrak{D}(X) = \{(x,G) \in X \times \mathfrak{F}(X) \mid x \notin G\}$. A map $\varphi : X \times \mathfrak{D}(X) \to [0,1]$ is called a **CR-operator** iff, for each $(x,G) \in \mathfrak{D}(X)$, it is true that $\varphi(x, (x,G)) = 0$ and $G \subset \{y \in X \mid \varphi(y, (x,G)) = 1\}$.

The following Theorems 18 and 19 are due to P. Zenor [63] and Theorems 20 and 21 to G. Gruenhage [16].

THEOREM 18. **X is metrizable iff it admits a continuous PN-operator φ such that, for every finite subset K of X and $x \in K$, $\varphi(y, \{x\}) \geq \varphi(y, K)$ for every $y \in X$.**

THEOREM 19. **X is metrizable iff it is the closed continuous image of a metric space (i.e. the image of a metric space by a closed continuous map) and admits a continuous CR-operator.**

THEOREM 20. **A separable space X is metrizable iff it admits a continuous CR-operator.**

THEOREM 21. **The following conditions are equivalent for a space X: i) X is a separable metrizable space; ii) X satisfies CCC and admits a continuous PN-operator; iii) X is a Lindelöf space and admits a continuous PN operator; iv) X is a separable space and admits a continuous PN-operator.**

Gruenhage also gave an example of a non-metrizable space which admits a continuous PN-operator.

3 IMAGES, PRE-IMAGES AND SUMS OF METRIZABLE SPACES

We shall discuss in this section results obtained about metrizability of images, pre-images and sums of metrizable spaces, as well as related topics. As for metrizability of closed continuous images of metrizable spaces, we should quote the classical theorem due to A.H. Stone, K. Morita and S. Hanai (p.216). Theorem 19 in the previous section also belongs to this category. E.A. Michael [37] and F.E. Siewiec [53] took Stone, Morita and Hanai's result further. Another remarkable result in this area was obtained by D.M. Hyman [29], who proved:

THEOREM 22. **Let X and Y be non-discrete spaces. If $X \times Y$ is the closed continuous image of a metric space, then $X \times Y$ is metrizable.**

A. Okuyama & Y. Yasui [45] generalized this theorem, in the case when $Y = [0,1]$, as follows.

DEFINITION 9. A pair (X,F) of a space X and its closed subset F is called **semi-canonical** iff there is an open cover $\mathcal{U}$ of $X - F$ such that, for each $x \in F$ and each nbd U of x in X, there is a nbd W of x in X satisfying $St(W, \mathcal{U}) \subset U$.

It is obvious that, for any closed subset F of a metric space X, (X,F) is semi-canonical. The same can be proved if X is the closed continuous image of a metric space.

THEOREM 23. **$(X\times[0,1], X\times\{0\})$ is a semi-canonical pair iff X is metrizable.**

On the other hand P.W. Harley [20] generalized Hyman's theorem as follows,

THEOREM 24. **Let X and Y be non-discrete spaces which are closed continuous images of metric spaces. Then $X\times Y$ is metrizable iff it is a Fréchet-Urysohn space; where a topological space Z is called Fréchet-Urysohn iff $x\in\overline{A}$ and $A\subset Z$ imply $x_n\to x$ for some $\{x_n \mid n=1,2,\ldots\}\subset A$.**

As is well-known, an M-space is characterized as the pre-image of a metric space by a **quasi-perfect map** (= a closed continuous map f such that $f^{-1}(y)$ is countably compact for each point y of the range of f). Thus metrization theorems on M-spaces stated in the first section give conditions for quasi-perfect pre-images of metrizable spaces to be metrizable. F.G. Slaughter & J.M. Atkins [54], G.M. Reed & P. Zenor [49], H.W. Martin [33, 34], A. Makkouk [32] and T. Przymusiński [47] also obtained interesting results on metrizability of images and pre-images of metric spaces by maps satisfying certain conditions.

As for metrizability of sums of closed metrizable subsets, we can quote classical theorems like Smirnov's theorem (p.214), Stone's theorem (p.215) for countable sums and Nagata's theorem (p.213) for a locally finite sum. In this area S.P. Franklin & B.V.S. Thomas [15] obtained a significant metrizability condition in terms of concrete spaces S_ω and S_2 (both defined below). Recently Y. Tanaka and Zhou Hao-xuan have taken Franklin & Thomas's investigation further to obtain the theorems which follow.

DEFINITION 8. A closed cover $\{F_\alpha \mid \alpha\in A\}$ of X is said to **determine** X iff $G\subset X$ is closed in X whenever $G\cap F_\alpha$ is closed in F_α for each $\alpha\in A$. The closed cover $\{F_\alpha \mid \alpha\in A\}$ is said to **dominate** X iff $G\subset X$ is closed in X whenever there is $A'\subset A$ satisfying: $G\subset\bigcup\{F_\alpha \mid \alpha\in A'\}$, and $G\cap F_\alpha$ is closed in F_α for each $\alpha\in A'$.

We denote by S_ω the space obtained from the topological sum of countably many converging sequences by identifying all the limit points and by S_2 the space consisting of the points $(N \times N) \cup N \cup \{0\}$ with the strongest topology such that $\{(m,n) \mid m = 1,2,\ldots\}$ converges to n for each $n \in N$ and $\{n \mid n = 1,2,\ldots\}$ converges to 0.

THEOREM 25 (Tanaka [57]). **Let X be a regular space determined by a closed cover $\mathcal{F} = \{F_\alpha \mid \alpha \in A\}$, where each F_α is metrizable. Provided i) or ii) below hold, then X is metrizable iff X contains closed copies of neither S_ω nor S_2 : i) $\mathcal{F}$ is star-countable (i.e. each element of $\mathcal{F}$ meets at most countably many elements of $\mathcal{F}$) ; ii) X is paracompact and $\mathcal{F}$ is point-countable.**

Franklin & Thomas proved this theorem earlier, in the case when A is countable and each F_α is compact. It is interesting that metrizability of an abstract space is characterized in terms of such concrete spaces as S_ω and S_2. Similar metrizability theorems were obtained by Tanaka for CW-complexes and other types of spaces. The following theorem, which may be regarded as a generalization of Nagata's theorem (p.213), is simpler and beautiful.

THEOREM 26 (Tanaka & Zhou Hao-xuan [58]). **Let X be a space dominated by a closed cover $\mathcal{F} = \{F_\alpha \mid \alpha \in A\}$, where each F_α is metrizable. Then X is metrizable iff it contains closed copies of neither S_ω nor S_2.**

Let us conclude this note with an extremely interesting theorem proved by A.V. Arhangelskii [3] on the metrizability of the sum of two metrizable (non-closed) subsets.

THEOREM 27. **Let X be a compact space such that $X = X_1 \cup X_2$, where X_1 and X_2 are metrizable. If X satisfies CCC, then X is separable metrizable. This result cannot be generalized to three summands.**

REFERENCES

[1] A.V. Arhangelskii, Mappings and spaces, Russ.Math.Surveys 21 (1966) no.4, 115-162.
[2] ____ , On herditary properties, Gen.Top.Appl. 3 (1973) 39-46.
[3] ____ , Bicompacta and unions of countable families of metrizable spaces, Sov.Math. 18 (1977) 165-169.
[4] Z. Balogh, On the metrizability of F_{pp}-spaces and its relations to the normal Moore space conjecture, Fund.Math. 113 (1981) 45-58.
[5] H.R. Bennett & D.J. Lutzer, Certain hereditary properties and metrizability in generalized ordered spaces, Fund.Math. 107 (1980) 71-84.
[6] D. Burke, R. Engelking & D.J. Lutzer, Hereditarily closure-preserving collections and metrization, Proc.Amer.Math.Soc. 51 (1975) 483-488.
[7] J. Chaber, Conditions which imply compactness in countably compact spaces, Bull.Acad.Pol.Sci.Ser.Math. 24 (1976) 993-998.
[8] ____ , On point-countable collections and monotonic properties, Fund.Math. 94 (1977) 209-219.
[9] C.H. Dowker, On countably paracompact spaces, Can.J.Math. 1 (1951) 219-224.
[10] R. Engelking, General Topology (Polish Scientific Publishers, Warsaw, 1977).
[11] M.J. Faber, Metrizability in Generalized Ordered Spaces, Math. Centre Tracts 53 (1974).
[12] V.V. Filippov, On feathered paracompacta, Sov.Math. 9 (1968) 161-164.
[13] W.G. Fleissner, Normal Moore spaces in the constructible universe, Proc.Amer.Math.Soc. 46 (1974) 294-298.
[14] ____ , If all normal Moore spaces are metrizable, then there is an inner model with a measurable cardinal, Trans.Amer.Math.Soc. 273 (1982) 365-373.
[15] S.P. Franklin & B.V.S. Thomas, On the metrizability of k_ω-spaces, Pacific J.Math. 72 (1977) 399-402.
[16] G. Gruenhage, Continuously perfectly normal spaces and some generalizations, Trans.Amer.Math.Soc. 224 (1976) 323-338.
[17] G. Gruenhage & P.J. Nyikos, Spaces with bases of countable rank, Gen.Top.Appl. 8 (1978) 233-257.
[18] G. Gruenhage & P. Zenor, Metrization of spaces with countable large basis dimension, Pacific J.Math. 59 (1975) 455-460.
[19] J.A. Guthrie & M. Henry, Metrization and paracompactness in terms of real functions, Bull.Amer.Math.Soc. 80 (1974) 720.
[20] P.W. Harley, Metrization of closed images of metric spaces, TOPO 72 (Proc. 2nd Pittsburgh Int.Conf. 1972), Springer Lecture Notes 378 (1974) 188-191.
[21] P.W. Harley & G.D. Faulkner, Metrization of symmetric spaces, Can. J.Math. 27 (1975) 986-990.
[22] R.W. Heath, D.J. Lutzer & P. Zenor, Monotonically normal spaces, Trans.Amer.Math.Soc. 178 (1973) 481-493.
[23] R.E. Hodel, Some results in metrization theory, 1950-1972, Topology Conf.Virginia Polytech.Inst. & State Univ. 1973, Springer Lecture Notes 375 (1974) 120-136.
[24] ____ , Metrizability of topological spaces, Pacific J.Math. 55 (1974) 441-459.

[25] R.E. Hodel, Extensions of metrization theorems to higher cardinality, Fund.Math. 87 (1975) 219-229.
[26] R.E. Hodel & A. Okuyama, Cardinal functions and a metrization theorem, Math.Japonica 21 (1976) 61-62.
[27] H.H. Hung, A contribution to the theory of metrization, Can.J.Math., 29 (1977) 1145-1151.
[28] ____ , A general metrization theorem, Top.Appl. 11 (1980) 275-279.
[29] D.M. Hyman, A note on closed maps and metrizability, Proc.Amer.Math. Soc. 21 (1969) 109-112.
[30] M. Katětov, Measures in fully normal spaces, Fund.Math. 38 (1951) 73-84.
[31] D.J. Lutzer, On generalized ordered spaces, Dissert.Math. 89 (1971) 1-36.
[32] A. Makkouk, On complete metrizability of some topological spaces, Acta Math.Acad.Sci.Hungar. 27 (1976) 243-246.
[33] H.W. Martin, Metrization of symmetric spaces and regular maps, Proc.Amer.Math.Soc. 35 (1972) 269-274.
[34] ____ , Perfect maps of symmetrizable spaces, Proc.Amer.Math.Soc.38 (1973) 410-412.
[35] ____ , Weak bases and metrization, Trans.Amer.Math.Soc. 222 (1976) 337-344.
[36] ____ , A note on the metrization of γ-spaces, Proc.Amer.Math.Soc. 57 (1976) 332-336.
[37] E.A. Michael, A quintuple quotient quest, Gen.Top.Appl. 2 (1972) 91-138.
[38] J. Nagata, On coverings and continuous functions, J.Inst.Polytech. Osaka City Univ.Ser.A 7 (1956) 29-38.
[39] ____ , A note on Filippov's theorem, Proc.Japan Acad. 45 (1969) 30-33.
[40] ____ , Modern General Topology: second, revised edition (North Holland Publ.Co., Amsterdam & London, 1974).
[41] ____ , Modern Dimension Theory: revised and extended edition (Heldermann Verlag, Berlin, 1983).
[42] P.J. Nyikos, A provisional solution to the normal Moore space problem, Proc.Amer.Math.Soc. 78 (1980) 429-435.
[43] ____ , Metrizability and the Fréchet-Urysohn property in topological groups, Proc.Amer.Math.Soc. 83 (1981) 793-801.
[44] P.J. Nyikos & H.C. Reichel, Some results on cardinal functions in metrization theory, Glas.Mat.Ser.III 15 (35) (1980) 183-202.
[45] A. Okuyama & Y. Yasui, On the semi-canonical property in the product space X×I, Proc.Amer.Math.Soc. 68 (1978) 229-234.
[46] V.V. Popov, Metrizability and the space of closed subsets, Russ. Math.Surveys 35 (1980) no.3, 268-273.
[47] T. Przymusińsky, Metrizability of inverse images of metric spaces, under open perfect and zero-dimensional mappings, Colloq. Math. 24 (1971/72) 175-180.
[48] T. Przymusińsky & F.D. Tall, The undecidability of the existence of a non-separable normal Moore space satisfying the countable chain condition, Fund.Math. 85 (1974) 291-297.
[49] G.M. Reed & P. Zenor, Pre-images of metric spaces, Bull.Amer.Math. Soc. 80 (1974) 879-880.
[50] M.E. Rudin, The metrizability of normal Moore spaces, Studies in Topology (Academic Press, New York, 1975), 507-516.
[51] E.V. Ščepin, A finite-dimensional compact absolute neighbourhood retract is metrizable, Sov.Math. 18 (1977) 402-406.

[52] T. Shiraki, M-spaces, their generalizations and metrization theorems, Sci.Rep.Tokyo Kyoiku Daigaku Sect.A 11 (1971) 57-67.

[53] F.E. Siewiec, On the theorem of Morita and Hanai and Stone, TOPO 72 (Proc. 2nd Pittsburgh Int.Conf. 1972), Springer Lecture Notes 378 (1974) 449-454.

[54] F.G. Slaughter & J.M. Atkins, On the metrizability of preimages of metric spaces under closed continuous functions, Proc.Univ. of Oklahoma Topology Conf. 1972, 13-22.

[55] F.D. Tall, Set-theoretic consistency results and topological theorems concerning the normal Moore space conjecture and related problems (Thesis, Univ.of Wisconsin 1969), Dissert.Math. 148 (1977).

[56] ____ , The normal Moore space problem, Topological Structures II (Part 2), Math.Centre Tracts 116 (1979) 243-261.

[57] Y. Tanaka, Metrizability of certain quotient spaces, Fund.Math. (to appear).

[58] Y. Tanaka & Zhou Hao-xuan, Spaces dominated by metric subsets (to appear).

[59] M.L. Wage, Countable paracompactness, normality and Mooore spaces, Proc.Amer.Math.Soc. 57 (1976) 183-188.

[60] J.M. van Wouwe, GO-spaces and Generalizations of Metrizability Math.Centre Tracts 104 (1979).

[61] P. Zenor, On continuously perfect normal spaces, Proc.Univ.of Oklahoma Topology Conf. 1972, 334-336.

[62] ____ , Extending continuous functions in compact metric spaces, Topology Conf.Virginia Polytech.Inst. & State Univ. 1973, Springer Lecture Notes 375 (1974) 277-283.

[63] ____ , Some continuous separation axioms, Fund.Math. 90 (1975/76) 143-158.

[64] P. Zenor & J. Chaber, On perfect subparacompactness and a metrization theorem for Moore spaces, Topology Proc. 2 (1977) 401-407.

SOME THOUGHTS ON LATTICE VALUED FUNCTIONS AND RELATIONS

M. W. Warner

Department of Mathematics,
The City University,
Northampton Square, London EC1V OHB.

1 INTRODUCTION

A lattice valued function is a function $f : X \to L$ from a set X to a lattice L. Both X and L may possess further structure. In fact, every real-valued function is lattice-valued by virtue of the usual max, min lattice on the ordered set of reals. It could not be our intention to discuss such a general situation. We concentrate, rather, on some areas where the actual lattice structure of L plays a major part in a topological theory. Continuous real-valued functions from a topological space are thus excluded per se, but feature within the context of fuzzy topological spaces. Some of the formal transition from ordinary to fuzzy spaces is largely mechanical. More interesting are the difficulties encountered, and on some of these we shall concentrate.

The term 'fuzzy' has been used by Poston (22) and Dodson (6) to describe a set with a reflexive, symmetric relation, elsewhere (30) called a tolerance space. We shall avoid confusion by adopting the latter term. Extending the tolerance relation to a fuzzy, (or L-fuzzy, Goguen (8)) relation, we review the topological analogues which can be introduced, with particular reference to homogeneity.

The main discussion, then, is on two topics, namely fuzzy topological spaces and sets with fuzzy relations. We conclude with a few general remarks on lattice-valued functions, topology and homology.

2 LATTICES

The following brief sketch of the elements of lattice theory which are needed in this paper may be pursued in any standard text (e.g. Birkhoff (3)).

Definition A lattice $(L, \leqslant)$ is a non-empty set L, partially ordered by a relation $\leqslant$ in which every pair of elements a,b has a greatest lower bound $a \wedge b$ and a least upper bound $a \vee b$.

These two bounds are called the meet and join, respectively, of a and b.

If every subset of L has g.l.b. and l.u.b., then L is said to be complete. A complete lattice then has a least element, often written 0, and a greatest element, 1.

Definition. A lattice is distributive when it satisfies both distributive laws ; namely

$$x \wedge (y \vee z) = (x \wedge y) \vee (x \wedge z)$$
$$x \vee (y \wedge z) = (x \vee y) \wedge (x \vee z)$$

for all x,y,z.

Definition. A complement of an element x of a complete lattice L is an element $y \in L$ such that $x \wedge y = 0$ and $x \vee y = 1$. The lattice L is complemented if all its elements have complements.

Definition A Boolean lattice is defined to be a complemented distributive lattice.

In such a lattice each element has exactly one complement x'. Moreover $(x')' = x$ and $(x \wedge y)' = x' \vee y'$.

The power set P(X) of a set X is a Boolean lattice where $\leq$ is set inclusion $\subseteq$, and meet and join are set intersection and union respectively.

The closed unit interval I = [0,1] is a complete distributive lattice with respect to the standard ordering. Here $a \wedge b = \min\{a,b\}$ and $a \vee b = \max\{a,b\}$. We can define for each x an element $x' = 1-x$. Then $(x')' = x$, but the requirements for x' to be a true complement are not satisfied, since in general, $x \vee x' \neq 1$ and $x \wedge x' \neq 0$.

Definition A morphism $f : L \to M$ of lattices is a function from L to M which preserves meet and join.

Definition. A sublattice S of a lattice L is a subset of L which is closed under meet and join.

3 SET-VALUED MAPPINGS

A set-valued mapping $f : X \to Y$ is a lattice-valued function, its codomain being the Boolean lattice $P(Y)$ of subsets of Y. A representative selection of papers devoted to the topological theory of set-valued mappings is contained in (26). We include here a few definitions of terms used in the ensuing discussion. These versions follow Berge (2) and differ slightly from those of Kuratowski (13) et al.

Let X,Y be topological spaces.

Definition. A set-valued function $f : X \to Y$ is upper semi-continuous at x_0 (u.s.c.) if, for every open set G containing $f(x_0)$, there exists a neighbourhood $U(x_0)$ of x_0 such that $x \in U \implies f(x) \subseteq G$.

f is upper semi-continuous if it is upper semi-continuous at every point of X, and for all $x \in X$, $f(x)$ is compact.

Definition. $f : X \to Y$ is lower semi-continuous at x_0 (l.s.c.) if for every open set G meeting $f(x_0)$, there exists a neighbourhood $U(x_0)$ such that $x \in U \implies f(x) \cap G \neq \phi$. f is lower semi-continuous if it is so at every point of X.

Definition. The union of two set-valued functions $f,g : X \to Y$ is the set-valued function $f \cup g : X \to Y$, $f \cup g(x) = f(x) \cup g(x)$.

Definition. The intersection $f \cap g : X \to Y$ is defined by $f \cap g(x) = f(x) \cap g(x)$.

Berge proves that finite unions and infinite intersections of u.s.c. functions are u.s.c.

A set-valued function $f : X \to Y$ is continuous if it is both upper and lower semi-continuous. We note that if the image sets of elements of X are all singletons then both upper and lower semi-continuity become ordinary classical continuity.

4 FUZZY SETS AND TOPOLOGY

A fuzzy subset of the set X was defined by Zadeh (28) to be a function A from X to the closed unit interval $[0,1] = I$. Then for $x \in X$, $A(x)$ can be called the grade of membership of x in A.

In the nineteen years since the appearance of Zadeh's paper

much has been written on fuzziness, which has been seized upon as a useful tool for treating mathematically the uncertainty and lack of clarity pervading the 'real' world. Attempts to fuzzify topology began to appear around 1968 (Chang (5)) and the work continues, for example by Wong (27), Goguen (8), Pu and Liu (23), Gottwald (9), Lowen (14), (15), Martin (16). Operations on fuzzy subsets are defined in terms of the corresponding lattice operations in I(§2). If $\mathcal{A} = \{A_\alpha | \alpha \in I\}$ is a family of fuzzy sets in X, the union $\cup\mathcal{A}$ and intersection $\cap\mathcal{A}$ are defined respectively by

$$(\cup\mathcal{A})(x) = \sup\{A_\alpha(x) | \alpha \in I\}, \; x \in X$$
$$(\cap\mathcal{A})(x) = \inf\{A_\alpha(x) | \alpha \in I\}, \; x \in X$$

The complement A' of A is given by $A'(x) = 1 - A(x)$, $x \in X$.

<u>Definition</u>. A family $\mathcal{T}$ of fuzzy sets in X is called a <u>fuzzy topology</u> iff (1) $X \in \mathcal{T}$, $\phi \in \mathcal{T}$, (2) $A \cap B \in \mathcal{T}$ whenever $A, B \in \mathcal{T}$, and (3) $\cup\{A_\alpha | \alpha \in I\} \in \mathcal{T}$ whenever each $A_\alpha \in \mathcal{T}$ $(\alpha \in I)$.

Here X is the 'crisp' set which takes the value 1 for all $x \in X$, and ϕ is the crisp set which takes the value 0 for all $x \in X$.

Thus far, the definitions are 'obvious' at least in that no reasonable alternatives present themselves. But difficulties arise with the concepts of fuzzy point and membership. A fuzzy point would be expected to have a singleton <u>support</u> (set of elements of X taking values $\neq 0$). A <u>fuzzy singleton</u> is defined to be a fuzzy set x_λ taking the value 0 for all $y \in X$ except one, say, $x \in X$. Its value at x is $\lambda (0 < \lambda \leq 1)$. This is taken to be the definition of a <u>fuzzy point</u> by Pu and Liu (23) and others, having been introduced by Goguen (8). It has the advantage that an ordinary point x_1 is a special case of a fuzzy point. This was not true in Wong's (27) definition where λ is not allowed to equal 1.

Now set inclusion is naturally given by $A \subseteq B$ iff $A(x) \leq B(x)$ for all $x \in X$. With the singleton set x_λ taken as a fuzzy point, it is possible to take set inclusion of the singleton to represent membership. Then $x_\lambda \in A$ iff $\lambda \leq A(x)$. Pu and Liu build up a reasonable neighbourhood theory from this approach.

There are arguments against identification of a singleton set with its single member and of being unable to distinguish between inclusion and membership. But no satisfactory alternative presents itself.

Wong defines $x_\lambda \varepsilon A$ iff $x_\lambda(y) < A(y)$ for all $y \varepsilon X$. Then $x_\lambda \varepsilon A$ implies that the support of A is the whole of X ; also if membership values are restricted to {0,1} we do not get the classical case. Perhaps the only advantage is that '$x_\lambda \varepsilon y_\mu$' is always false.

In any case Gottwald (9) has pointed out that Wong's proofs of three theorems on local properties of fuzzy topology are incorrect and shows that an adjustment of the membership definition is required, namely, $x_\lambda \varepsilon A$ iff $\lambda < A(x)$ for x the support of x_λ. With this definition Wong's theorems are correct, although it is now possible for x_λ to be a member of x_μ $(\lambda < \mu)$.

But the theorems are still not <u>good generalisations</u> in the sense of Lowen (14) who stipulates that a property P' of a fuzzy topological space is a good generalisation of P iff (a) when a topological space (X,τ) has property P, then the fuzzy topological space $(X,\mathcal{T})$ has property P', and (b) if (X,τ) does not have P then $(X,\mathcal{T})$ does not have P'. Here $\mathcal{T} = \{\hat{z} | z \varepsilon \tau\}$, $\hat{z}$ the characteristic function of z.

The Pu and Liu development then is so far the most successful, the authors' avowed intention of generalising Kelley (12) to fuzzy being well under way.

In all attempts at developing a fuzzy topological theory complementation produces a stumbling block. The lattice I is not complemented, and in general $(A \cap A')(x) \neq 0$. Pu and Liu cope with this by introducing the concept of quasi-coincidence. Two sets A,B are quasi-coincident iff there exists $x \varepsilon X$ such that $A(x) > B'(x)$. Then A and A' cannot be quasi-coincident. The above authors build up quasi-neighbourhood systems (in which a fuzzy point is not necessarily a member of its q-neighbourhood) and develop a 'q-generalisation' of Kelley. While this can be considered satisfactory from a formal view-point, we feel that the complementation problem merits further thought. Lowen (14) has considered which maps to I could replace Zadeh's complement but no real improvement emerges.

Now Zadeh (29) and others e.g. Natvig (21) have suggested that, for a fuzzy set A, the value A(x) can be interpreted as the possibility of x lying in A. Then $(A \cap A')x$ measures the possibility of x both lying in A and simultaneously not doing so.

And $(A \cup A')x$ represents the possibility of x lying either in A or A'. If this does not have value 1 then there is an alternative to x either lying in A or not lying in A.

This failure of the excluded middle law has been claimed as a positive virtue of fuzzy theory, being a valid mathematical expression of ambiguity. Muir (18) argues against this view and suggests that the failure is not only undesirable but is a result of the theory's shortcomings rather than being one of its original purposes. If we accept Natvig's (21) conclusion that the difference between possibility and probability is illusory an alternative form of fuzzy theory immediately presents itself, namely that set membership be represented by a Boolean lattice B of subsets of I. Goguen (8) in the early days of fuzzy drew attention to the fact that properties of the lattice L of truth values are reflected immediately in L-fuzzy theory. Thus, defining a fuzzy subset A of X to be a function $A : X \to B$ produces a theory in which $(A \cap A')(x) = A(x) \cap A'(x) = A(x) \cap (A(x))' = \phi$ for all $x \in X$. We can then recapture numerical values by taking B to be the collection of Borel sets of I and assigning to each set A(x) its measure, as in probability theory.

It would appear that a theory of B-fuzzy topology could be constructed along the lines of the existing classical numerical-valued fuzzy topology. In particular, if X is a topological space an obvious B-fuzzy topology τ for X is obtained by defining closed fuzzy subsets to be upper semi-continuous functions (§3).

<u>Definition</u>. For a B-fuzzy topological space $(X, \mathcal{T})$ a subfamily $\mathcal{B}$ of $\mathcal{T}$ is a base for $\mathcal{T}$ iff for each $A \in \mathcal{T}$, there exists $\mathcal{B}_A \subseteq \mathcal{B}$ such that $A = \cup \mathcal{B}_A$.

Attempts to show that this definition of base is equivalent to the classical 'open neighbourhood' definition fail because if a B-set λ is contained in a union $\cup A_\alpha$ it does not necessarily lie in one of the A_α. Thus if the fuzzy point x_λ is a member of $\cup A_\alpha$ we cannot deduce that $x_\lambda \in A_\alpha$ for some value of α.

This difficulty arises for similar reasons in ordinary fuzzy, and is overcome by Pu and Liu using quasi-coincidence. A form of quasi-coincidence would therefore seem appropriate in B-fuzzy. Pu and Liu take a fuzzy point x_λ to be quasi-coincident with the fuzzy subset A, $(x q A)$ iff $\lambda > A'(x)$. The formal generalisation to B-fuzzy is the condition $\lambda \supset A'(x)$. Then it is true that if a fuzzy point x_λ is quasi-coincident with UA_α , there exists α such that $x_\lambda \, q \, A_\alpha$.

Definition. A B-fuzzy set A in $(X,\mathcal{T})$ is called a Q-neighbourhood of x_λ iff there exists $C \in \mathcal{T}$ such that $x_\lambda \, q \, C$, $C \subseteq A$. It is straightforward now to prove that a subfamily $\mathcal{B}$ of a B-fuzzy topology $\mathcal{T}$ on X is a base for $\mathcal{T}$ iff for each B-fuzzy point x_λ of X and each open Q-neighbourhood U of x_λ there exists a member $C \in \mathcal{B}$ such that $x_\lambda \, q \, C$, $C \subseteq U$.

Definitions of upper and lower semi-continuity exist for ordinary (not set-valued) functions and are used in classical fuzzy to induce (sometimes called weakly) a fuzzy topology for a given topological space (Pu and Liu (23), Martin (16)). Taking lower semi-continuous functions as open sets, there is a theorem by Pu and Liu which states that if X is a completely regular topological space the family of continuous functions from X to I forms a basis for this topology.

Although semi-continuity in I^X is not a special case of the set-valued semi-continuity of §3, there still seem to be parallel results in B-fuzzy. Taking open fuzzy subsets to be the complements of upper semi-continuous functions, we conjecture that, for a space X satisfying suitable separation axioms, continuous set-valued functions from X to I form a basis for the B-fuzzy topology τ ; that is, if A is the complement of an upper semi-continuous function, and $x_\lambda \, q \, A$, there exists a continuous function A_* such that $x_\lambda \, q \, A_*$, $A_* \subseteq A$.

Further B-fuzzification of Kelley would proceed along the lines of Pu and Liu (23).

Problems next arise in finding an appropriate definition for a fuzzy function from a set X to a set Y. A fuzzy relation between sets X,Y, (see later), is simply a function $\lambda : X \times Y \to I$. But how are we to specialise this by excluding the fuzzy equivalent of one-many relations ? There is no simple answer to this. Butnariu (4) defines a fuzzy map R over a set X to be a function $R : X \to$ {Fuzzy subsets of X} , and thus obtains some fixed point theorems. A discussion of fuzzy switching functions in automata appears in Kandel and Lee (10).

However, in the theory of fuzzy topology fuzzy functions are not necessary. Rather an ordinary function f from the 'crisp' set X to the 'crisp' set Y is required to be fuzzy continuous with respect to the fuzzy topologies $\mathcal{T},\mathcal{U}$ respectively, the condition being that if $A \varepsilon \mathcal{U}$ then $f^{-1}(A) \varepsilon \mathcal{T}$. (The fuzzy set $f^{-1}(A)$ is defined by $f^{-1}(A)(x)=A(f(x))$. The formal topological theory then continues with definitions of closure, separation axioms, connectedness, products and quotients. Pu and Liu (24) find that a judicious use of the quasi-versions of these definitions sharpens the theorems of other authors, e.g. Chang (5) Goguen (8) Wong (27).

For instance :

Definitions. $(X,\mathcal{T})$ is a fuzzy quasi-T_0-space iff for every $x \varepsilon X$, $\lambda, \mu \varepsilon [0,1]$, $\lambda \neq \mu$, either $x_\lambda \notin \overline{x}_\mu$ or $x_\mu \notin \overline{x}_\lambda$.

$(X,\mathcal{T})$ is a fuzzy T_0-space iff for any two fuzzy points e, d, $e \neq d$, either $e \notin \overline{d}$ or $d \notin \overline{e}$.

$(X,\mathcal{T})$ is a fuzzy T_1-space iff all fuzzy points are closed sets.

$(X,\mathcal{T})$ is a fuzzy Hausdorff (T_2) space iff for any two fuzzy points e,d such that $\sup e \neq \sup d$, there exist quasi-neighbourhoods B,C of e,d respectively such that $B \cap C = \phi$.

Theorem : If $(X,\mathcal{T})$ is both T_2 and quasi-T_0, then it is also T_1 .

There are at least seven (different) current definitions of compactness, an account of which is given by Lowen (14), who points out which are the 'good' generalisations of the classical theory. We have nothing to add to this account.

5 LATTICE-VALUED RELATIONS

If L is a lattice and X,Y sets, then a lattice-valued relation (l.v.r.) from X to Y is a function $\lambda : X \times Y \to L$.

When L = I, this is the usual fuzzy relation of fuzzy set theory. In general it is Goguen's (8) fuzzy L-relation, as distinct from his L-fuzzy L-relation (which is a function from $L^{X \times Y} \to L$). Goguen develops an algebraic theory of fuzzy relations now adopted in classical text books e.g. (11).

For example, composition of fuzzy L-relations $\lambda : X \times Y \to L$, $\mu : Y \times Z \to L$, is given by $(\mu \circ \lambda)(x,z) = \bigvee_y \{(x,y) \wedge \mu(y,z)\}$.

Definitions of symmetry, reflexivity and transitivity emerge naturally, (e.g. see (11)).

Definition. A fuzzy (I)-relation $R : X \times X \to I$ is symmetric iff (i) $R(x,x) = 1$ for all $x \in X$, reflexive iff (ii) $R(x,y) = R(y,x)$ for all $x,y \in X$, and transitive iff
(ii) $R(x,z) = \max_y(\min(R(x,y)), R(y,z))$ for all $x,y,z \in X$.

A weakening of transitivity gives a fuzzy equivalence relation which looks startlingly like a pseudo metric. This replaces condition (iii) by (iv) $\min(R(x,y)), R(y,z)) \leq R(x,z)$.

Some of the most interesting l.v.r. spaces use the lattice of subsets of a given set ; e.g. (i) An automaton transition function $\delta : X \times Q \to Q$ (input set X, state space Q) may be recast as a function $\lambda : Q \times Q \to P(X)$ where $\lambda(q,q') = \{x \in X ; q' = \delta(x,q)\}$, (20).
(ii) If L is the lattice of Borel sets in a probability space Ω, an l.v.r. may be interpreted as assigning, via any measure on Ω, a probability of the relation between two events, (20).

Definition. A morphism from an l.v.r. space (Q,L,λ) to an l.v.r. space (Q',L',λ') is a pair of functions $\beta : Q \to Q'$, $\gamma : L \to L'$ where γ is a lattice morphism, and $\lambda' : (\beta(q),\beta(q')) \geq \gamma(\lambda(q,q'))$ for all $q,q' \in Q$. If β,γ are bijective and (β^{-1},γ^{-1}) is a morphism, we have an isomorphism of l.v.r. spaces for which $\lambda'(\beta(q),\beta(q')) = \gamma(\lambda(q,q'))$.

When $L = L'$ and $\gamma = 1$, the morphism is referred to as β. Muir and Warner (20) have built up a theory of homogeneity for l.v.r. spaces, the main theorem of which produces a necessary and sufficient condition for an l.v.r. space Q to be isomorphic to a quotient $\frac{G}{H}$ of its isomorphism group G by a stabiliser subgroup H. The condition obtained is that the l.v.r. space (Q,L,λ) should be very homogeneous (v.h.), that is, that for all $q', q'' \in Q$, there exists $g \in G$ such that $g(q') = q''$ and if e is the identity in G,

$$\lambda(q',q'') \leq \bigvee_{\substack{g \text{ s.t.} \\ g(q')=q''}} \mu(e,g) \tag{1}$$

$$\text{where } \mu(e,g) = \bigwedge_{q \in Q} (q,g(q))$$

For the simple case when $L = \{0,1\}$, and λ is a reflexive symmetric relation, (tolerance), the theorem appeared in (19). We use it to illustrate the 'physical meaning' of the somewhat obscure theoretical definition provided by (1). A topological analogy assists the illumination.

In a tolerance space (X,ρ) the condition 'very homogeneous' reads 'For all x', x" there exists an isomorphism g such that g(x') = x", and if $x' \rho x''$ then $\min_{\overline{x} \in X} \rho(\overline{x},g(\overline{x})) = 1$, so $\overline{x} \rho g(\overline{x})$ for $\overline{x} \varepsilon X$. If the tolerance relation is thought of as representing 'nearness', the condition on g requires g to move all points of X to nearby points ; that is, the mapping which transfers x' to the nearby point x" can be achieved without disrupting the rest of X.

There exists in the topological category an analogous group quotient theorem (e.g. McCarty (17)). In this case the homeomorphism group of a topological space is not always the appropriate action group G. Attempts to make this so by mimicking the l.v.r. theory produce a transformation group G which is not a topological group. However, if we stipulate that there should exist a topological transformation group G of the space X acting transitively on X, (the analogue of homogeneity), the following theorem is classical (17).

<u>Theorem</u>. If G is a transitive topological transformation group acting on the space X, and x_0 is a given (base)-point in X, the projection $p : G \to X$, $p(g) = g(x_0)$, induces a continuous bijection $\overline{p} : {}^G/_H \to X$ where H is the subgroup of G which fixes x_0 . If, further, p has a local cross-section at x_0, then $\overline{p}$ is a homeomorphism.

We recall that a local cross-section to p at x_0 is a continuous function $s : S \to G$, where S is a neighbourhood of x_0, such that ps = identity. So $\overline{p}^{-1}|_S(x) = [s(x)]$ since $\overline{p}([s(x)]) = ps(x) = x$, and $\overline{p}$ is bijective.

For tolerance it is natural to replace topological neighbourhoods by their tolerance analogues, namely the set of points 'near to ' the base-point x_0. This set $\mu(x_0) = \{x \varepsilon X ; x \rho x_0\}$ is called the monad of x. (20).

Let G be a tolerance group (20) acting transitively on X, for example, the set of isomorphisms on X (tolerance preserving bijections with tolerance preserving inverses). Again let $p : G \to X$ be given by $p(g) = g(x_0)$ for some x_0, whereupon a local cross-section to p at x_0 will be a tolerance map $s : \mu(x_0) \to G$ such that ps = 1. Thus

$s(x_0) = h \in H_{x_0}$ since $s(x_0)$ maps x_0 to itself. Since s is a tolerance map it maps any $x \in \mu(x_0)$ to an element $g \in G$ which is near to h, that is $g(\bar{x}) \rho h(\bar{x})$ for all $\bar{x} \in X$.

Thus the v.h. condition for tolerance spaces seems to have been replaced by 'for $x \rho x_0$, there exist $g \in G$, $h \in H_{x_0}$ such that $g(\bar{x}) \rho h(\bar{x})$ for all $\bar{x} \in X'$. But clearly we can modify the cross-section by right multiplication by h^{-1}. For $s' : \mu(x_0) \to G$ defined by $s'(x) = s(x)h^{-1}$ is still a cross-section since $ps'(x) = s'(x)(x_0) = s(x)h^{-1}(x_0) = s(x)(x_0) = ps(x) = x$, and now $s'(x)$ is near the identity of G as required. Then for $x \rho x_0$ there exists $\bar{g} \in G$ such that $\bar{g}(x_0) = x$ and $\bar{g}(\bar{x}) \rho \bar{x}$ for all $\bar{x} \in X$. Homogeneity extends this immediately to all $x' \rho x''$, so that the cross-section condition is identical with the tolerance version of very homogeneous.

6 LATTICE VALUED FUNCTIONS

A pretopological space is a set X with a function $\lambda : P(X) \to P(X)$ such that $A \subseteq \lambda(A)$ for all $A \subseteq X$. (e.g. Badard (1)). This is a weakening of Kuratowski's (13) definition of a topological space in terms of closure. The closure $\lambda(A)$ of A is written $\bar{A}$ and satisfies the three closure axioms. Both kinds of spaces are thus defined using a lattice valued function (l.v.f.). Neither of these functions can be deduced from a (closure) point function $\lambda : q \longmapsto \bar{q}$, $q \in X$, $\bar{q} \in P(X)$. In fact for a T_1-space $q = \bar{q}$ and λ is the identity.

Definition. A morphism of lattice valued functions $\lambda : Q \to L$, $\lambda' : Q' \to L'$ is a pair $\beta : Q \to Q'$, $\gamma : L \to L'$ (γ a lattice morphism), such that $\lambda'(\beta(q)) \geq \gamma\lambda(q)$. Thus for $L = L' = I$ and γ the identity, the fuzzy subset λ is required to be a subset of $\lambda'\beta (\lambda \subseteq \lambda'\beta)$ (Zadeh (28)). Adapting this to the closure operation, let $\gamma = \beta: P(X) \to P(X')$ be induced by the function $f : X \to X'$. Then f is a morphism iff $\overline{f(A)} \supseteq f(\bar{A})$ for all $A \subseteq X$, the classical requirement for f to be a continuous function from the topological space X to the topological space X'.

For a fuzzy subset $\lambda : X \to I$ and a function $f : X \to Q$, the induced fuzzy set in X is $\lambda f : X \to I$, written $f^{-1}(\lambda)$ by Pu and Liu. For topological closure, given $f : X \to Q$, $B \subseteq X$, we must have $\bar{B} = f^{-1}\overline{(f(B))}$ which is the classical induced topology on X.

Thus closure treated as an l.v.f. gives rise naturally to classical topological definitions.

7 TOPOLOGY AND HOMOLOGY

A natural attempt to put a topology on an l.v.r. space could proceed as follows. Let $\lambda : X \times X - \Delta \to L$ be a lattice valued relation not defined on the diagonal Δ of $X \times X$. For $x \in X$, $\ell \in L$, define $B_\ell(x) = \{x' \in X - \{x\} \ ; \ \lambda(x,x') \geq \ell\} \cup \{x\}$. Then a subset V of X is open if for all $x \in V$, $\exists\, \ell \in L$ such that $B_\ell(x) \subseteq V$. It follows that the open subsets of X so defined form a topology , $\mathcal{T}$ on X. And it is straightforward to show that l.v.r. morphisms are continuous with respect to this topology. In fact the condition for l.v.r. morphisms is stronger than required, since the following statements hold true. If (X,L,λ), (X',L',λ') are l.v.r. spaces, a function $f : X \to X'$ is <u>nice</u> if for all $x \in X$, $\ell' \in L$, there exists $\ell \in L$ such that $\lambda(x,x') \geq \ell \implies \lambda'(f(x),f(x')) \geq \ell'$. Then nice functions are continuous with respect to the topology $\mathcal{T}$.

Dowker (7) defines an n-simplex of a relation $\rho : Y \times Z \to \{0,1\}$ from a set Y to a set Z as $(n+1)$ elements of Y which are ρ-related to some element of Z. Once simplexes are defined, a simplicial homology theory results by standard calculations.

We can alternatively define an n-simplex as $(n+1)$ elements of Z which are ρ-related to some element of Y. Dowker proves that the corresponding homology modules H(Y) and H(Z) are isomorphic. In l.v.r. terms the following definition suggests itself.

<u>Definition</u>. Let $\lambda : Q \times Q' \to L$ be a lattice valued relation from Q to Q' and let $\ell \in L$. An ℓ-n-simplex of λ is an ordered set $[q_0 \ldots q_n]$ of elements of Q such that $\lambda(q_i,q') > \ell$ for some $q' \in Q'$, $i = 0, \ldots, n$.

With $L = \{0,1\}$, $\ell = 0$, we restore Dowker's original definition. Clearly the requirement that $\lambda(q,q') > \ell$ is an ordinary relation from Q to Q', so for given ℓ , Dowker's homology isomorphism theorem still holds. Moreover if $\ell \geq \ell'$ every ℓ-n-simplex is an ℓ'-n-simplex so any sequence of lattice elements $\ell_0 \leq \ell_1 \leq \ell_2 \leq \ldots$ yields a filtration on the chain complex C of ℓ_0-simplexes which is compatible with the gradation and differential of C (25).

The apparatus of spectral sequence theory is tailor made for computing homologies from filtrations, so one may expect to find intriguing relationships between the exact homology sequences for each $\ell \in L$ and the lattice structure of L.

Reference List.

1 Badard, R. 'Fuzzy pretopological spaces and their representation'. J. Math. Anal. & App., 81 (2), (1981), 378-390.

2 Berge, C. 'Topological spaces'. Oliver and Boyd (1963).

3 Birkhoff, G. 'Lattice theory', Providence, Amer. Math. Soc. (1966).

4 Butnariu, D. 'Fixed points for fuzzy mappings'. Fuzzy sets and systems. 7 (1982), 191-207.

5 Chang, C. L. 'Fuzzy topological spaces'. J. Math. Anal. and App. 24, (1968), 182-190.

6 Dodson, C. T. J. 'Hazy spaces and fuzzy spaces'. Bull. Lond. Math. Soc. 6, (1974), 191-197.

7 Dowker, C. H. 'Homology groups of relations'. Annals of Math. 56, (1952), 84-95.

8 Goguen, J. A. 'L-fuzzy sets'. J. Math. Anal. and App. 18, (1967), 145-174.

9 Gottwald, S. 'Fuzzy points and local properties of fuzzy topological spaces'. Fuzzy sets and systems. 5, (1981), 199-201.

10 Kandel, A. and Lee, S. C. 'Fuzzy switching and automata'. Arnold, London (1979).

11 Kaufmann, A. 'Theory of fuzzy subsets', Vol. 1. Academic Press, New York (1975).

12 Kelley, J. L. 'General Topology'. Van Nostrand, (1955).

13 Kuratowski, K. 'Topology'. Academic Press, (1966).

14 Lowen, R. 'A comparison of different compactness notions in fuzzy topological spaces'. J. Math. Anal. App. 64 (1978), 446-454.

15 Lowen, R. 'On fuzzy complements'. Inform. Sci. 14 (1978), 2, 107-113.

16 Martin, H. W. 'Weakly induced fuzzy topological spaces'. J. Math. Annal and App. 78 (1980), 634-639.

17 McCarty, G. 'Topology'. McGraw Hill (1967).

18 Muir, A. 'Fuzzy sets and probability'. Kybernetes, 10, (1981), 197-200.

19 Muir, A. and Warner, M. W. 'Homogeneous tolerance spaces'. Czech. Math. Journal 30 (105) (1980), 118-126.

20 Muir, A. and Warner, M. W. 'Lattice-valued relations and automata'. Discrete App. Math. 7 (1984) 65-78.

21 Natvig, B. 'Possibility v. Probability'. Fuzzy sets and systems 10 (1983), 31-37.

22 Poston, T. 'Fuzzy geometry'. Thesis, Univ. of Warwick (1968).

23 Pu, P. M. and Liu, Y. M. 'Fuzzy topology I. Neighbourhood structure of a fuzzy point and Moore-Smith convergence'. J. Math. Anal and App. 76 (2) (1980), 571-599.

24 Pu, P. M. and Liu, Y. M. 'Fuzzy topology II. Product and quotient spaces'. J. Math. Anal. and App. 77 (1) (1980), 20-37.

25 Spanier, E. H. 'Algebraic Topology'. McGraw-Hill, (1966).

26 Springer Lecture Notes in Mathematics 171. 'Set-valued mappings, selections and topological properties of 2^X'. (1970).

27 Wong, C. K. 'Fuzzy points and local properties of fuzzy topology'. J. Math. Anal. and App. 46 (1974). 316-328.

28 Zadeh, L. A. 'Fuzzy sets'. Information and Control 8 (1965), 338-353.

29 Zadeh, L. A. 'Fuzzy sets as a basis for a theory of possibility'. Fuzzy sets and systems 1 (1978) 3 - 28.

30 Zeeman, E. C. 'The topology of the brain and visual perception'. In 'The topology of 3-manifolds'. Ed. M. K. Fort, 240-56. Prentice Hall (1961).

General topology over a base

I.M. James

1. Introduction

In ordinary topology one is concerned with the category Top of spaces and maps, i.e. continuous functions. In this article, however, I wish to consider rather the category Top_B of spaces and maps over a given space B. My aim is to show that many of the familiar definitions and theorems of ordinary topology can be generalized, in a natural way, so that one can develop a theory of topology over a base. In fact, once the definitions have been suitably formulated the proofs of the theorems are mostly just fairly routine generalizations (see [4], Chapter 3) of those used in ordinary topology. There are, however, certain results which have no counterpart in the ordinary theory and for these, of course, I will give proofs.

A space over B, I recall, is a space X together with a map $p : X \to B$, called the projection. Usually X alone is sufficient notation. If X is a space over B then any subspace of X may be regarded as a space over B by restriction of the projection. Also B itself is regarded as a space over B with the identity map as projection.

If X,Y are spaces over B with projections p,q, respectively, then a map $X \to Y$ over B is a map $\phi : X \to Y$ of spaces such that $q\phi = p$. The category Top_B of spaces and maps over B has various features which are relevant. First of all B, as a space over itself, constitutes a final object of the category; the maps $B \to X$ over B are called sections. Next the fibre sum $X \underset{B}{+} Y$ and fibre product $X \underset{B}{\times} Y$ are defined, for spaces X,Y over B. The fibre sum is simply the disjoint union, as in Top, with the projection given by p on X, by q on Y. The fibre product is the subset of the topological product $X \times Y$ consisting of pairs (x,y) such that $px = qy$; the projection is given by $(x,y) \to px = qy$. Infinite fibre sums and products can also be defined but we shall not be concerned with these here.

For each point b of B a functor

$$\mathrm{Top}_B \to \mathrm{Top}$$

is defined which associates with each space X over B the fibre $X_b = p^{-1}b$, and similarly for maps over B. Note that for spaces X,Y over B a map $\phi : X \to Y$ of spaces satisfies the condition $p = q\phi$ if and only if $\phi X_b \subset Y_b$ for all points b of B. For this reason the term "fibre-preserving map" is often used instead of map over B, when it is clear which base space B is intended.

For each space B' and map $\xi : B' \to B$ a functor

$$\xi^* : \mathrm{Top}_B \to \mathrm{Top}_{B'}$$

is defined, where $\xi^*X = X \underset{B}{\times} B'$ for each space X over B. Here B' is regarded as a space over B using ξ as projection. The projection $\xi^*X \to B'$ is given by pr_2, while a canonical map $\xi^*X \to X$ is given by pr_1. The functor id*, where $id : B \to B$ is the identity, is naturally equivalent to the identity functor on $\underline{Top}_B$. Also if $\xi : B' \to B$ and $\xi' : B'' \to B'$ are maps then $(\xi \circ \xi')^*$ is naturally equivalent to $\xi'^* \circ \xi^*$.

Note that if X,Y are spaces over B then $\xi^*(X \underset{B}{+} Y)$ is naturally equivalent to $\xi^*Y \underset{B}{+} \xi^*Y$, and $\xi^*(X \underset{B}{\times} Y)$ is naturally equivalent to $\xi^*X \underset{B}{\times} \xi^*Y$.

If B' is a subspace of B and ξ the inclusion then ξ^* is naturally equivalent to the functor

$$\underline{Top}_B \to \underline{Top}_{B'}$$

which associates with each space X over B the space $X_{B'} = p^{-1}B'$ over B', with the projection given by p, and similarly with maps over B.

We also have a functor

$$B\times \; : \underline{Top} \to \underline{Top}_B$$

which associates with each space T the space $B \times T$ over B, with projection pr_1, and similarly with maps.

To illustrate the use of this language I reformulate two well-known results of ordinary topology (see [2], for example).

<u>Proposition</u> (1.1). Let $\phi : X \to Y$ be a map over B. If ϕ is open (resp. closed) then the map $\phi' : X_{B'} \to Y_{B'}$ determined by ϕ is open (resp. closed) for each subset B' of B.

<u>Proposition</u> (1.2). Let $\phi : X \to Y$ be a map over B. Let $\{B_j\}(j \in J)$ be a family of subsets of B such that either

(i) $\{\text{Int } B_j\}$ is an open covering of B, or

(ii) $\{B_j\}$ is a locally finite closed covering of B.

If each of the maps $\phi_j : X_{B_j} \to Y_{B_j}$ determined by ϕ is open (resp. closed) then ϕ is open (resp. closed).

The standard functors and binary functors of Top, such as the join, can be extended to Top_B as follows. Let X and Y be spaces over B. The fibre-join $X \underset{B}{*} Y$ is defined as the quotient space of the fibre-sum

$$I \times X \underset{B}{\times} Y \underset{B}{+} X \underset{B}{+} Y$$

with respect to the relations

$$(0,x,y) \sim x,\ (1,x,y) \sim y \qquad (x \in X_b,\ y \in Y_b,\ b \in B),$$

and similarly for maps over B. In particular the fibre-cone $\Gamma_B X$ and fibre-suspension $\Sigma_B X$ of the space X over B are given by

$$\Gamma_B X = X \underset{B}{*} (B \times \{0\}),\quad \Sigma_B X = X \underset{B}{*} (B \times \{0,1\}).$$

In general the fibre-join operation is not associative.

2. Topologizing a set over a base

Let B be a space and let X be a set over B, with projection $p : X \to B$. If we wish to topologize X as a space over B then we have at least to ensure that p is continuous. Thus the open sets for any such topology must include the sets $X_U = p^{-1}U$ for each open set U of B. I refer to these as the special open sets.

Let Γ be any collection of subsets of X. By the topology generated by Γ _after supplementation_ I mean the topology generated, in the usual sense, by the collection consisting of Γ supplemented by the special open sets.

If Γ satisfies the usual basis condition, i.e. if the intersection of two members of Γ is a union of members of Γ, then I say that Γ is a basis for the topology of X after supplementation. Otherwise I say that Γ is a subbasis for the topology of X after supplementation.

For example, take Γ to be the empty collection. Then the topology generated by Γ after supplementation is just the topology induced by the projection.

Proposition (2.1). Let X be a space over B and, for any space B' and map $\xi : B' \to B$, let ξ^*X be the induced space over B', with canonical map $\tilde{\xi} : \xi^*X \to X$. If ξ is a basis (resp. subbasis) for the topology of X after supplementation then $\tilde{\xi}^{-1}\Gamma$ is a basis (resp. subbasis) for the topology of ξ^*X after supplementation.

With these definitions some of the most basic results of ordinary topology can be generalized, as follows:

Proposition (2.2). Let $\phi : X \to Y$ be a function over B, where X and Y are spaces over B. Let Γ be a subbasis for the topology of Y after supplementation. Then ϕ is continuous if and only if $\phi^{-1}V$ is open for each member V of Γ.

Proposition (2.3). Let $\phi : X \to Y$ be a function over B, where X and Y are spaces over B. Let Γ be a basis for the topology of X after supplementation. Then ϕ is open if and only if ϕU is open for each member U of Γ.

Proposition (2.4). Let $\phi : X \to Y$ be an injective function over B, where X and Y are spaces over B. Let Γ be a subbasis for the topology of X after supplementation. Then ϕ is open if and only if ϕU is open for each member U of Γ.

If X is a set over B and X has a topology we can always refine the topology so as to make the projection continuous. Specifically the open sets in the new topology are the unions

$$\bigcup_{j \in J} (V_j \cap X_{W_j})$$

for arbitrary J, where each V_j is an open set of X in the old topology, and each W_j is an open set of B. To distinguish between the two topologies, we use $\check{X}$ when we refer to the old topology and $\hat{X}$ when we refer to the new one.

Of course the open sets of the old topology satisfy the basis condition, in a particularly strong way, and the new topology is generated by this basis after supplementation. Thus the procedure we have just described is natural in the sense of (2.1). In particular the fibres of X have the same topology whether the old or the new topology is used.

Proposition (2.5). Let $\phi : X \to Y$ be a function over B, where X is a set over B and Y is a space over B. Suppose that ϕ is proper, in some topology $\check{X}$ for the set X. Then ϕ is proper in the refined topology $\hat{X}$.

The first step is to show that each fibre $\phi^{-1}y$ $(y \in Y)$ is compact in the new topology as well as the old. Suppose therefore that we have a covering of $\phi^{-1}y$ by sets $\{U_k\}$ $(k \in K)$ which are open in the new topology. We have to extract a finite subcovering. Now each U_k is the union of a family

$$\{V_{j,k} \cap X_{W_{j,k}}\} \qquad (j \in J_k),$$

where each $V_{j,k}$ is open in the old topology and each $W_{j,k}$ is open in B. Without real loss of generality we may assume that each $W_{j,k}$ contains the point $qy = b$, say. Now the family

$$\left\{\bigcup_{j \in J_k} V_{j,k}\right\} \qquad (k \in K)$$

is a covering of $\phi^{-1}y$ by sets which are open in the old topology. Since $\phi^{-1}y$ is compact in the old topology we can extract a finite subcovering, indexed by a finite subset $L \subset K$. Since each set $X_{W_{j,\ell}}$ $(\ell \in L)$ contains X_b and hence $\phi^{-1}y$ it follows at once that $\{U_\ell\}$ $(\ell \in L)$ covers $\phi^{-1}y$. Thus we have extracted a finite subcovering of the original covering as required.

It remains to be shown that the function $\phi : \hat{X} \to Y$ is closed. So let U be an (open) neighbourhood of $\phi^{-1}y$ in the new topology. We have to show that there exists a neighbourhood N of y in Y such that $\phi^{-1}N \subset U$. Now U is the union of the members of a covering

$$\{V_j \cap X_{W_j}\} \qquad (j \in J)$$

of $\phi^{-1}y$, where each V_j is open in the old topology and each W_j is open in B. Since $\phi^{-1}y$ is compact, as we have seen, we can assume without real loss of generality that J is finite. We can also assume, without real loss of generality, that each W_j contains the point $qy = b$, Now

$$V = \bigcup_{j \in J} V_j$$

is a neighbourhood of $\phi^{-1}y$ in the old topology and so $N' = Y - \phi(X - V)$ is a neighbourhood of y such that $\phi^{-1}N' \subset V$. Also

$$W = \bigcap_{j \in J} W_j$$

is a neighbourhood of b, and $N = N' \cap Y_W$ is a neighbourhood of y such that $\phi^{-1}N \subset U$ as required.

3. Special classes of space over a base

In ordinary topology one considers various classes of space, for example discrete spaces, compact spaces, Hausdorff spaces and so forth. Some of these classes are closed under the sum operation +, in which case they are called additive, or the product operation ×, in which case they are called multiplicative. Some are hereditary, in the sense that the properties in question are inherited by closed subspaces, or by open subspaces, or by all subspaces.

Given a property P of spaces we seek to define a property P_B of spaces over B, for each space B, so that the following three conditions are satisfied:

<u>Condition</u> (3.1). Let X,Y be equivalent spaces over B. If X has property P_B then so does Y.

<u>Condition</u> (3.2). The space X has property P_*, as a space over the point-space $*$, if and only if the space X has property P.

<u>Condition</u> (3.3). Let X be a space over B. If X has property P_B then ξ^*X has property $P_{B'}$ for each space B' and map $\xi : B' \to B$.

These conditions imply that $B \times T$ has property P_B, as a space over B, if and only if T has property P.

It is convenient to refer to P_B as either "P over B" or "fibrewise P", as seems most natural in a given situation. In general there is no relation between X having property P_B, as a space over B, and X having property P, as a space. If X has property P_B then each fibre X_b has property P, from the above conditions; however, this necessary condition is not, in general, sufficient, as we shall see.

We will also aim to define P_B so that if P is additive, multiplicative or hereditary, in any sense, then so is P_B.

I do not suggest that there is any canonical extension of P to P_B. In fact this is not possible. For example, take P to be the universal property, which is satisfied by every space. Then we can take P_B to be the universal property, which is satisfied by every space over B. Or we can take P_B to be the following property.

Definition (3.4). The space X over B is open if the projection $X \to B$ is an open map.

Here it is understood that every space is an open space. I give one result, for use later, where this property is involved.

Proposition (3.5). Let $\phi : X \to Y$ be a map over B, where Y is a space over B and X is an open space over B. If

$$\phi \times id : X \times_B X \to Y \times_B X$$

an open map then so is ϕ.

Definition (3.6). The space X over B is indiscrete over B if for each open set U of X there exists an open set V of B such that $U = p^{-1}V$.

In other words the indiscrete topology is the topology induced by the projection. Note that a fibre preserving surjection with a fibrewise indiscrete space as codomain is necessarily continuous.

Definition (3.7). The space X over B is discrete over B if for each point x of X there exists a neighbourhood U of x and a neighbourhood V of px such that p maps U homeomorphically onto V.

In other words the condition is that the projection is locally homeomorphic. A fibre preserving function with a fibrewise discrete space as domain is necessarily continuous if the projection of the codomain is an open map.

The classes of indiscrete and discrete spaces over a base are both additive and multiplicative. The class of indiscrete spaces is hereditary for all subspaces while the class of discrete spaces is hereditary for open subspaces.

Proposition (3.8). Let X be a space over B and let Y be a discrete space over B. Let $\theta, \phi : X \to Y$ be maps over B. Then the coincidence set of (θ, ϕ) is open in X.

If θ and ϕ have no coincidence points there is nothing further to prove. So suppose that $\theta x = \phi x$ for some point x of X. There exists a neighbourhood U of $\theta x = \phi x$ in Y and a neighbourhood V of $px = q\theta x$ in B such that q maps U homeomorphically onto V. Then θ coincides with ϕ on the neighbourhood

$$\theta^{-1}U \cap \phi^{-1}U \cap p^{-1}V$$

of x. Therefore the set of coincidence points is open, as asserted.

Definition (3.9). The space X over B is disconnected over B if there exists a partition $X = U \cup V$, where U,V are disjoint open sets such that $pU = B = pV$.

Here too the class of spaces is additive and multiplicative, but not hereditary. The negation of non-connected over B; however the negation does not satisfy our third condition. We have

Proposition (3.10). Let $\phi : X \to Y$ be a continuous surjection over B. If X is connected over B then so is Y.

To illustrate this concept consider, in the real plane, the subsets

$$X = \{t,i\}, \quad B = \{t,it\} \qquad (-1 \le t \le 1,\ i = \pm 1)$$

with the projection p given by $p(t,i) = (t,it)$. Here X is connected over B while, as spaces, B is connected but X is not.

4. Compact spaces over a base

Definition (4.1). The space X over B is compact over B if the projection $p : X \to B$ is proper, i.e. if the projection is closed and each fibre X_b is a compact space.

This satisfies our three conditions and moreover is both additive and multiplicative. Furthermore the property is hereditary so far as closed subspaces are concerned.

The most familiar examples of compact spaces are closed and bounded subspaces of the euclidean $\mathbb{R}^n$ $(n = 0,1,\dots)$. For spaces over a base we have

Proposition (4.2). Let X be a closed subspace of $B \times \mathbb{R}^n$. Then X is compact over B if there exists a map $\alpha : B \to \mathbb{R}$ such that X_b is bounded by $\alpha(b)$ for each point b of B.

For suppose that the condition is satisfied. Since the euclidean norm $D : \mathbb{R}^n \to \mathbb{R}$ is a proper map, so is the product

$$\mathrm{id} \times D : B \times \mathbb{R}^n \to B \times \mathbb{R} .$$

Consider the graph Γ of α in $B \times \mathbb{R}$. The projection $\Gamma \to B$ is a homeomorphism and so proper. Also the map

$$(\mathrm{id} \times D)^{-1}\Gamma \to \Gamma$$

determined by $\mathrm{id} \times D$ is proper. Therefore $(\mathrm{id} \times D)^{-1}\Gamma$ is compact over B. But X is closed in $B \times \mathbb{R}^n$, by hypothesis, and so closed in $(\mathrm{id} \times D)^{-1}\Gamma$. Thus X is compact over B, as asserted.

Incidentally the converse of (4.2) holds provided we place some restriction on B, for example metrizability; I owe this remark to Albrecht Dold.

Many of the standard results about proper maps (see Bourbaki [2], for example) become easier to understand in the present language. I give three illustrations of this.

Proposition (4.3). Let $\phi : X \to Y$ be a surjective map over B. If X is compact over B then so is Y.

Proposition (4.4). Let X be a space over B. Let $\{X_j\}$ be a finite covering of X such that X_j is compact over B for each index j. Then X is compact over B.

Proposition (4.5). Let X be a space over B. Let $\{B_j\}$ $(j \in J)$ be a family of subsets of B such that either

(i) the family $\{\mathrm{Int}\, B_j\}$ covers B, or

(ii) the family $\{B_j\}$ is a locally finite closed covering of B.

Suppose that X_{B_j} is compact over B_j for each index j. Then X is compact over B.

Locally compact spaces over a base are defined in the obvious way:

Definition (4.6). The space X over B is locally compact over B if each point x of X admits a neighbourhood N such that $C\ell\, N$ is compact over B.

This satisfies our three conditions and moreover is both additive and multiplicative. Furthermore the property is hereditary in the sense of closed subspaces.

Proposition (4.7). Let $\phi : X \to Y$ be a continuous open surjection over B. If X is locally compact over B then so is Y.

5. Separation axioms over a base

I will discuss the Hausdorff and regularity axioms although the other separation axioms can be dealt with similarly.

Definition (5.1). The space X over B is Hausdorff over B if for each point b of B and each distinct pair of points x, x' of X_b there exist neighbourhoods U of x, U' of x' in X which are disjoint.

Thus the only difference from the Hausdorff axiom for X as a space is that both points x, x' have to lie in the same fibre. However, to require each fibre X_b to be a Hausdorff space is insufficient.

Clearly the fibrewise Hausdorff property satisfies the three conditions, is additive and multiplicative, and is hereditary for all subspaces.

Discrete spaces over B are Hausdorff over B, indiscrete spaces over B, in general, are not.

<u>Proposition</u> (5.2). The space X over B is Hausdorff over B if and only if the diagonal embedding

$$\Delta : X \to X \times_B X$$

is closed.

<u>Proposition</u> (5.3). Let Y be a Hausdorff space over B. Then for each space X over B and each map $\phi : X \to Y$ over B the graph

$$\Gamma_\phi : X \to X \times_B Y$$

is a closed embedding.

<u>Proposition</u> (5.4). Let Y be a Hausdorff space over B. Then for each space X over B and each pair of maps $\theta, \phi : X \to Y$ over B the coincidence set of (θ, ϕ) is closed in X.

Comparing (5.4) with (3.8) above we obtain

<u>Corollary</u> (5.5). Let X be a connected space over B and let Y be a discrete space over B. Let $\theta, \phi : X \to Y$ be maps over B. Then either there exists a fibre of X throughout which $\theta = \phi$, or there exists a fibre of X throughout which $\theta \neq \phi$.

<u>Proposition</u> (5.6). Let $\phi : X \to Y$ be a map over B, where X is compact over B and Y is Hausdorff over B. Then ϕ is proper.

<u>Proposition</u> (5.7). Let $\phi : X \to Y$ be a proper surjection over B. If X is Hausdorff over B then so is Y.

Let us turn now to the regularity axiom.

Definition (5.8). The space X over B is regular over B if for each point x of X and each neighbourhood V of x there exists a neighbourhood U of x such that $X_b \cap C\ell\, U \subset V$, where $b = px$.

It is important to notice that if the condition holds for all V in a subbasis for the supplemented topology of X then it holds for all V and so X is regular over B.

Another way to formulate the definition is as follows. Let E be a subset of a fibre X_b, with E closed in X, and let x be a point of X_b which is not contained in E. Then X is regular over B if, in this situation, there exist neighbourhoods U of x and V of E in X which are disjoint.

The property satisfies our three conditions and moreover is additive and multiplicative. It is also hereditary for closed subspaces. Discrete spaces over B are regular over B and so are indiscrete spaces over B.

Proposition (5.9). If X is compact Hausdorff over B then X is regular over B.

Proposition (5.10). Let $\phi : X \to Y$ be a proper surjection over B. If X is regular over B then so is Y.

The familiar shrinking theorems of ordinary topology can also be generalized, as in

Proposition (5.11). Let X be regular over B. Let C be a compact subspace of some fibre X_b of X and let V be a neighbourhood of C in X. Then there exists a neighbourhood U of C in X such that $X_b \cap C\ell\, U \subset V$. Further, if X is also locally compact over B then U can be chosen so that $C\ell\, U$ is also compact over B.

Proposition (5.12). If $\phi : X \to Y$ is a quotient map over B then so is

$$\phi \times id_T : X \underset{B}{\times} T \to Y \underset{B}{\times} T$$

for each locally compact regular space T over B.

Corollary (5.13). Let X,Y,Z be spaces over B. Suppose either that (i) X and Z are locally compact regular over B or that (ii) X and Y (or Y and Z) are compact regular over B. Then

$$(X \underset{B}{*} Y) \underset{B}{*} Z \equiv X \underset{B}{*} (Y \underset{B}{*} Z).$$

6. The adjoint of the fibre product

Consider, for spaces X,Y over B, the set

$$\mathrm{map}_B(X,Y) = \coprod_{b \in B} \mathrm{map}(X_b, Y_b)$$

over B, where map (X_b, Y_b) denotes the set of maps of X_b into Y_b. There are various ways in which this set might be topologized so that the space over B thus obtained would, subject to certain restrictions, serve as an adjoint to the fibre product. The topology I am going to propose coincides with the one studied by Booth [1] when B is compact Hausdorff but is finer in general. It also coincides with the one considered by Niefield [5] when X is Hausdorff over B but is coarser in general. For each pair of subsets $K \subset X$ and $V \subset Y$ let (K,V) denote the subset of $\text{map}_B(X,Y)$ consisting, for each point $b \in B$, of the maps $\phi : X_b \to Y_b$ such that $\phi K_b \subset V_b$. Note that if K_b is empty, for some b, then (K,V) contains all the maps $X_b \to Y_b$. In particular $\text{map}_B(X,Y) = (\emptyset,Y)$.

<u>Definition</u> (6.1). The fibrewise compact-open topology for the set $\text{map}_B(X,Y)$ over B is the topology generated, after supplementation, by the family of subsets (K,V), where $K \subset X$ is compact over B and $V \subset Y$ is open.

We refer to the subbasic sets (K,V) as the fibrewise compact-open sets. When $B = *$ the fibrewise compact-open topology on $\text{map}_*(X,Y)$ reduces to the compact-open topology on $\text{map}(X,Y)$. It is easy to check that $\text{map}_B(X,B)$ is equivalent to B and that $\text{map}_B(B,X)$ is equivalent to X, for all spaces X over B. If Y is indiscrete over B then so is $\text{map}_B(X,Y)$, for all spaces X over B.

Let X,Y,Z be spaces over B. Precomposition with a map $\theta : X \to Y$ over B determines a map

$$\theta^* : \text{map}_B(Y,Z) \to \text{map}_B(X,Z)$$

over B, while postcomposition with a map $\phi : Y \to Z$ over B determines a map

$$\phi_* : \text{map}_B(X,Y) \to \text{map}_B(X,Z)$$

over B. If θ is a proper surjection then θ^* is an embedding, while if ϕ is an embedding then ϕ_* is an embedding.

<u>Proposition</u> (6.2). For spaces X,Y,Z over B there is a natural equivalence

$$\mathrm{map}_B(X \underset{B}{+} Y, Z) \equiv \mathrm{map}_B(X,Z) \underset{B}{\times} \mathrm{map}_B(Y,Z)$$

of spaces over B.

<u>Proposition</u> (6.3). If Y is regular over B then so is $\mathrm{map}_B(X,Y)$, for all spaces X over B.

This result is no longer true if we replace "regular" by "Hausdorff".

<u>Proposition</u> (6.4). For spaces X,Y,Z over B, with X regular over B, there is a natural equivalence

$$\mathrm{map}_B(X, Y \underset{B}{\times} Z) \equiv \mathrm{map}_B(X,Y) \underset{B}{\times} \mathrm{map}_B(X,Z)$$

of spaces over B.

After these preliminaries we can begin to show that $\mathrm{map}_B(\ ,\ )$ has the properties required for an adjoint functor to the fibre product $\underset{B}{\times}$. The first step is

<u>Proposition</u> (6.5). Let X,Y,Z be spaces over B. If the function $h : X \underset{B}{\times} Y \to Z$ over B is continuous then so is the function $\hat{h} : X \to \mathrm{map}_B(Y,Z)$ over B given by

$$\hat{h}(x)(y) = h(x,y) \qquad (x \in X_b,\ y \in Y_b,\ b \in B).$$

Composition of maps determines a function

$$\mathrm{map}_B(Y,Z) \times_B \mathrm{map}_B(X,Y) \to \mathrm{map}_B(X,Z)$$

over B, and we have

Proposition (6.6). If Y is locally compact regular over B then the function

$$\mathrm{map}_B(Y,Z) \times_B \mathrm{map}_B(X,Y) \to \mathrm{map}_B(X,Z)$$

over B is continuous for all spaces X,Z over B.

Corollary (6.7). Let $\hat{h} : X \to \mathrm{map}_B(Y,Z)$ be a map over B, where Y is locally compact regular over B. Then $h : X \times_B Y \to Z$ is a map over B, where

$$h(x,y) = \hat{h}(x)(y) \qquad (x \in X_b,\ y \in Y_b,\ b \in B).$$

We now come to the exponential law itself.

Proposition (6.8). Let X,Y,Z be spaces over B and let

$$\xi : \mathrm{map}_B(X \times_B Y,Z) \to \mathrm{map}_B(X,\mathrm{map}_B(Y,Z))$$

be the injection defined by taking adjoints as in (6.5). If X is regular over B then ξ is continuous. If Y is also regular over B then ξ is an open embedding. If, further, Y is also locally compact over B then ξ is an equivalence of spaces over B.

Let X,Y be spaces over B. For each space B' and map $\xi : B' \to B$ we have a continuous bijection

$$\xi_\# : \mathrm{map}_{B'}(\xi^*X,\xi^*Y) \to \xi^*\mathrm{map}_B(X,Y)$$

over B'. In general this is not an equivalence but we prove

Proposition (6.9). Let X be locally compact regular over B. Then $\xi_\#$ is an equivalence of spaces over B', for all spaces Y over B.

This can be proved by a formal argument as follows. By (6.7) the evaluation function

$$\mathrm{map}_B(X,Y) \underset{B}{\times} X \to Y$$

is continuous, hence so is the pull-back

$$\xi^*(\mathrm{map}_B(X,Y) \underset{B}{\times} X) \to \xi_*Y.$$

Since ξ^* commutes with the fibre product we can rewrite this as

$$\xi^*\mathrm{map}_B(X,Y) \underset{B'}{\times} \xi^*X \to \xi_*Y.$$

Taking the adjoint, as in (6.5), we obtain a map

$$\xi^*\mathrm{map}_B(X,Y) \to \mathrm{map}_{B'}(\xi^*X,\ \xi^*Y)$$

which is obviously inverse to $\xi_\#$.

The space $\mathrm{map}_B(X,Y)$ over B should not be confused with the space $\mathrm{MAP}_B(X,Y)$ of maps $X \to Y$ over B. The former has the fibrewise compact-open topology while the latter has the ordinary compact-open topology. Now a continuous injection

$$\sigma : \mathrm{MAP}_B(X,Y) \to \mathrm{sec}_B\mathrm{map}_B(X,Y)$$

is defined, where sec_B denotes the space of sections. Here σ transforms the map $\phi : X \to Y$ over B into the section $\sigma(\phi)$ which sends the point $b \in B$ into the map $\phi_b : X_b \to Y_b$. We prove

<u>Proposition</u> (6.10). Let X be compact regular over B. Then σ is an equivalence of spaces, for all spaces Y over B.

First observe that σ is bijective. For an inverse function is defined by associating with each section $s : B \to \mathrm{map}_B(X,Y)$ the composition

$$X \equiv B \times_B X \to \mathrm{map}_B(X,Y) \times_B X \to Y$$

of $s \times_B \mathrm{id}_X$ with the evaluating function. To see that σ^{-1} is continuous, i.e. that σ is open, take a compact-open subset (C,V) of $\mathrm{MAP}_B(X,Y)$. Then $\sigma(C,V) = (pC,(C,V))$, which is open since pC is compact and since C is closed in X and so compact over B.

Proposition (6.11). Let Y be a fibre space over B and let X be a locally compact regular fibre space over B. Then $\mathrm{map}_B(X,Y)$ is a fibre space over B.

We need to establish that the projection $\pi : \mathrm{map}_B(X,Y) \to B$ has the homotopy lifting property for maps with arbitrary domain A. So let $\phi : I \times A \to B$ be a homotopy. We regard $I \times A$ as a space over B through ϕ and A as a space over B through ϕ_0. Since X is a fibre space over B there exists a map (in fact a fibre homotopy equivalence)

$$\eta : (I \times A) \times_B X = \phi^*X \to I \times \phi_0^*X = I \times (A \times_B X)$$

over $I \times A$, such that

$$\eta((0,a),x) = (0,(a,x)) \qquad (\phi_0 a = px).$$

Suppose that ϕ_0 can be lifted to a map

$$\hat{f} : A \to \mathrm{map}_B(X,Y).$$

Since X is locally compact regular over B the adjoint

$$f : A \times_B X \to Y$$

is defined, so that the following diagram is commutative.

$$\begin{array}{ccc} A \times_B X & \xrightarrow{f} & Y \\ {\scriptstyle pr_1}\downarrow & & \downarrow{\scriptstyle q} \\ A & \xrightarrow[\phi_0]{} & B \end{array}$$

Since q is a fibration we can lift the projection

$$I \times (A \times_B X) \to B,$$

regarded as a homotopy of qf, to a homotopy

$$\psi : I \times (A \times_B X) \to Y$$

of f. Thus we obtain a map

$$h = \psi\eta : (I \times A) \times_B X \to Y,$$

of which the adjoint

$$\hat{h} : I \times A \to map_B(X,Y)$$

is a homotopy of $\hat{f}$ lifting the given homotopy ϕ of $\pi\hat{f}$. Thus π has the homotopy lifting property and the proof of (6.11) is complete. An analogous result is given by Booth in [1].

7. Sectioned spaces over a base

A sectioned space over B is a triple consisting of a space X and maps

$$B \xrightarrow{s} X \xrightarrow{p} B$$

such that $ps = id_B$; usually X alone is sufficient notation. The map p is called the projection and the map s the section. We regard B as a sectioned space over itself using the identity as section and projection.

Let X,Y be sectioned spaces over B with projections

p,q and sections s,t, respectively. By a map of sectioned spaces over B we mean a map $\phi : X \to Y$ of spaces such that $q\phi = p$ and $\phi s = t$. The pointed set of such maps is denoted by $MAP_B^B(X,Y)$, the basepoint being the nul-map tp. With this definition of morphism the category Top_B^B of sectioned spaces over B is defined.

If X,Y are sectioned spaces over B the push-out of the cotriad

$$X \xleftarrow{s} B \xrightarrow{t} Y$$

of sections is denoted by $X \vee_B Y$ and called the fibrewedge sum. The sections s,t also define a triad

$$X \xrightarrow[u]{} X \times_B Y \xleftarrow[v]{} Y,$$

where the components of u are (id_X, tp) and the components of v are (sq, id_Y). The push-out of the cotriad

$$X \times_B Y \xleftarrow[u \vee v]{} X \vee_B Y \longrightarrow B$$

is denoted by $X \wedge_B Y$ and called the fibre-smash product. Although the fibre-wedge sum is associative the fibre-smash product in general is not, nor does it distribute over the fibre-wedge sum. However we have

<u>Proposition</u> (7.1). Let X,Y,Z be sectioned spaces over B with closed sections. Suppose either that (i) X and Z are locally compact regular over B or that (ii) X and Y (or Y and Z) are compact regular over B. Then

$$(X \wedge_B Y) \wedge_B Z \equiv X \wedge_B (Y \wedge_B Z).$$

<u>Proposition</u> (7.2). Let X,Y,Z be sectioned spaces over B with closed sections. Then

$$(X \underset{B}{\wedge} X) \underset{B}{\vee} (Y \underset{B}{\wedge} Z) \equiv (X \underset{B}{\vee} Y) \underset{B}{\wedge} Z.$$

For any space X over B, a compact space X^+_B over B can be constructed as follows. Take X^+_B, as a set, to be the disjoint union X + B, and give X^+_B the following topology. The generating open sets, before supplementation, are of two kinds. The first kind are the open sets of X, regarded as subsets of X^+_B. The second kind are the complements in X^+_B of the closed subsets of X which are compact over B. Clearly the inclusion $X \to X^+_B$ is an open embedding. Moreover X^+_B reduces to the topological sum X + B when X is compact over B.

To see that X^+_B is compact over B, we use (2.5). We have given X^+_B the topology generated, after supplementation, by the two families of subsets described above. Both families are closed under finite intersection and unrestricted union. Hence a neighbourhood U of a fibre X^+_b, in the old topology, is a union $V \cup (X^+_B - K)$, where $V \subset X$ is open and $K \subset X$ is closed and compact over B. Now the section $s : B \to X^+_B$ does not meet X and so does not meet V. Therefore $s(b) \in X^+_B - K$, hence $b \notin pK$. Since pK is closed in B there exists a neighbourhood N of b which does not meet pK. Then $(p^+)^{-1}N$ does not meet K and so is contained in $X^+_B - K$ and hence in U. Thus p^+ is closed in the old topology. Also X^+_b is the compactification of X_b and so compact.

Thus X_B^+ is compact over B in the old topology and so compact over B in the new.

Suppose that X is locally compact Hausdorff over B. Then X_B^+ is Hausdorff over B. Certainly each pair of distinct points in a fibre of X can be separated in X and so can be separated in X_B^+. It remains to be shown that each point x of X can be separated from its projection px = b in the other summand of X_B^+. Since X is locally compact over B there exists a neighbourhood U of x such that $C\ell\ U$ is compact over B. Then U and $X_B^+ - C\ell\ U$ are disjoint neighbourhoods of x and b, respectively. Thus X_B^+ is Hausdorff over B, as asserted.

Clearly each function $\phi : X \to Y$ over B determines a section-preserving function $\phi_B^+ : X_B^+ \to Y_B^+$ over B, and vice versa. I assert that ϕ_B^+ is continuous if and only if ϕ is proper. For suppose that ϕ is proper. If $K \subset Y$ is closed and compact over B then $\phi^{-1}K \subset X$ is closed and compact over B. Hence if $Y_B^+ - K$ is a neighbourhood of some point of the section of Y_B^+ then $X_B^+ - \phi^{-1}K$ is a neighbourhood of the corresponding point of the section of X_B^+. This proves continuity at points of the section; continuity away from such points is obvious.

Conversely suppose that ϕ_B^+ is continuous. Then $\phi^{-1}(Y_B^+ - K) = X_B^+ - \phi^{-1}K$ is open for each closed and compact K over B, and hence $\phi^{-1}K$ is closed and compact over B. So let $\{U_j\}$ $(j \in J)$ be an open covering of Y such that $C\ell\, U_j$ is compact over B for each index j. Then each of the maps $\phi^{-1}C\ell\, U_j \to C\ell\, U_j$ determined by ϕ is closed and so compact. Therefore ϕ is compact, by (4.5).

Fibrewise compactification has satisfactory naturality properties. For let X be a locally compact Hausdorff space over B. Then ξ^*X is a locally compact Hausdorff space over B' for each space B' and map $\xi : B' \to B$. Hence the fibrewise compactifications of X and ξ^*X are defined, as sectioned spaces over B and B' respectively. As we have seen there is a continuous bijection

$$\theta : (\xi^*X)^+_B \to \xi^*(X^+_B)$$

of sectioned spaces over B'. Now X^+_B is Hausdorff over B, hence $\xi^*(X^+_B)$ is Hausdorff over B', and $(\xi^*X)^+_{B'}$ is compact over B'. Therefore ξ is an equivalence of sectioned spaces over B', by (5.6).

<u>Proposition</u> (7.3). Let X,Y be locally compact Hausdorff spaces over B. Then

$$X^+_B \vee_B Y^+_B \equiv (X +_B Y)^+_B,\quad X^+_B \wedge_B Y^+_B \equiv (X \times_B Y)^+_B$$

as sectioned spaces over B.

The proof is purely formal.

8. <u>The adjoint of the fibre-smash</u>

If X,Y are sectioned spaces over B we denote by $\mathrm{map}^B_B(X,Y)$ the subspace of $\mathrm{map}_B(X,Y)$ consisting of pointed maps, where the basepoints in the fibres are determined by the sections. Here $\mathrm{map}_B^B(X,Y)$ itself is regarded as a sectioned space over B, the section being that which sends each point $b \in B$ into the nul-map $X_b \to Y_b$. It is easy to see that $\mathrm{map}_B^B(B \times \dot{I}, X)$ is naturally equivalent to X for all sectioned spaces X over B.

<u>Proposition</u> (8.1). Let X,Y be sectioned spaces over B, with closed sections. Then

$$\mathrm{map}_B^B(X \vee_B Y, Z) \equiv \mathrm{map}_B^B(X,Z) \times_B \mathrm{map}_B^B(Y,Z)$$

for all sectioned spaces Z over B.

<u>Proposition</u> (8.2). Let X,Y,Z be sectioned spaces over B, with Y regular over B. Then

$$\mathrm{map}_B^B(X, Y \times_B Z) \equiv \mathrm{map}_B^B(X,Y) \times_B \mathrm{map}_B^B(X,Z),$$

as sectioned spaces over B.

<u>Proposition</u> (8.3). Let X,Y,Z be sectioned spaces over B. If the function $\hat{h} : X \wedge_B Y \to Z$ of sectioned spaces over B is continuous then so is the function $\hat{h} : X \to \mathrm{map}_B^B(Y,Z)$, where

$$\hat{h}(x)(y) = h(x,y) \qquad (x \in X_b,\ y \in Y_b,\ b \in B).$$

In the theory of Top_B^B composition of pointed maps determines a section-preserving function

$$\mathrm{map}_B^B(Y,Z) \wedge_B \mathrm{map}_B^B(X,Y) \to \mathrm{map}_B^B(X,Z),$$

for all sectioned spaces X,Y,Z over B. We have

Proposition (8.4). Suppose that the sectioned space Y is locally compact regular over B. Then the composition function

$$\mathrm{map}_B^B(Y,Z) \wedge_B \mathrm{map}_B^B(X,Y) \to \mathrm{map}_B^B(X,Z)$$

is continuous for all sectioned spaces X,Z over B.

Corollary (8.5). Let $\hat{h} : X \to \mathrm{map}_B^B(Y,Z)$ be a map of sectioned spaces, where X,Y,Z are sectioned spaces with Y locally compact regular over B. Then $h : X \wedge_B Y \to Z$ is continuous, where

$$h(x,y) = \hat{h}(x)(y) \qquad (x \in X_b,\ y \in Y_b,\ b \in B).$$

Proposition (8.6). Consider the continuous bijection

$$\xi : \mathrm{map}_B^B(X \wedge_B Y,Z) \to \mathrm{map}_B^B(X,\mathrm{map}_B^B(Y,Z)).$$

given as in (8.3), where X,Y,Z are sectioned spaces over B. If X is regular over B then ξ is continuous. If Y is also regular over B then ξ is an open embedding. If, further, Y is also locally compact over B then ξ is an equivalence of sectioned spaces over B.

9. Topological transformation groups

Next let us turn to the category of G-spaces, where G is a topological group. We can always regard a G-space X as a space over its orbit space X/G. However it is more significant to consider the shearing map

$$\theta : G \times X \to X \times_{X/G} X$$

which is given by

$$(g,x) = (gx,x) \qquad (g \in G,\ x \in X).$$

The shearing map is always surjective, of course, and is injective when the action of free. The point of view I am going to adopt is to regard $G \times X$ as a space over $X \underset{X/G}{\times} X$ with projection θ; the fibres are then essentially the stabilizers of the action. Consider now a property of spaces over a base satisfying the three conditions of §3.

<u>Definition</u> (9.1). The action of G on X has property P if $G \times X$ has property P as a space $X \underset{X/G}{\times} X$.

We have at once

<u>Proposition</u> (9.2). If the action of G on X has property P then each of the stabilizers of the action has property P.

It follows, for example, that the action of G on X is connected if at least one of the stabilizers is connected, in particular if there exists a free orbit.

Although the necessary condition of (9.2) will not be sufficient, in general, there are some properties for which it is, for example the Hausdorff property:

<u>Proposition</u> (9.3). The action of G on X is Hausdorff if (and only if) each stabilizer is Hausdorff.

For let (g,x) and (g',x) be distinct points in the same fibre as θ. Then $x = x'$ and $gx = g'x$, so that $h = g^{-1}g' \in G_x$, with $h \neq e$, the neutral element. If G_x is Hausdorff there exists a neighbourhood W of e in G such that $h \in W^{-1}W$; then $Wg \times X$ and $Wg' \times X$ are disjoint neighbourhoods of (g,x) and (g',x), respectively. Thus the action is Hausdorff, as asserted.

By a principal G-space over a space B I mean an open space X over B, with surjective projection, together with a fibre-preserving free action of G on X. In this situation the projection induces a homeomorphism between X/G and B.

<u>Proposition</u> (9.4). Let $\phi : X \to Y$ be a G-map over B, where X and Y are principal G-spaces over B. If the action of G on Y is compact then ϕ is an equivalence.

By (3.5) it is sufficient to show that

$$\phi \times \mathrm{id} : X \times_B X \to Y \times_B X$$

is an equivalence. Now the shearing map

$$G \times Y \to Y \times_B Y$$

is a homeomorphism, since the action is compact; let $\psi : Y \times_B Y \to G$ be the first projection of the inverse homeomorphism. Then the composition

$$Y \times_B X \xrightarrow{\mathrm{id}\times\Delta} Y \times_B X \times_B X \xrightarrow{\mathrm{id}\times\phi\times\mathrm{id}} Y \times_B Y \times_B X \xrightarrow{\psi\times\mathrm{id}} G \times X \xrightarrow{\theta} X \times_B X$$

is an inverse of $\phi \times \mathrm{id}$, and so the result follows.

10. Uniform maps over a base

Throughout this final section we work over a base B which satisfies the T_1 condition, i.e. points are closed. I do not know whether this restriction can be avoided.

Definition (10.1). A uniform space over the T_1 space B is a uniform space X together with a function $p : X \to B$ such that for each point b of B and neighbourhood N of b there exists a member U of the uniform structure such that $U[X_b] \subset X_N$.

We refer to the condition on the projection p as vertical uniform continuity. It implies that p is continuous. The opposite implication holds when each of the fibres X_b is compact, in particular when p is a proper map.

Example (10.2). Let G be a topological group, with the right uniform structure. Then G is a uniform space over the left factor space G/H for each closed subgroup H of G.

For each uniform space X the completion $\hat{X}$ is defined, as a uniform space. We regard X as a subspace of $\hat{X}$, in the usual way, and recall that for any subspace A of X the completion $\hat{A}$ of A may be identified with the closure $C\ell A$ of A in $\hat{X}$. When X is a uniform space over B let us write

$$\hat{X}_B = \bigcup_{b \in B} C\ell X_b,$$

where the closures are taken in $\hat{X}$. I assert that $\hat{X}_B$, with the uniform structure obtained from $\hat{X}$ by resriction, constitutes a uniform space over B.

The first step is to show that if b,b' are distinct points of B then $C\ell X_b$ and $C\ell X_{b'}$ are disjoint. To see this, let N be a neighbourhood of b which does not contain b'. Since the projection $p : X \to B$ is vertically uniformly continuous there exists a member U of the uniform structure of X such that $U[X_b] \subset X_N$. Now U is the restriction of a member U* of the uniform structure of $\hat{X}$. Let V* be a symmetric member of the uniform structure of $\hat{X}$ such that $V^* \circ V^* \subset U^*$. Suppose, to obtain a contradiction, that there exists a point $x \in \hat{X}$ such that $x \in C\ell X_b$ and $x \in C\ell X_{b'}$. Then the neighbourhood $V^*[x]$ of x in $\hat{X}$ contains a point $y \in X_b$ and a point $y' \in X_{b'}$. Hence $(y,y') \in U$ and so $y' \in U[y] \subset U[X_b] \subset X_N$. But $X_{b'}$ is disjoint from X_N, since $b' \notin N$, and so we have the required contradiction.

We conclude, therefore, that a function $p^* : \hat{X}_B \to B$ can be defined, such that $p^{*-1}b = \hat{X}_b = C\ell X_b$. The next step is to show that p* is vertically uniformly continuous. Again let b be a point of B, let N be a neighbourhood of b, and let U* be a member of the uniform structure of $\hat{X}$ such that $U[X_b] \subset X_N$, where U is the restriction of U* to X. I assert that $V^*[\hat{X}_b] \subset \hat{X}_N$, where V* is a symmetric member of the uniform structure of $\hat{X}$ such that $V^* \circ V^* \circ V^* \subset U^*$. For suppose, to obtain a contradiction, that there exists a point $x' \in V^*[\hat{X}_b]$ such that $p^*x' = b' \notin N$. Then $x' \in V^*[x]$, for some point $x \in \hat{X}_b$.

Also the neighbourhood $V^*[x]$ of x in $\hat{X}$ contains a point y of X_b and a point y' of $X_{b'}$. Thus V^* contains (x,x'), (x,y) and (x,y'), hence U contains (y,y'). Thus $y' \in U[y] \subset U[X_b] \subset X_N$. But $X_{b'}$ is disjoint from X_N, since $b' \notin N$, and so we have the required contradiction.

Thus we have shown that for each uniform space X over B a uniform space $\hat{X}_B$ over B is defined. We may refer to $\hat{X}_B$ as the fibrewise completion of X. The fibrewise completion of uniformly continuous fibre-preserving maps may be defined in a similar manner.

Following Niefield [5] one may consider the special case when B is itself a uniform space and the projection $p : X \to B$ is uniformly continuous. In that case $\hat{X}$ may be regarded as a space over $\hat{B}$ through the completion $\hat{p} : \hat{X} \to \hat{B}$ of p, and then $\hat{X}_B$ may be identified with the restriction of $\hat{X}$ to B.

References

1. P.I. Booth, The exponential law for maps I, Proc. London Math. Soc. (3) 20, 179-192, 1970.

2. N. Bourbaki, Topologie générale, Hermann, Paris 1965.

3. D.E. Cohen, Products and carrier theory, Proc. London Math. Soc. (3), 26 (1957), 219-248.

4. I.M. James, General topology and homotopy theory, Springer, New York 1984.

5. S.B. Niefield, Cartesianness; topological spaces, uniform spaces and affine schemes, J. Pure and Applied Alg. 23 (1982), 147-168.

Mathematical Institute,
Oxford.

κ-DOWKER SPACES

M.E. Rudin
Department of Mathematics, University of Wisconsin, Madison
Wisconsin, U.S.A. 53706

INTRODUCTION

By a <u>space</u> we mean a Hausdorff topological space.

In one of the most referred to papers in all of mathematics, Dowker (1951), we have:

Theorem 1

The following are equivalent for a normal space X:

(1) X is countably paracompact.

(2) $X \times (\omega + 1)$ is normal.

(3) Every countable open cover of X can be shrunk.

Part of Dowker's theorem was that $X \times (\omega + 1)$ is normal if and only if $X \times I$ is normal for the closed unit interval I, in fact, if and only if $X \times C$ is normal for all compact metric C. At that time a space was said to be <u>binormal</u> if not only X but also $X \times I$ was normal, a hypothesis used in a number of homotopy extension theorems (see Borsuk (1937), Morita (1975), Starbird (1975)). The normality of products was a subject full of mysteries and a well known problem was to try to find a normal nonbinormal space. After Dowker's theorem such a space, if any, became known as a <u>Dowker space</u>.

Paracompactness was then a rather new concept, and the idea of using cardinal functions in topology, for instance studying nonparacompact spaces in terms of the minimal cardinality of an open cover having no locally finite refinement, was not so common as it is today.

As Dowker recognized, condition (3), although not the countable case of some clearly important concept, is the <u>useful</u> condition. In order to check that conditions (1) and (2) hold, one most easily checks (3); in constructing a counterexample to (1) or (2) one aims at having (3) fail. While trivially equivalent to (3), Dowker's original (3) was stated in a different way. If κ is a cardinal, we say today that $\{V_\alpha | \alpha < \kappa\}$ is a shrinking of an open cover $\{U_\alpha | \alpha < \kappa\}$ if $\{V_\alpha | \alpha < \kappa\}$ is also an open cover and $\overline{V}_\alpha \subset U_\alpha$ for all $\alpha < \kappa$; we say that $\{U_\alpha | \alpha < \kappa\}$ is monotone if $U_\alpha \subset U_\beta$ for all $\alpha < \beta < \kappa$. Part of Dowker's theorem was that, in normal spaces, every countable open cover could be shrunk if and only if every monotone countable open cover had a monotone shrinking. Even today shrinking is not a common term, however, I use it to state (3) because I think the concept is important (and the clue to generalizing Dowker spaces.)

General topologists study pathologies. The fact that normality is not always preserved in products and that normal spaces may not be paracompact are two of the most common and basic pathologies. Dowker's theorem is beautiful because it shows that the "countable case" of these pathologies is the same; it is useful for it reminds us of the ever present threat that even a normal space may have a countable open cover which cannot be shrunk, but constructing such spaces is a fragile operation. Searching for Dowker spaces of various kinds has been a recurring problem.

Since (Rudin, 1971) we have known that there is a Dowker space. The space given is a subspace of a box product of a countable increasing sequence of regular cardinals. Modulo your choice of cardinals and other alterations (such as Dow & van Mill, 1982), this is still the only known example of a Dowker space which does not depend on some special model for set theory. It is not a nice space because its cardinal functions such as cardinality, weight, and character, all seem larger than necessary. However, if one makes various set theoretic assumptions such as the existence of a Souslin line (Rudin, 1955 and 1973), or $\diamond$ (Ostaszewski (1976), Weiss (1981), Rudin (1983 a) Handbook (1984)), or $\clubsuit$(see de Caux 1976), the continuum Hypothesis (Juhasz et al., 1976), or Martin's Axiom (Bell (1981), Weiss (1981)), or the existence of a compact cardinal (Watson, 1984), one can construct Dowker spaces which have a remarkable variety of additional properties. (See Handbook (1984) for a survey article on Dowker spaces.) In fact, it is a bit of a game: Whenever one works with some set of topological properties which do not obviously deny the existence of a Dowker space, one asks, could there exist such a space which is also Dowker? Quite often it is consistent with Zermelo Frankel Set Theory that the answer is yes. However, it is a difficult game and it is not a joke. If often turns out to be necessary in solving apparently unrelated problems. For instance, we have no idea, with or without set theoretic assumptions, whether there is a normal nonparacompact space with a σ-disjoint base (see Rudin, 1983 a). But we know that if there is one, it must be a Dowker space (Nagami, 1955).

Despite our continuing interest in Dowker spaces and the fact that looking at countable problems in a more general setting has sometimes been useful, we have not been especially successful in our attempts to generalize Dowker spaces. A Dowker space is clearly the countable (and hence most vital) case of something. Throughout the paper we shall use κ to stand for an arbitrary infinite regular cardinal. How should we define a κ-Dowker space?

POSSIBLE κ-COVERING PROPERTIES

As we shall see, there are some nice "κ-Dowker" theorems, i.e. generalizations of Dowker's theorem which prove that in normal spaces some κ-covering property is equivalent to some κ-product being normal. But the associated "κ-Dowker" spaces, i.e. normal spaces not having these properties, have not proved very interesting. What we want is some difficult-to-fail-in-the-presence-of-normality covering property whose failure leads to the kind of basic, delicate pathology found in ordinary Dowker spaces.

Our plan is to present the obvious covering properties, to prove some theorems and present some examples indicating their interrelationships, and to define a κ-Dowker space to be a normal space which fails to have the weakest of our properties. Some of the theorems are (very) old. The ideas have all been known to me for a long time and some of the "new" theorems have been previously stated by me without proof (see Rudin, 1978 and 1983 c). I will star previously unpublished theorems. But I present it all as one topic to accentuate and perhaps throw some light on the complexity.

We consider the following properties of a normal space X:

(1) <u>X is κ-paracompact</u>: Every open cover of X of cardinality $\leq \kappa$ has a locally finite refinement.

(2) <u>X is κ-$\mathcal{B}$</u>: Every monotone open cover of X of cardinality $\leq \kappa$ has a monotone shrinking.

(3) <u>X is κ-$\mathcal{D}$</u> (or weak κ-$\mathcal{B}$ or weak κ-shrinking): Every

monotone open cover of X of cardinality $\leq \kappa$ can be shrunk.

(4) X is κ-shrinking: Every open cover of X of cardinality $\leq \kappa$ can be shrunk.

The dreadful names are unfortunately historical. Previously the shrinking property has meant that every open cover can be shrunk (see Smith (1979), Rudin (1983 b), Le Donne (1984).) The $\mathcal{B}$-property has meant that every monotone open cover has a monotone shrinking (see Zenor (1970), Chiba (1982), Yasui (1983).) I have chosen to call the formerly-called-weak version of these properties the $\mathcal{D}$-property primarily because Dowker begins with a D, but also to avoid the double names and any implication that weak means nonbasic. Observe however that every κ-$\mathcal{B}$ and κ-shrinking space is trivially κ-$\mathcal{D}$.

Also observe that if κ were singular, a space would have one of our κ covering properties if and only if it had the associated λ property for all regular $\lambda < \kappa$.

By the proof of Dowkers' theorem, all of these properties are the same if $k = \omega$. For uncountable κ the situation is more complicated.

κ-PARACOMPACTNESS

Theorem 2

The following are equivalent for a normal space X:

(1) X is κ-paracompact.

(2) $X \times (\kappa+1)$ is normal.

(3) $X \times C$ is normal for all compact spaces C of weight $\leq \kappa$.

Proof

That (3) implies (2) is trivial.

That (1) implies (3) is now classic: Suppose C is a compact space of weight $\leq \kappa$ and X is normal and paracompact. Let H and K be closed and disjoint in $X \times C$.

Let $\mathcal{B}$ be an open basis for C of cardinality at most κ, and let $F = \{f : \mathcal{D} \to \{H,K\} \mid \mathcal{D}$ is a finite subset of $\mathcal{B}$ covering $C\}$. For $f \in F$ let

$$U_f = \bigcup\left\{ U \text{ open in } X \;\middle|\; \begin{array}{l} \text{if } B \in \text{domain } f \\ f(B) = H \text{ implies } \overline{U \times B} \cap K = \emptyset \\ \text{and } f(B) = K \text{ implies } \overline{U \times B} \cap H = \emptyset. \end{array} \right\}$$

Then $\{U_f | f \in F\}$ is an open cover of X of cardinality $\leq \kappa$ and it has a locally finite refinement $\{V_f | f \in F\}$ with $V_f \subset U_f$. Let $V_H = \cup\{(V_f \times B) | f \in F,\ B \in \text{domain } f, \text{ and } f(B) = H\}$ and define V_K similarly interchanging H and K. Clearly $(V_H - \overline{V}_K)$ is an open set containing H whose closure misses K as desired.

To prove that (2) implies (1), suppose κ is minimal for this to fail and that $\{0_\alpha | \alpha < \kappa\}$ is an open cover of X. For $\alpha < \kappa$, let $U_\alpha = \cup\{0_\beta | \beta < \alpha\}$.

Observe that we lose nothing by indexing with κ since our cover has a cofinal subsequence of cofinality λ for some $\lambda \leq \kappa$ and $(\lambda + 1)$ is a closed subspace of $(\kappa + 1)$.

Let $K = \cup\{(X - U_\alpha) \times \{\alpha\} | \alpha < \kappa\}$ and $H = X \times \{\kappa\}$. Since $X \times (\kappa + 1)$ is normal and H and K are disjoint and closed, there is an open V_0 in $X \times (\kappa + 1)$ with $H \subset V_0$ and $\overline{V}_0 \cap K = \emptyset$. For $n \in \omega$ choose open V_{n+1} with $\overline{V}_n \subset V_{n+1}$ and $\overline{V}_{n+1} \cap K = \emptyset$.

For each $\alpha < \kappa$ and $n < \omega$, define $V_{\alpha n} = \{x \in X | \langle x,\beta\rangle \in V_n$ for some $\beta \leq \alpha\}$. Then for each $n \in \omega$, $\overline{V}_{\alpha n} \subset V_{\alpha(n+1)} \subset U_\alpha$ and $\{V_{\alpha n} | \alpha < \kappa\}$ is a monotone open cover of X with $\overline{\cup\{V_{\beta n} | \beta < \alpha\}} \subset \cup\{V_{\beta(n+1)} | \beta < \alpha\}$. Let $W_{\alpha n} = V_{\alpha n} - \overline{\cup\{V_{\beta m} | \beta < \alpha,\ m \in \omega\}}$ and $W_n = \cup\{W_{\alpha n} | \alpha < \kappa\}$. Then $\{W_n | n \in \omega\}$ is an open cover of X (since for each $x \in X$ there is a minimal α with $x \in V_{\alpha n}$ for some n).

Since X is countably paracompact, there is a locally finite open cover $\{Z_n | n \in \omega\}$ of X with $\overline{Z}_n \subset W_n$. For $\alpha < \kappa$ and $n < \omega$, let $Z_{\alpha n} = Z_n \cap W_{\alpha n}$. Observe that if $x \in \overline{Z}_n \subset W_n$, there is an $\alpha < \kappa$ with $x \in W_{\alpha n}$. But $\{W_{\alpha n} | \alpha < \kappa\}$ are disjoint. So $\{Z_{\alpha n} | \alpha < \kappa,\ n \in \omega\}$ is a locally finite open refinement of $\{U_\alpha | \alpha < \kappa\}$ with $\overline{Z}_{\alpha n} \subset U_\alpha$.

Since each $\overline{Z}_{\alpha n} \times (\alpha + 1)$ is a closed subset of $X \times (\kappa + 1)$ and $\overline{Z}_{\alpha n} \subset U_\alpha = \cup\{0_\beta | < \alpha\}$, by the minimality of κ there is a locally finite open cover $\mathcal{Z}_{\alpha n}$ of $Z_{\alpha n}$ refining $\{Z_{\alpha n} \cap 0_\beta | \beta < \alpha\}$. Thus $\{Z \cap Z_{\alpha n} | \alpha < \kappa,\ n < \omega, \text{ and } Z \in \mathcal{Z}_{\alpha n}\}$ is a locally finite open refinement of $\{0_\alpha | \alpha < \kappa\}$.

This completes the proof of Morita's elegant theorem which so nicely generalizes Dowker's theorem. A simple theorem we proved enroute (Rudin, 1983 c) is that a normal space is κ-paracompact if and only if every monotone open cover of cardinality $\leq \kappa$ has a locally finite refinement, and the refinement can be assumed to be a monotone shrinking.

But the delicate pathology of Dowker spaces is totally lost. For uncountable κ there is no problem in finding spaces which are normal but not κ-paracompact; for uncountable regular κ, κ itself (with the order topology) is normal and λ-paracompact for all $\lambda < \kappa$, but not κ-paracompact. Think of your favorite normal non-paracompact space; by definition it is not κ-paracompact for some κ. There is nothing subtle going on here; the failure of κ-paracompactness for uncountable κ is a gross property.

We next prove that κ-paracompactness is the strongest of our covering properties. I hesitate to credit anyone with the following theorem, but it can be found in Mack (1967), Morita (1975), Rudin (1978):

Theorem 3

If a normal space X is κ-paracompact, then it is κ-shrinking, $\kappa - \mathcal{B}$, and $\kappa - \mathcal{D}$.

Proof

Let $\{U_\alpha | \alpha < \kappa\}$ be an open cover of X and let $\{W_\alpha | \alpha < \kappa\}$ be a locally finite open cover of X with $W_\alpha \subset U_\alpha$. Let $K_\alpha = \{x \in X | \alpha = \max\{\beta < \alpha | x \in W_\beta\}\}$; observe that $\overline{K}_\alpha \subset \bigcup_{\beta<\alpha} K_\beta$. By induction on κ choose an open V_α with $(K_\alpha - \bigcup_{\beta<\alpha} V_\beta) \subset V_\alpha \subset \overline{V}_\alpha \subset W_\alpha$. Then $\{V_\alpha | \alpha < \kappa\}$ is a shrinking of $\{U_\alpha | \alpha < \kappa\}$. If $\{U_\alpha | \alpha < \kappa\}$ is monotone, let $\{Z_\alpha | \alpha < \kappa\}$ be a locally finite refinement of $\{V_\alpha | \alpha < \kappa\}$ with $Z_\alpha \subset V_\alpha$. Then $\{\bigcup_{\beta<\alpha} Z_\beta | \alpha < \kappa\}$ is a monotone shrinking of $\{U_\alpha | \alpha < \kappa\}$. The proofs are the same for all $\lambda < \kappa$.

Theorem 4

[*] Caryn Navy's space S (1981) is normal, countably paracompact, shrinking, $\mathcal{B}$, and $\mathcal{D}$, but not ω_1-paracompact. It is also a P-space in the sense of Morita, i.e. its product with every metric space is normal.

Navy's S was constructed as an example of a normal paralindelöf space which is not paracompact. Navy proves that S is normal and countably paracompact but not ω_1-paracompact. So it remains for us to show that S is shrinking and $\mathcal{B}$. To do this we give:

Definition of S

Let F be the set of all functions from ω into ω_1. For

$n \in \mathbb{N}$, let $\Sigma_n = \{f \upharpoonright n | f \in F\}$ and let P_n be the set of all subsets of Σ_n. Let $\Sigma = \cup\{\Sigma_n | n \in \mathbb{N}\}$ and P be the set of all finite subsets of $\cup\{P_n | n \in \mathbb{N}\}$. Let

$$\Delta = \left\{ \langle \sigma, \tau, A \rangle \;\middle|\; \begin{array}{l} (1)\quad A \in P;\ \sigma \text{ and } \tau \in \Sigma_n \text{ for some } n. \\ (2)\quad \sigma(0) < \tau(0) < \sigma(1) < \tau(1) < \ldots < \tau(n-1). \end{array} \right\}$$

If $\rho \in \Sigma$ and $\mathcal{B} \in P$, define

$$B(\rho, \mathcal{B}) = \{f \in F | f \text{ extends } \rho\}$$
$$\cup \left\{ \langle \sigma, \tau, A \rangle \in \Delta \;\middle|\; \begin{array}{l} \text{One of } \sigma \text{ and } \tau \text{ extends } \rho, \\ \mathcal{B} \subsetneq A, \text{ and if } A \in A \cap P_m \\ \sigma \upharpoonright m \in A \text{ if and only if } \tau \upharpoonright m \in A. \end{array} \right\}$$

Navy's space S is $F \cup \Delta$ topologized by having $\{B(\rho, \mathcal{B}) | \rho \in \Sigma \text{ and } \mathcal{B} \in P\} \cup \{\text{singletons from } \Delta\}$ as an open basis.

Proof that S *is shrinking and* B

Suppose that $\{U_\alpha | \alpha < \kappa\}$ is an open cover of S.

For $n \in \mathbb{N}$, let

$$\Sigma_n^* = \left\{ \sigma \in \Sigma_n \;\middle|\; \begin{array}{l} \exists \alpha < \kappa \text{ and } \mathcal{B} \in P \text{ with } B(\sigma, \mathcal{B}) \subset U_\alpha \\ \text{but } \nexists \alpha' < \kappa \text{ and } \mathcal{B}' \in P \text{ and} \\ m < n \text{ with } B(\sigma' \upharpoonright m, \mathcal{B}') \subset U_{\alpha'}. \end{array} \right\}$$

For $\sigma \in \Sigma_n^*$ choose $\alpha_\sigma < \kappa$ and $\mathcal{B}_\sigma \in P$ with $B(\sigma, \mathcal{B}_\sigma) \subset U_{\alpha_\sigma}$. Define $\mathcal{C}_\sigma = \mathcal{B}_\sigma \cup \{\Sigma_m^* | m \leq n\} \cup \{\{\tau \in \Sigma_n^* | \alpha_\tau = \alpha_\sigma\}\}$.

Then $\{V_\alpha | \alpha < \kappa\}$ is a shrinking of $\{U_\alpha | \alpha < \kappa\}$ where $V_\alpha = \cup\{B(\sigma, \mathcal{C}_\sigma) | \alpha_\sigma = \alpha\}$. To see that $\overline{V}_\alpha \subset U_\alpha$ suppose that $f \in \overline{V}_\alpha - U_\alpha$; $f \in F$ since the points of $S - F$ isolated. There must be an $n \in \mathbb{N}$ such that $f \upharpoonright n = \sigma \in \Sigma_n^*$. Since $f \notin U_\alpha, \alpha_\sigma \neq \alpha$. Let $E = \{\tau \in \Sigma_n^* | \alpha_\tau = \alpha\}$. We claim that $B(\sigma, \mathcal{C}_\sigma \cup \{E\}) \cap V_\alpha = \emptyset$ contradicting $f \in \overline{V}_\alpha$. For suppose $\tau \in \Sigma_m^*$ and $\alpha_\tau = \alpha$. If $m < n, \Sigma_m^* \in \mathcal{C}_\sigma \cap \mathcal{C}_\tau$ and $\tau \in \Sigma_m^*$ while $\sigma \notin \Sigma_m^*$, hence $B(\sigma, \mathcal{C}_\sigma) \cap B(\tau, \mathcal{C}_\tau) = \emptyset$. If $m > n$, $\Sigma_n^* \in \mathcal{C}_\sigma \cap \mathcal{C}_\tau$ and $\sigma \in \Sigma_n^*$ while $\tau \notin \Sigma_n^*$ so again $B(\sigma, \mathcal{C}_\sigma) \cap B(\tau, \mathcal{C}_\tau) = \emptyset$. If $n = m$, $E \in \mathcal{C}_\sigma, \sigma \in E$ and $\tau \notin E$, so $B(\sigma, \mathcal{C}_\sigma) \cap B(\tau, \mathcal{C}_\tau \cup \{E\}) = \emptyset$ in any case. Thus S is shrinking.

To prove that S is $\mathcal{B}$, assume that $\{U_\alpha | \alpha < \kappa\}$ is monotone. Since S is contably paracompact, we can assume that κ has uncountable cofinality; and since S has weight ω_1, we can assume that $\kappa = \omega_1$.

Reread the definitions of Σ_n^*, α_σ, $\mathcal{B}_\sigma$, and $\mathcal{C}_\gamma$, and let $\Sigma^* = \cup\{\Sigma_n^* | n \in \mathbb{N}\}$. Choose a closed unbounded subset Γ of ω_1 such that $\gamma \in \Gamma$ and $\sigma \in \Sigma^*$ and (range σ) $\subset \gamma$ imply that $\alpha_\sigma < \gamma$. Without loss of generality $\Gamma = \omega_1$.

For $\gamma \in \omega_1$, define

$$T_\gamma = \left\{\tau_\gamma \in \Sigma_{n+1} \;\middle|\; \begin{array}{l} \exists n \in \mathbb{N} \text{ with } \tau \restriction n \in \Sigma^* \\ \text{and } \tau(n) < \gamma . \end{array}\right\}$$

Let $W_\gamma = \cup\{B(\tau,\mathcal{C}_\sigma) | \tau \in T_\gamma,\ \sigma \in \Sigma^*, \text{ and } \tau \text{ extends } \sigma\}$. We claim $\{W_\gamma | \gamma < \omega_1\}$ is a monotone shrinking of $\{U_\gamma | \gamma < \omega_1\}$. It is monotone since $\{T_\gamma | \gamma < \omega_1\}$ is.

To prove that $\overline{W}_\gamma \subset U_\gamma$, suppose that $f \in \overline{W}_\gamma - U_\gamma$. For some $m \in \mathbb{N}$ there is $\rho \in \Sigma_m^*$ extended by f; let $\mu = f \restriction (m+1)$. We prove that $B(\mu,\mathcal{C}_\rho)$ is a neighborhood of f missing W_γ, contradicting $f \in \overline{W}_\gamma$. For suppose $n \in \mathbb{N}$, $\tau \in T_\gamma \cap \Sigma_{n+1}$, $\sigma = \tau \restriction n$ and $B(\mu,\mathcal{C}_\rho) \cap B(\tau,\mathcal{C}_\sigma) = \emptyset$. As in our proof that S is shrinking, $B(\rho,\mathcal{C}_\rho) \cap B(\sigma,\mathcal{C}_\sigma) = \emptyset$ unless $n = m$. Since $B(\mu,\mathcal{C}_\rho) \subset B(\rho,\mathcal{C}_\rho)$ and $B(\tau,\mathcal{C}_\sigma) \subset B(\sigma,\mathcal{C}_\sigma)$ we can assume $n = m$. Since $\{U_\delta | \delta < \omega_1\}$ is monotone and $f \notin U_\gamma$, $\alpha_\rho > \gamma$. So by our definition of $\Gamma = \omega_1$, $\mu(n-1) = \rho(n-1) > \gamma$. Since $\tau \in T_\gamma$, $\tau(n) < \gamma$ and there is no term of Δ in both $B(\mu,\mathcal{C}_\rho)$ and $B(\tau,\mathcal{C}_\sigma)$. Since nothing can extend both μ and τ which are distinct and have n as their domain, $B(\mu,\mathcal{C}_\rho) \cap B(\tau,\mathcal{C}_\sigma) = \emptyset$.

THE κ-$\mathcal{B}$ PROPERTY

If λ is an infinite cardinal, let $(\lambda+1)'$ be $(\lambda+1)$ topologized by using the singletons from λ together with intervals of the form $(\alpha,\lambda]$ for $\alpha < \lambda$, as a basis. Yasui (1983) proves:

Theorem 5

The following are equivalent for normal spaces X.

(1) X is κ - $\mathcal{B}$.

(2) $X \times (\lambda+1)'$ is normal for all $\lambda \leq \kappa$.

Proof

To see that (2) implies (1) let $\{U_\alpha | \alpha < \kappa\}$ be a monotone open cover of X. Since $K = \cup\{(X-U_\alpha) \times \{\alpha\} | \alpha < \kappa\}$ and $H = X \times \{\kappa\}$ are closed and disjoint in $X \times (\kappa+1)'$, there is an open V with $H \subset V$ and $\overline{V} \cap K = \emptyset$. If $V_\alpha = \cup\{\text{open } W \text{ in } X | (W \times [\alpha,\kappa]) \subset V\}$, then $\{V_\alpha | \alpha < \kappa\}$ is trivially a monotone shrinking of $\{U_\alpha | \alpha < \kappa\}$. A similar proof handles $\lambda < \kappa$.

To see that (1) implies (2), suppose that H and K are closed and disjoint in $X \times (\kappa+1)'$.

Choose an open set A in X such that $\{x \in X | \langle x,\kappa\rangle \in H\} \subset A$ but $\{x \in \overline{A} | \langle x,\kappa\rangle \in K\} = \emptyset$. For each $\alpha < \kappa$, define $U_\alpha = (X-\overline{A}) \cup \cup\{\text{open } W \text{ in } X | (W \times [\alpha,\kappa]) \cap K = \emptyset\}$. Since $\{U_\alpha | \alpha < \kappa\}$ is a monotone open cover of X which is κ - $\mathcal{B}$, there is a monotone shrinking $\{V_\alpha | \alpha < \kappa\}$ of $\{U_\alpha | \alpha < \kappa\}$. Then $V = \cup\{(V_\alpha \cap A) \times (\alpha,\kappa] | \alpha < \kappa\}$ is an open set in $X \times (\kappa+1)'$ containing $(X \times \{\kappa\}) \cap H$ such that $\overline{V} \cap K = \emptyset$. Similarly we can find an open set W in $X \times (\kappa+1)'$ containing $(X \times \{\kappa\}) - V$ such that $\overline{W} \cap H = \emptyset$.

For each $\alpha < \kappa$ choose an open set A_α in X such that $\{x \in X | \langle x,\alpha\rangle \in H\} \subset A_\alpha$ and $\{x \in X | \langle x,\alpha\rangle \in K\} \cap \overline{A}_\alpha = \emptyset$. Then $(V \cup \cup\{A_\alpha \times \{\alpha\} | \alpha < \kappa\}) - \overline{W}$ is an open set in $X \times (\kappa+1)'$ containing H whose closure misses K.

Again we have certainly proved a generalization of Dowker's theorem for $(\omega+1)' = (\omega+1)$ and being ω - $\mathcal{B}$ is equivalent to Dowker's third condition. We know that Navy's S is $\mathcal{B}$ but not paracompact so we cannot hope for all three conditions in the uncountable case. But how many of the consequences of κ-paracompactness does κ - $\mathcal{B}$ carry? Yasui (1983) asks, for instance, if closed continuous maps preserve the $\mathcal{B}$ property in normal spaces and if there is a class $\mathcal{C}$ of compact spaces such that $X \times C$ is normal for all $C \in \mathcal{C}$ if and only if X is a normal $\mathcal{B}$ space? We prove that the answer to both questions is <u>no</u>.

Theorem 6

[*] Closed continuous maps do not necessarily preserve the $\mathcal{B}$-property.

Proof

Reread the definition of Navy's S. For $\alpha \in \omega_1$, let $F_\alpha = \{f \in F | f(0) = \alpha\}$. If $T = \Delta \cup \{F_\alpha | \alpha < \omega_1\}$, define $p : S \to T$ by

$p(x) = x$ for $x \in \Delta$ and $p(x) = F_\alpha$ for $x \in F_\alpha$. Topologize T by having U be open in T if and only if $p^{-1}(U)$ is open in S. The identification map p is continuous and closed; S is a $\mathcal{B}$ space and T is not.

To see that T is not ω_1-$\mathcal{B}$ we use Navy's proof that S is not collectionwise normal. In particular she proves that there do not exist uncountably many disjoint open sets in S each containing an F_α.

For $\alpha < \omega_1$, let $U_\alpha = \Delta \cup \{F_\beta | \beta < \alpha\}$. Suppose that $\{V_\alpha | \alpha < \omega_1\}$ were a monotone shrinking of the monotone open cover $\{U_\alpha | \alpha < \omega_1\}$ of T. For each $\alpha < \omega_1$ there is an $\alpha^* \in \omega_1$ such that $F_\alpha \in F_{\alpha^*}$. Choose a closed unbounded subset Γ of ω_1 such that $\gamma \in \Gamma$ and $\alpha < \gamma$ imply that $\alpha^* < \gamma$. Without loss of generality we assume that $\Gamma = \omega_1$. Thus for all $\gamma \in \omega_1$, $F_\gamma \in V_{\gamma+1}$. Since $F_\gamma \notin U_\gamma, F_\gamma \notin \overline{V}_\gamma$. Thus $W_\gamma = p^{-1}(V_{\gamma+1} - \overline{V}_\gamma)$ is an open set in S containing F_γ and $\{W_\gamma | \gamma \in \omega_1\}$ are disjoint which contradicts Navy.

Several old theorems about products with a compact factor seem worth mentioning here.

Theorem 7

(Rudin, 1975) If C is compact and Y is the image of X under a continuous closed map, then $X \times C$ is normal implies $Y \times C$ is normal.

Theorem 8

(Alas 1971) If X^x is the one point compactification of X with the discrete topology, then $X \times X^x$ is normal if and only if X is countably paracompact and collectionwise normal.

Theorem 9

(Rudin 1975) If C is a compact space and $X \times C$ is normal, then X is (weight of C) - collectionwise normal.

Theorem 10

[*] There is no class $\mathcal{C}$ of compact spaces such that $X \times C$ is normal for all $C \in \mathcal{C}$ if and only if X is $\mathcal{B}$. Nor is there such a class distinguishing shrinking or $\mathcal{D}$ spaces.

Proof

Recall the S and T from the proof of Theorem 6. Since T is the closed continuous image of S, by Theorem 7, for compact C,

$T \times C$ is normal whenever $S \times C$ is. But S has property $\mathcal{B}$ while T does not. For the last statement note that S and T and Bing's G [31] are all examples of normal, countably paracompact, not ω_1-collectionwise normal spaces which are shrinking and $\mathcal{D}$. By Theorems 1 and 9, $G \times C$ (or $S \times C$ or $T \times C$) is normal for a compact C if and only if C is metric. But in the last section we construct a normal, countably paracompact space X_{ω_1} which is neither ω_1-shrinking nor ω_1 - $\mathcal{D}$. By Theorem (1), however, $X_{\omega_1} \times C$ is normal for all compact metric C. Thus the normality of products with a compact factor does not distinguish the shrinking, $\mathcal{B}$, or $\mathcal{D}$ properties.

Indeed normal $\mathcal{B}$-spaces are (surprisingly) weaker than paracompact spaces in spite of the current scarcity of spaces which fall in this gap. But are spaces which are normal and countably paracompact but not κ - $\mathcal{B}$ for some uncountable κ, interesting? Not really for the reasons already given, namely, we know hundreds of them. To find a property whose denial gives us some of the delicate pathology found in the ordinary Dowker space, we need something much weaker than the κ - $\mathcal{B}$ property.

THE κ-SHRINKING AND κ - $\mathcal{D}$ PROPERTIES

Amazingly enough, we do not know at this time if the classes of normal spaces satisfying these properties are different in all models of set theory, but we can now show that if $V = L$ then they are different (Beslagic & Rudin, 1984).

My preference is for the shrinking property because it is so marvelously simple: no well ordering of covers, no special mention of normality.

All spaces are 0- and 1-shrinking.

If $1 < n < \omega$, a space is n-shrinking if and only if it is normal.

A space is ω-shrinking if and only if it is normal and countably paracompact.

That every open cover of a space should be shrinkable seems like a tremendously more powerful characteristic of a space than that every open cover of cardinality 2 should be shrinkable. Yet we know precisely one class of (very wierd) spaces which are normal and not shrinking. (We describe these spaces later in this section.) A very basic totally unsolved problem is to find some "nice" normal spaces which have an open cover which cannot be shrunk.

A Σ-product of metric spaces, or compact or p-spaces with countable tightness, are examples of nice, natural, but not paracompact spaces. To prove that these spaces are normal is not trivial (Kambarov (1978), Chiba (1982), Yajima (1983)) and even knowing this, it is not clear that all open covers can be shrunk. However, it is true in these cases (see Rudin (1983 b) and Le Donne (1984)); although the proofs are considerably harder than the normality proofs. It seems unreasonable to me, however, that being able to shrink covers of cardinality 2 should imply that all covers shrink in all half way decent spaces.

Since $\mathcal{D}$ is clearly the weakest of our properties, we define a space X to be a <u>κ-Dowker space</u> if X is normal but <u>not</u> a κ - $\mathcal{D}$ space. This seems to carry with it the fragile pathology of an ordinary Dowker space which by this definition is an ω-Dowker space.

Is there some product theorem to go with this definition of a κ-Dowker space? We present a rather weak one; but it indicates the continued involvement of the normality of products. (See Atsuji (1976) and Rudin (1978).)

Theorem 11

If $X \times Y$ is normal and the minimal cardinality of a subset of Y with a limit point is κ, then every monotone open cover of X indexed by κ can be shrunk.

Proof

Suppose $\{U_\alpha | \alpha < \kappa\}$ is a monotone open cover of X. Let $(\kappa + 1)$ index a subset of Y with $\{\kappa\}$ being a limit point of the set. Then $K = \cup\{(X - U_\alpha) \times \{\alpha\} | \alpha < \kappa\}$ and $H = X \times \{\kappa\}$ are closed and disjoint and there is an open set V in $X \times Y$ with $H \subset V$ and $\overline{V} \cap K = \emptyset$. For $\alpha < \kappa$, let $V_\alpha = \{x \in X | \langle x, \alpha\rangle \in V\}$. Then $\{V_\alpha | \alpha < \kappa\}$ shrinks $\{U_\alpha | \alpha < \kappa\}$.

We define below a normal space X_κ having a monotone open cover indexed by κ which cannot be shrunk. Thus (See Atsuji (1976) and Rudin (1978)):

Corollary 12

If $X \times Y$ is normal for all normal Y, then X is discrete.

For regular κ, $\kappa = \min\{\lambda < \kappa | X_\kappa$ fails to have a λ-covering property$\}$. For all κ, although X_κ is κ-Dowker, namely normal and not $\kappa - \mathcal{D}$ (and thus not κ-shrinking or $\kappa - \mathcal{B}$ or κ-paracompact), X_κ is $\lambda - \mathcal{D}$ and λ-shrinking and $\lambda - \mathcal{B}$ and λ-paracompact and even λ-ultraparacompact for all $\lambda < cf(\kappa)$. (A space is ultraparacompact if every open cover has a disjoint open refinement. A κ-ultraparacompact space has all of our other covering properties trivially.) This answers questions found in Zenor (1970), Atsuji (1976), Chiba (1982), and Yasui (1983) in an especially general way. Unfortunately the spaces are not what I would call "nice".

Definition

Choose an increasing sequence $\{\lambda_\alpha | \alpha < \kappa\}$ of regular cardinals for which $\lambda_\alpha = \lambda_\alpha^\kappa$. If κ^+ is the smallest cardinal greater than κ, choose $\lambda_0 > \kappa^+$. Then X_κ (Morita (1975), Ostaszewski (1976)) is the subspace of the box product $\square_{\alpha<\kappa} (\lambda_\alpha + 1)$ consisting of precisely those functions f for which there is a $\beta < \kappa$ such that $\kappa <$ cofinality of $f(\alpha) < \lambda_\beta$ for all $\alpha < \kappa$.

Notation

We use $X = X_\kappa$, $F = \prod_{\alpha<\kappa} (\lambda_\alpha + 1)$, $C = \{f \in F | f(\alpha) < \lambda_\alpha$ for all $\alpha < \kappa\}$ and cf (δ) = cofinality of δ. For f and g in F we say $f < g$ if $f(\alpha) < g(\alpha)$ for all $\alpha < \kappa$, and $f \leq g$ if $f(\alpha) \leq g(\alpha)$ for $\alpha < \kappa$. For $g < f$ in F, let $B_{gf} = \{h \in X | g < h \leq f\}$; if $f \in X, B_{gf}$ is a basic open neighborhood of f. Trivially X is Hausdorff.

Theorem 12

[*] X_κ does not have the κ-$\mathcal{D}$-property (nor any of our other κ-covering properties.)

Proof

For $\alpha < \kappa$ let $U_\alpha = \{f \in X | f(\beta) < \lambda_\beta$ for all $\alpha \leq \beta < \kappa\}$. Suppose, contrary to the theorem, that there were a shrinking $\{V_\alpha | \alpha < \kappa\}$ of the monotone open cover $\{U_\alpha | \alpha < \kappa\}$ of X.

Lemma 1

For all $\alpha < \kappa$ there is $f_\alpha \in C$ such that

$\{g \in C \cap V_\alpha | f_\alpha < g\} = \emptyset$.

Assuming Lemma 1, let $g \in F$ be defined by $g(\beta) = \sup\{f_\alpha(\beta) | \alpha < \kappa\} + \kappa^+$. Since λ_β is regular and $\kappa < \kappa^+ < \lambda_\beta$, $g(\beta) < \lambda_\beta$ and $cf(g(\beta)) = \kappa^+$. Hence $g \in C \subset X$. But $g \notin \bigcup_{\alpha<\kappa} V_\alpha = X$. This contradiction proves our theorem.

Proof of Lemma (1)

Fix $\alpha < \kappa$. Index $\{p:(\alpha+1) \to \lambda_\alpha | p(\beta) < \lambda_\beta$ for all $\beta \leq \alpha\}$ as $\{p_\sigma | \sigma < \lambda_\alpha\}$, making sure that each p is p_σ for λ_α-many σs. Since $\alpha < \kappa$ and $\lambda_\alpha = \lambda_\alpha^\kappa$, this is possible. Assuming our lemma is false, for each $\sigma < \lambda_\alpha$ we choose $g_\sigma \in C \cap V_\alpha$ such that $p_\sigma(\beta) < g_\sigma(\beta)$ if $\beta \leq \alpha$ and $\sup\{g_\tau(\beta) | \tau < \sigma\} < g_\sigma(\beta)$ if $\alpha < \beta < \kappa$. Define $f \in F$ by $f(\beta) = \lambda_\beta$ if $\beta \leq \alpha$ and $f(\beta) = \sup\{g_\sigma(\beta) | \sigma < \lambda_\alpha\}$ if $\alpha < \beta < \kappa$. Since $f \notin U_\alpha$ and $\overline{V}_\alpha \subset U_\alpha$, $f \notin \overline{V}_\alpha$. So there is $g < f$ in F such that $B_{gf} \cap V_\alpha = \emptyset$. For all $\alpha < \beta < \kappa$ there is $\sigma_\beta < \lambda_\alpha$ such that $g(\beta) < g_{\sigma_\beta}(\beta)$. There is $\sigma < \lambda_\alpha$ such that $\sigma > \sup\{\sigma_\beta | \alpha < \beta < \kappa\}$ and $p_\sigma(\beta) = g(\beta)$ for all $\beta \leq \alpha$. Then $g_\sigma \in B_{gf} \cap V_\alpha$ contradicting the fact that this set is empty; our lemma and theorem are thus proved.

Theorem 13

[*] If $\lambda < cf(\kappa)$ then X is λ-ultraparacompact (and thus has all of our λ-covering properties.)

Proof

Assume $\lambda < cf(\kappa)$ and that $\{U_\mu | \mu < \lambda\}$ is an open cover of X; we prove that $\{U_\mu | \mu < \lambda\}$ has an open disjoint refinement $\mathcal{W}$.

If $A \subset F$, we define "the top of A" to be the term t_A of F defined by $t_A(\alpha) = \sup\{f(\alpha) | f \in A\}$ for all $\alpha < \kappa$.

By induction, for each $\sigma < \kappa^+$, we define an open cover $\mathcal{W}_\sigma$ of X by disjoint open sets such that $\tau < \sigma < \kappa^+$ and $W \in \mathcal{W}_\sigma$ imply there is $V \in \mathcal{W}_\tau$ such that:

(1) $W \subset V$

(2) $W \not\subset U_\mu$ for any $\mu < \lambda$, then $t_W \neq t_V$.

(3) If $V \subset U_\mu$ for some $\mu < \lambda$, then $W = V$.

Assuming such $\mathcal{W}_\sigma$s exist, define $\mathcal{W} = \{W \in \mathcal{W}_\sigma | \sigma < \kappa^+$ and $W \subset U_\mu$ for some $\mu < \lambda\}$. To see that $\mathcal{W}$ covers X, suppose $f \in X$. For every $\sigma < \kappa^+$ there is $W_\sigma \in \mathcal{W}_\sigma$ such that $f \in W_\sigma$ and, by (1), $W_\tau \subset W_\sigma$ whenever $\tau < \sigma$. For each $\alpha < \kappa$, if $\tau < \sigma < \kappa^+$,

$t_{W_\tau}(\alpha) \geq t_{W_\sigma}(\alpha)$. So there is $\tau_\alpha < \kappa^+$ such that $t_{W_{\tau_\alpha}}(\alpha) = t_{W_\sigma}(\alpha)$ for all $\tau_\alpha < \sigma < \kappa^+$. Since $\sup\{\tau_\alpha | \alpha < \kappa\} = \tau < \kappa^+$, $W_\tau = W_\sigma$ for all $\tau < \sigma < \kappa$, and, by (2), $W_\tau \in \mathcal{W}$. Thus $\mathcal{W}$ is a refinement of $\{U_\mu | \mu < \lambda\}$ which, by (3), is made up of disjoint open sets, as desired.

To choose such $\mathcal{W}_\sigma$s, we define $\mathcal{W}_0 = \{X\}$. Then we assume that $0 < \sigma < \kappa^+$ and that $\mathcal{W}_\tau$ has been defined for all $\tau < \sigma$.

If σ is a limit, we define $\mathcal{W}_\sigma = \{\bigcap_{\tau<\sigma} W_\tau | W_\tau \in \mathcal{W}_\tau$ for $\tau < \sigma$ and $W_\rho \supset W_\tau$ for $\rho < \tau < \sigma\}$. Clearly $\mathcal{W}_\sigma$ then has the desired properties when one observes:

<u>The intersection of $\leq \kappa$ open sets is open</u>. To see this, suppose that $\{Z_\alpha | \alpha < \kappa\}$ is a family of open sets with $f \in \bigcap_{\alpha<\kappa} Z_\alpha$. For each $\alpha < \kappa$ choose $g_\alpha < f$ in F such that $B_{g_\alpha f} \subset Z_\alpha$. Define $g \in F$ by $g(\beta) = \sup\{g_\alpha(\beta) | \alpha < \kappa\}$ for all $\beta < \kappa$. Since $cf(f(\beta)) > \kappa, g < f$. Thus $f \in B_{gf} \subset \bigcap_{\alpha<\kappa} B_{g_\alpha f} \subset \bigcap_{\alpha<\kappa} Z_\alpha$, and $\bigcap_{\alpha<\kappa} Z_\alpha$ is proved open.

If $\sigma = \tau + 1$, for each $V \in \mathcal{W}_\tau$ we define $\mathcal{W}_V$. If $V \subset U_\mu$ for some $\mu < \lambda$, we define $\mathcal{W}_V = \{V\}$. Otherwise $\mathcal{W}_V$ is an open cover of V by disjoint subsets of V which do not have t_V as their top unless they are contained in U_μ for some $\mu < \lambda$. Then $\mathcal{W}_\sigma = \cup\{\mathcal{W}_V | V \in \mathcal{W}_\tau\}$ has the desired properties.

So suppose V is open, $V \not\subset U_\mu$ for any $\mu < \lambda$, and $t = t_V$. Our remaining task is to define $\mathcal{W}_V$. We consider several cases.

Case (1)

There is $\alpha < \kappa$ with $cf(t(\alpha)) \leq \kappa$.

Choose a closed increasing sequence $\{\alpha_\delta | \delta < cf(t(\alpha))\}$ of ordinals cofinal with $t(\alpha)$; choose $\alpha_0 = 0$. For each $\gamma < cf(t(\alpha))$ let $W_\gamma \{f \in V | \alpha_\gamma < f(\alpha) \leq \alpha_{\gamma+1}\}$. Since $f \in X$ implies $cf(f(\alpha)) > \kappa$ and $cf(\alpha_\gamma) \leq \kappa$ for limit γ_s, $\mathcal{W}_V = \{W_\gamma | \gamma < cf(t(\alpha))\}$ is an open cover of V by disjoint open sets whose members do not have t as their top, as desired.

Case (2)

For all $\alpha < \kappa$, $cf(t(\alpha)) > \kappa$.

Lemma 2

In case (2) there is an $f < t$ in F and $\mu < \lambda$ such that $\{g \in V | f < g\} \subset U_\mu$.

Assuming Lemma 2 , if $S \subset \kappa$, define

$W_S = \{g \in V | f(\alpha) < g(\alpha)$ if $\alpha \in S$, and $g(\alpha) \leq f(\alpha)$ if $\alpha \in \kappa - S\}$.

Then, since the only W_S having t as its top is $W_\kappa = \{g \in V | f < g\}$, $\mathcal{W}_V = \{W_S | S \subset \kappa\}$ has the desired properties. All that remains is:

Proof of Lemma (2)

If $\gamma < \kappa$, let $V_\gamma = \{g \in V | cf(g(\alpha)) < \lambda_\gamma$ for all $\alpha < \kappa\}$. We prove:

Lemma 3

If $\gamma < \kappa$, there are $f_\gamma < t$ in F and $\mu_\gamma < \lambda$ such that $\{g \in V_\gamma | f_\gamma < g\} \subset U_{\mu_\gamma}$.

Assuming Lemma 3, define f by $f(\alpha) = \sup\{f_\gamma(\alpha) | \gamma < \kappa\}$ for all $\alpha < \kappa$. Since we have Case (2) with $cf(t(\alpha)) > \kappa$ for all α and $f_\gamma < t$ for all $\gamma, f < t$. Since $\lambda < cf(\kappa)$ there is some $\mu < \lambda$ which is μ_γ for κ many $\gamma < \kappa$. This f and μ satisfy Lemma 2. For suppose $g \in V$ and $f < g$. Since $V = \bigcup_{\gamma<\kappa} V_\gamma$, $g \in V_{\gamma*}$ for some $\gamma^* < \kappa$. There is $\gamma > \gamma^*$ in κ with $\mu_\gamma = \mu$; and $g \in V_{\gamma*} \subset V_\gamma$. Since $f_\gamma \leq f < g$, $g \in U_{\mu_\gamma} = U_\mu$. Which is exactly what Lemma 2 requires. So it suffices to have:

Proof of Lemma (3)

Fix $\gamma < \kappa$ and assume that, contrary to Lemma 3, for all $f < t$ in F and $\mu < \lambda$, there is $g \in (U_\gamma - U_\mu)$ with $f < g$.

Define $A = \{\alpha < \kappa | cf(t(\alpha)) \leq \lambda_\gamma\}$. For all $\alpha \in A$ choose $T_\alpha \subset t(\alpha)$ of cardinality $cf(t(\alpha))$ and cofinal with $t(\alpha)$. Since $\lambda_\gamma = \lambda_\gamma^\kappa$ and $\lambda < \kappa < \lambda_\gamma$, we can index $\{p: A \to \bigcup_{\alpha \in A} T_\alpha | p(\alpha) \in T_\alpha$ for all $\alpha \in A\}$ as $\{p_{\delta\mu} | \delta < \lambda_\gamma$ and $\mu < \lambda\}$ making sure each p is $p_{\delta\mu}$ for λ_γ many δs for each fixed $\mu < \lambda$.

By double induction we now choose $g_{\delta\mu} \in (V_\gamma - U_\mu)$ for all $\delta < \lambda_\gamma$ and $\mu < \lambda$. Assume $\delta < \lambda_\gamma$ and $\mu < \lambda$ and that $g_{\beta\nu}$ has been chosen for $\beta < \delta$ or $\beta = \delta$ and $\nu < \mu$. Let $f \in F$ be defined by $f(\alpha) = p_{\delta\mu}(\alpha)$ if $\alpha \in A$, and $f(\alpha) = \sup\{g_{\beta\nu}(\alpha) | \beta < \delta$ or $\beta = \delta$ and $\nu < \mu\}$ if $\alpha \in (\kappa - A)$.

If $\alpha \in (\kappa - A)$ and $g \in V_\gamma$, then $g(\alpha) \leq t(\alpha)$ and $cf(g(\alpha)) < \lambda_\gamma < cf(t(\alpha))$ so $g(\alpha) < t(\alpha)$. Since $g_{\beta\nu} \in V_\gamma$ and the number of $\beta\nu$'s is less than λ_γ, $f(\alpha) < t(\alpha)$ for all $\alpha \in (\kappa - A)$. Also $p_{\delta\mu}(\alpha) < t(\alpha)$ for $\alpha \in A$, so $f < t$. By our assumption that Lemma 3 fails, there is $g_{\delta\mu} \in (V_\gamma - U_\mu)$ with $f < g_{\delta\mu}$.

Define $g \in F$ by $g(\alpha) = t(\alpha)$ for $\alpha \in A$ and $g(\alpha) = \sup\{g_{\delta\mu}(\alpha) | \delta < \lambda_\gamma$ and $\mu < \lambda\}$ if $\alpha \in (\kappa - A)$. Since $cf(g(\alpha)) = \lambda_\gamma$ if $\alpha \in (\kappa - A)$ and $cf(g(\ )) \leq \lambda_\gamma$ if $\alpha \in A$, $g \in X$. Thus $g \in U_\mu$ for some $\mu < \lambda$ and there is $h < g$ in F with $B_{hg} \subset U_\mu$. For $\alpha \in (\kappa - A)$, by our choice of $g(\alpha)$, there is $\delta_\alpha < \lambda_\gamma$ such that $g_{\delta_\alpha\mu}(\alpha) > h(\alpha)$. By our choice of p there is $\delta > \sup\{\delta_\alpha | \alpha \in (\kappa - A)\}$ such that $p_{\delta\mu}(\alpha) = h(\alpha)$ for all $\alpha \in A$. By our definition of $g_{\delta\mu}$, $h < g_{\delta\mu} < g$. So $g_{\delta\mu} \in B_{hg} \subset U_\mu$ which contradicts $g_{\delta\mu} \notin U_\mu$.

This proof can also be used to prove:

Theorem 14

If $\omega < cf(\kappa)$ and $\lambda < cf(\kappa)$ and M is a metric space of weight $\leq cf(\kappa)$, then $X \times M$ is normal and λ-paracompact.

Proof

Let $X_\kappa = X$ and let $\mathcal{O}$ be an open basis for M of cardinality $\leq \kappa$.

By induction, for each $\sigma < \kappa^+$, we define a cover $\mathcal{W}_\sigma$ of X by disjoint open sets such that $\tau < \sigma < \kappa^+$ and $W \in \mathcal{W}_\sigma$ imply there is $V \in \mathcal{W}_\tau$ such that:

(1) $W \subset V$.

(2) If there is $p \in M$ such that $V \times O$ is not in any U_μ for any $O \in \mathcal{O}$, then $t_W \neq t_V$.

(3) If $\{O \in \mathcal{O} | V \times O \subset U_\mu$ for some $\mu\}$ covers M, then $W = V$.

If such $\mathcal{W}_\sigma$ exist, then $\mathcal{W} = \{W \in \mathcal{W}_\sigma | \sigma < \kappa^+$ and $W \in \mathcal{W}_{\sigma+1}\}$ is an open cover of X by disjoint sets. If $W \in \mathcal{W}$ and $\mu < \lambda$, let $O_{\mu W} = \{O \in \mathcal{O} | W \times O \subset U_\mu\}$. Since $\{O_{\mu W} | \mu < \lambda\}$ covers M which is metric, it has a shrinking $\{O_{\mu W'} | \mu < \lambda\}$. Thus $\{\cup\{W \times O_{\mu W'} | W \in \mathcal{W}\} | \mu < \lambda\}$ is a shrinking of $\{U_\mu | \mu < \lambda\}$.

To show that such $\mathcal{W}_\sigma$s exist, define $\mathcal{W}_0 = \{X\}$ and for limit σs let $\mathcal{W}_\sigma = \{\cap_{\tau<\sigma} W_\tau | W_\tau \in \mathcal{W}_\tau\}$.

Suppose $\sigma = \tau + 1$, $V \in \mathcal{W}_\tau$, and $t = t_V$. We can assume that $cf(t(\alpha)) > \kappa$ for all $\alpha < \kappa$; it will suffice to find an $f < t$ such that $\{O | B_{ft} \times O \subset U_\mu\}$ for some $\mu < \lambda$ covers M.

Suppose $p \in M$ and choose a basis $\{O_n | n \in \omega\}$ for p. If $\mu < \lambda$ and for each $n \in \omega$ and $f < t$ in F there is $v \in V$ and

$o \in O_n$ with $\langle v,o\rangle \notin U_\mu$, then, as in the proof of Lemma 3, there is a point $g \in X$ with $\langle g,p\rangle \notin U_\mu$ for any μ. Thus, for some $n \in \omega$ we can find $f < t$ in F with $B_{ft} \times O_n \subset U_\mu$.

For each $O \in \mathcal{O}$ and $\mu < \lambda$ for which there is $f < t$ in F with $B_{ft} \times O \subset U_\mu$, choose one such f, called $f_{O\mu}$. Let $f \in F$ be defined by $f(\alpha) = \sup\{f_{O\mu}(\alpha) \mid O \in \mathcal{O}, \mu < \lambda\}$. Since the cardinality of $\mathcal{O}$ is $\leq \kappa$, $f < t$. Since $\{O \in \mathcal{O} \mid$ there is $\mu < \lambda$ for which $f_{O\mu}$ is defined$\}$ covers M, our proof is complete.

REFERENCES

Alas, O.T. (1971). On a characterization of collectionwise normality. Can. Math. Bull., 14, 13-15.

Atsuji, M. (1976). On normality of the product of two spaces. Gen. Top. IV Proceedings Fourth Prague Top. Conf., pp. 25-27. Prague.

Bell, M. (1981). On the combinatorial Principal P(c). Fund Math. CXIV, 137-145.

Beslagic, A. & Rudin, M.E. (1984). Set theoretic constructions of nonshrinking open covers. Top. Appl., to appear.

Bing, R.H. (1951). Metrization of topological spaces. Canad. J. Math., 3, 175-186.

Borsuk, K. (1937). Sur les prolongments des transformations continues. Fund. Math., 28, 203.

de Caux, P. (1976). A collectionwise normal, weakly θ-refinable Dowker space which is neither irreducible nor realcompact. Top. Proc., 1, 66-77.

Chiba, K. (1982). On the weak $\mathcal{B}$-property and Σ-products. Math. Japon., 27, 737-746.

Le Donne, A. (1984). Shrinking property in Σ-products of paracompact p-spaces. Top. Appl., to appear.

Dow, A. & van Mill, J. (1982). An extremally disconnected Dowker space. Proc. of AMS, 86, 669-672.

Dowker, C.H. (1951). On countably paracompact spaces. Canad. J. Math., 3, 219-224.

Handbook for Set Theoretic Topology. (1984). North Holland Press.

Juhasz, I., et al. (1976). Two more hereditarily separable non-Lindelöff spaces. Canad. J. Math., 28, 998-1005.

Kambarov, A.P. (1978). On tightness and normality of Σ-products. Soviet Math. Dokl., 19, 403-407.

Mack, J. (1967). Directed covers and paracompact spaces. Canad. J. Math., 19, 649-654.

Morita, K. (1961). Paracompactness and product spaces. Fund. Math., 53, 223-236.

Morita, K. (1975). On generalizations of Borsuk's homotopy extension theorem. Fund. Math. LXXXVIII, 1-6.

Nagami, K. (1955). Paracompactness and strong screenability. Nagoya Math. J., 88, 83-88.

Navy, C. (1981). Paracompactness in para-Lindelöf spaces. Thesis. Madison, Wis.: Univ. of Wis.

Ostaszewski, A. (1976). On countably compact, perfectly normal spaces. J. London Math. Soc., 14, 505-516.

Rudin, M.E. (1955). Countable paracompactness and Souslin's problem. Canad. J. Math., 7, 543-547.

Rudin, M.E. (1971). A normal space X for which $X \times I$ is not normal. Fund. Math. LXXIII, 179-186.

Rudin, M.E. (1973). A separable Dowker space. Symp. Math. Isti. Maz. di Alta Math., 125-313.

Rudin, M.E. (1975). The normality of products with one compact factor. Gen. Top. Appl., 5, 45-59.

Rudin, M.E. (1978). κ-Dowker spaces. Czech. Math. J., 28, no.103, 324-326. Praha.

Rudin, M.E. (1983 a). A normal screenable nonparacompact space. Top. Appl., 15, 313-322.

Rudin, M.E. (1983 b). The shrinking property. Can. Bull.

Rudin, M.E. (1983 c). Yasui's questions. Questions and Answers in Gen. Top., Osaka Kyoiku Univ.

Smith, J.C. (1979). Applications of shrinkable covers. Proc. Amer. Math. Soc., 73, no.3, 379-387.

Starbird, M. (1975). The Borsuk homotopy extension theorem without the binormality condition. Fund. Math. LXXXVII, 207-211.

Watson, S. A Dowker space from a compact cardinal. to appear.

Weiss, W. (1981). A Dowker space from $MA+\diamond\Gamma_C$. Pacific J. Math., 94, 485-492.

Yajima, Y. (1983). On Σ-products and Σ-spaces. Fund. Math.

Yasui, Y. (1983). Some questions on the $\mathcal{B}$- and weak $\mathcal{B}$-property. Questions and Answers in Gen. Top., 1, no.1, 18-24.

Zenor, P. (1970). A class of countably paracompact spaces. Proc. Amer. Math. Soc., 42, 258-262.

GRADUATION AND DIMENSION IN LOCALES

J. Isbell
State University of New York at Buffalo, Buffalo,
New York 14214, USA, and Wesleyan University,
Middletown, Connecticut 06457, USA

INTRODUCTION

Among many ways of defining or characterizing the dimension of a topological space, three are the inductive (Menger-Urysohn), the covering (Lebesgue), and the chain (Krull). Krull dimension, one less than the maximum length of a chain of completely prime filters of the topology, is spectacularly wrong for the most popular spaces, vanishing for all non-empty Hausdorff spaces; but it seems to be the only dimension of interest for the Zariski spaces of algebraic geometry.

The definition proposed here is by Lebesgue <u>or</u> Krull, out of Menger-Urysohn. The <u>graduated dimension</u> gdim X of a T_0 space or locale X is the minimum n such that some sublattice of the topology (respectively, frame) T(X) which is a basis is a directed union of finite topologies of Krull dimension n. The idea of minimizing over bases is from Menger-Urysohn.

A first view of gdim is that it is a modified Krull dimension designed to be reasonable for a wider class of spaces. How reasonable is it? For metrizable spaces or locales X, ind X $\leq$ gdim X $\leq$ dim X. There is essentially only one known example for ind $\neq$ dim in metrizable spaces, Roy's (1968) example, and for it gdim agrees with ind. A second handy class of "geometric" spaces is the compact Hausdorff spaces Y, for which the lineup of familiar dimension functions is dim Y $\leq$ ind Y $\leq$ Ind Y (and dim Y = 1, ind Y = 2, Ind Y = 3 is possible (Filippov 1970)). Here gdim Y $\geq$ Ind Y, but whether ever strictly I do not know. (Pasynkov has a dimension function Ind_c defined only for compact Hausdorff spaces, greater than or equal to gdim, and he asks whether it can differ from Ind (Pasynkov 1965).) A third important class, I would urge, is the finite T_0 spaces. For these, Krull dimension K dim = ind = gdim = dimension of the geometric realization; the geometric realization is familiar for those finite spaces Z which are simplicial complexes, and the first barycentric subdivision $Z_{(1)}$ of any finite space is a simplicial complex. Observe that dim and Ind are as ridiculous (though not as quiet) for finite spaces as Kdim for Hausdorff spaces.

For arbitrary locales, gdim $\geq$ ind. In fact, if gdim $X \leq n$ then X has a basis $\mathcal{B}$ the boundaries of whose elements, and also finite unions H of boundaries, have ind $H \leq$ gdim $H \leq n - 1$ in virtue of the basis $\{U \cap H | U \in \mathcal{B}\}$. One may say, gdim includes quite a bit of inductiveness. I do not know if this condition characterizes gdim. It yields gdim $Y \geq$ Ind Y for all compact locales Y. For all X, gdim X is the minimum Krull dimension of a coherent space containing X. (Coherent = spectral = profinite T_0; cf. Johnstone 1982.)

It is more complicated to explain how gdim is a covering dimension. It is not a refinement dimension, and could not be; there are arbitrarily large finite T_0 spaces (in the sense of inclusion) Z, every open covering of which is refined by {Z}. dim X, according to Čech (1933) (note, not Lebesgue (1921)), is one less than the smallest integer which is the order of enough finite open covers to include refinements of all finite open covers of X. To get a definition of gdim we make three changes, only one of which will be fully specified at this point. (1) Not all finite covers. (There is precedent for this in Katětov's (1950) modification of dim, which has been widely adopted, and in my modification mindim (Isbell 1964), which has not.) (2) Not refinement, but a smaller preordering relation. (3) Not order. Precisely, for spaces, not locales, given a finite open cover $\{U_i\}$, define the <u>carrier</u> of each point x as the intersection of all U_i that contain x. The <u>height</u> of the cover is the maximum length of a chain of carriers. Use height in place of order. This can be reformulated lattice-theoretically, giving the definition of height in locales.

The height of a cover is less than or equal to its order. Therefore if hdim is defined like dim except for (3), height for order, one has hdim $X \leq$ dim X. For normal locales, though, we get hdim X = dim X. Since there are no theorems about dim in non-normal spaces (or locales), this shows that the success of dim provides no argument against height. The same change (3) in the definition of Katětov's modified dim does not change it at all. For mindim, I do not know.

Graduation is the structure that determines embeddings in coherent spaces. It can be given by a basis $\mathcal{B}$, or by a suitable family of coverings.

I am indebted to Jean Benabou and to Peter Johnstone for

helpful conversations about this material.

1 BASIC CONCEPTS

I shall assume the reader has some acquaintance with locales. Isbell (1972) provides enough; or Chapter II of Johnstone (1982), supplemented by pages 24-25 of Isbell (1972).

Sober spaces X are dual to topologies T(X), the lattice of all open sets; continuous maps $f : X \to Y$ correspond to morphisms $f^*: T(Y) \to T(X)$ ($f^*(U) = f^{-1}(U)$), which are lattice homomorphisms preserving 0, 1, and infinite joins.

N.B. We shall be concerned with mere lattices sometimes. We adopt Johnstone's (1982) convention that 0 and 1 are part of the structure: sublattices must contain them, lattice morphisms must preserve them.

Frames are certain complete lattices, the smallest algebraic category containing the topologies. Locales X are objects of the (abstract) dual category; we write T(X) for the frame that defines X. Its elements are <u>open parts</u> or <u>open sublocales</u> of X. There are also general sublocales $S \subset X$, defined by the quotient frames T(S) of T(X).

We shall need easy generalizations of fifteen or more results from spaces to locales. First, three more general points: (1) Every locale is a sublocale of a space. In fact the free frames are topologies, and of course every frame is a quotient of a free frame.

(2) Many simple statements in a locale, such as inclusions $\bar{A} \leq B$, can be proved by proving them locally, i.e., relatively on each element of an open cover. An open cover is just a set of open parts U_i with join 1. I do not find this localization principle explicitly stated anywhere. Call it

1.0. If $\{U_i\}$ is an open cover of a locale X then every sublocale S is the join, in the lattice of sublocales, of all $S \wedge U_i$.

Proof. In the sublocales of S, this says that nothing but 1 contains all $V_i = S \wedge U_i$. Now V_i is the open sublocale of S corresponding to the open part $V_i' \in T(S)$ which is the image of $U_i \in T(X)$. (This is true since both descriptions give the sublocale whose frame is the quotient of T(X) by the smallest congruence relation containing the S-congruence and the "$U_i = 1$" congruence.) So the proposition reduces to "If $\{V_i'\}$ is an open cover of S then the join of the corresponding sublocales V_i is 1". Joining sublocales (to get 1) means intersecting congruence relations (to get the diagonal). If (Y,Z) is any off-diagonal pair in $T(S) \times T(S)$, then since $Y = \vee(Y \wedge V_i')$, $Z = \vee(Z \wedge V_i')$, there is i such that $Y \wedge V_i' \neq Z \wedge V_i'$; so (Y,Z) is not in the i-th congruence relation.

An important addition: joins localize, i.e., among sublocales the open ones U satisfy $U \wedge (\vee S_i) = \vee(U \wedge S_i)$ (Isbell 1972).

(3) For finite calculations in frames, we may speak substantially as if they were topologies. For they are distributive lattices, and every distributive lattice is isomorphic with a ring of sets which can be taken as a basis of a topology (Johnstone 1982).

Now, most of the things we need to generalize are in my book (Isbell 1964; call it US). I shall go somewhat beyond need, but not wildly. For instance, the first theorem on complete uniform spaces (US II.10) is false for complete uniform locales. Why it is false and what happens instead are clear from results in US, but I won't go into that. We have (4) everything before II.10 generalizes, with five exceptions which have points or "one-to-one" in their statements. There is no difficulty in any of the proofs, with one large exception: Weil's theorem that every uniform cover is realized by a uniformly continuous map into a metric space. ("Realized" means refined by the inverse image of some uniform cover of the metric space.)

For Weil's theorem in a uniform locale, begin as in (US I.14) with the given uniform cover $\mathcal{U}_0$ and a sequence of uniform covers $\mathcal{U}_{n+1} <^* \mathcal{U}_n$. Introduce Cauchy filters (relative to this sequence) of open parts, defined as proper filters $\mathcal{F}$ that meet all $\mathcal{U}_n$. Check that if $\mathcal{F}$ is Cauchy so is $\mathcal{F}^* = \{St(F,\mathcal{U}_n) : F \in \mathcal{F}, n \in \omega\}$. Let Y be the set of Cauchy filters $\mathcal{F}^*$. ($\mathcal{F}^* = \mathcal{F}^{**}$.) Transfer the covers $\mathcal{U}_n$ to coverings $\mathcal{T}_n$ of the set Y, by transferring elements U of $\mathcal{U}_n$ to $t(U) = \{\mathcal{F} \in Y : U \in \mathcal{F}\}$. Check $\mathcal{T}_{n+1} <^* \mathcal{T}_n$. (If t(U) meets t(U') in a proper filter then U meets U'; similarly for inclusion.) Then we have a countably-based uniformity on the set Y; topologize Y in the obvious way, and we have a metric uniform space. It is now routine to define (in frames) the natural map $g : X \to Y$ and check that $\mathcal{U}_0$ is realized.

As a corollary, to every uniform cover of a uniform locale X one can subordinate an equiuniformly continuous partition of unity. Just realize the cover by a map $q : X \to Y$ to a (metric) space and make the following adjustment. If the given cover of X is $\{U_\alpha\}$, we have some uniform cover $\{V_\beta\}$ of Y with $\{q^{-1}(V_\beta)\}$ refining $\{U_\alpha\}$; thus for each β one can choose $\alpha = c(\beta)$ so that $q^{-1}(V_\beta) \subset U_\alpha$. Then define $W_\alpha = \cup[V_\beta : c(\beta) = \alpha]$. We have $\{V_\beta\} < \{W_\alpha\}$, so $\{W_\alpha\}$ is a uniform cover of Y. If $\{f_\alpha\}$ is an equiuniformly continuous partition of unity subordinated to it (US IV.11), $\{f_\alpha q\}$ is one subordinated to $\{U_\alpha\}$. (Remark. If $\{V_\beta\}$ is finite-dimensional one can partition 1 first and then clump; all partial sums are equiuniformly continuous. I do not know if that extra condition can be imposed in general.)

(5) Next, a bunch of results in Chapters IV - V of US, to be used later. IV. 20 - IV. 25, V. 1 - V. 5, V. 13, and V. 19 - V. 23 generalize to locales. The proof of IV. 20 mentions points, but these can be taken in the metric space given by Weil's theorem. The proofs through V. 5 are locale proofs already. Both V. 13 and V. 23 begin with reductions to the compact case, which generalize at once (using V. 20 - V. 22, which have locale proofs), and a compact uniform locale is a space. Finally, V. 19 simply uses new terminology to restate V. 13.

(6) For separable metrizable locales X, ind X = dim X; this is also the minimum mindim X of dim Y over Hausdorff compactifications Y of X, and the minimum of dim Z over metrizable compactifications Z of X. Seven propositions in US (generalized) prove this. VI. 3, VI. 4, VI.19, VI. 20 already have locale proofs. In VI. 23 there is a typographical error, but "Each point x of X" was meant, and is illegitimate for locales. However, here the uniformizable Lindelöf locale X is embedded in a compact Hausdorff space Y, and we are avoiding a closed set K where Y - K contains X. Put "Each point x of Y - K", and the proof works. (dim X = mindim X.) Sixth, the proof of VIII. 7 says "Consider any point p ..." and proceeds by cases. But there are only finitely many cases, and it is trivial to rephrase the proof in terms of open sublocales and their closures. Finally, the proof of VIII. 16 appears to use three pointwise arguments. The first merely shows that X is covered by basic open parts whose closures do not meet both of the disjoint closed parts A, B (true in regular locales), and the second is as easily handled. The last one proves an inclusion of sublocales; read it with a view to localization, (2) above, and it localizes. (Combined with VI. 23, we have dim X = mindim X $\leq$ ind X for uniformizable Lindelöf locales.)

The next generalization we need is:

(7) For a metrizable locale X, dim X = Ind X, and this is also the minimum of dim Y over (completely) metrizable spaces Y containing X.

VIII. 15 of US says this for spaces, except that it ends "... the minimum of $\Delta d\ \mu X$ for all compatible metric uniformities μ on X". Here Δd is the covering dimension defined by orders of (infinite) uniform covers; it is preserved by completion ((5) above), and a complete metric locale is a space (Isbell 1972), so (7) will follow from VIII. 15 for locales. For that, we need (more of (5) and) VIII. 1 - VIII. 4, VIII. 10, VIII. 14. Those already have locale proofs (rephrasing in the proof of VIII. 2, from V "containing f(x)" to $f^*(V)$). Finally, in the proof of the Dowker-Hurewicz theorem VIII. 15, the pointwise arguments for $E \subset M$, $F \subset N$ are seen on examination to be local arguments, justified by (2).

(8) We shall need

If every open sublocale of X is normal, so is every sublocale.

Urysohn's Lemma: Disjoint closed parts of a normal locale are separated by a map to the real line.

An F_σ sublocale of a normal locale is normal.

We may as well add

A regular locale with a σ-locally finite basis is metrizable.

For the first of these, Dowker's (1953) proof is a locale proof. This is also true of the usual proofs of Urysohn's Lemma, which has been done for locales by Papert and Papert (1958). The quick proof of normality of F_σ's, using Urysohn's Lemma, is then a locale proof, as follows. Let A and B be disjoint relatively closed in $\vee H_i$ where $H_1, H_2, \ldots$ are closed in the normal locale X. Then each $A \wedge H_i$ is closed and disjoint from B^-, so there is real-valued f_i on X taking $A \wedge H_i$ to 2^{-i}, B^- to 0, and X to the interval between. Similarly there are g_i vanishing on A^- and taking maximum value 2^{-i} constantly on $B \wedge H_i$. Then $\Sigma f_i - \Sigma g_i$ separates A and B in disjoint open parts.

Finally, for the metrization theorem note that the closures of a locally finite family of sublocales form a locally finite family, and their join is closed, by localization (including localizing the join). Then the usual proofs (as in Engelking (1968) or Kelley (1955)) may speak of neighborhoods, but use bases and they generalize.

To begin, then: a grade of a locale X is a finite epi-quotient. One reason for the new name is a wish to avoid the red herring of quotient topologies. $f : X \to Y$ is epic if and only if $f^* : T(Y) \to T(X)$ is monic, which means (in frames) one-to-one. The epi-quotient given by an epic is an equivalence class, uncomfortably large but simple enough; it corresponds exactly to the subframe $f^*(T(Y))$ of $T(X)$. If f represents a grade (or loosely speaking, f is a grade), that subframe is a finite sublattice. In particular, it is an open cover of X. The grade can be recovered from that cover and its nerve.

Let us begin again. Let X be a T_0 space, and C a finite open cover of X. The abstract nerve N(C) of C is the partially ordered set of all non-empty subsets D of C that have non-empty intersection, ordered by inclusion. We identify finite T_0 spaces with finite partially ordered sets: open sets of the topology are upper sets of the ordering. There is a canonical continuous mapping p_C from X to N(C), $p_C(x)$ being $\{U \in C | x \in U\}$. (Verification: the set $p_C(x)$ is non-empty since C is a covering, and its intersection is non-empty since x is in it. In the finite space N(C), an open set is precisely an upper set; the inverse image of a principal upper set $\uparrow(D)$ is the open intersection of D, so the inverse image of any open set is open.)

The <u>grade</u> G(C) of a covering C of a space X is the partially ordered set $p_C(X) \subset N(C)$. In other words, its elements are those non-empty $D \subset C$ that not only have a common point but have a generic common point x, belonging to no other element of C. There is another canonical function p^* from N(C) to T(X), $p^*(D) = \cap[U : U \in D]$. The <u>grade cover</u> $G^*(C)$ is $p^*(G(C))$.

1.1. A cover C and $G^*(C)$ generate the same sublattice of T(X); $G^*(C)$ consists of its join-irreducible elements. The lattice determines G(C) up to isomorphism and is the isomorphic image of the topology of G(C) under p_C^*, where $p_C^*(V) = p_C^{-1}(V)$. For $D \in N(C)$, $p^*(D)$ is p_C^* of the principal upper set $\uparrow(D)$. $p^* : G(C) \to G^*(C)$ is bijective and order-reversing.

Proof. Officially, p_C has codomain not G(C) but N(C); the domain of p_C^* is the topology of N(C), which has a sub-basis indexed by the non-empty elements U of C, $\{\uparrow(\{U\})\}$. $p_C^*(\uparrow(\{U\})) = U$, so p_C^* takes the generated sublattice T(N(C)) to the sublattice of T(X) generated by those U. Adding an empty U, if there is one, does not change the sublattice. Since p_C^* maps T(N(C)) surjectively to T(G(C)) which then maps injectively, the second sentence of 1.1 is true of the sublattice of T(X) generated by C. The third sentence is true by inspection (p_C takes a point x to a superset of D iff $x \in \cap[U : U \in D]$). But the principal upper sets form a basis for T(G(C)); hence $G^*(C)$ generates the same sublattice as does C, and generates it by union alone. So its join-irreducibles must be in $G^*(C)$. In the isomorphic lattice T(G(C)), the principal upper sets are join-irreducible. This proves the first sentence and bijectiveness of

p^*; as for order-reversingness of $p^* = p_C^*\uparrow$, $\uparrow$ is order-reversing and p_C^* order-preserving.

1.2. Corollary. $G^*(G^*(C)) = G^*(C)$.

A cover (finite open) is a grade cover if and only if it is the smallest basis for the topology it generates. All this extends to locales by (3), finite calculations, including the preceding sentence; finite frames are topologies. In spaces, C is a grade cover if and only if (a) each element of C has a generic point, and (b) each point is generic in some element of C.

Among finite covers, C is a subdivision of D if C is a refinement of D and each element of D is the union of the elements of C that it contains.

1.3. If C is a subdivision of D then the lattice generated by C contains the lattice generated by D; for grade covers C, the converse is true. In particular, if C is a grade cover and each element of D is a union of elements of C, then C is a subdivision of D.

Proof. If C generates all the elements of D (by union) then it generates what they generate. A grade cover C, by 1.1, generates its whole lattice L by union. If D is a cover contained in L, C generates it. As for refinement, each $A \in C$ has (we may assume by (3)) a generic point x; A is the smallest element of L containing x. Some $B \in D \subset L$ contains x since D is a cover, so $A \subset B$.

The order of a finite (open) cover C is the maximum size of an element of N(C); we define the height of C as the maximum size of a chain in G(C). Order is the maximum size of a chain in N(C), and is greater than or equal to height.

Let us define hdim $X \leq n$ if every finite open cover of X has a finite open refinement of height at most $n + 1$. Evidently hdim $\leq$ dim; what is more interesting is

1.4. For normal locales hdim = dim.

The proof involves geometric realization and barycentric subdivision. Both constructions work perfectly well for arbitrary finite T_0 spaces, though they are more familiar for simplicial complexes. First, then, for any simplicial complex P, the (traditional) geometric realization gP is the set of all probability measures on the vertices of P whose support is a simplex, topologized by pointwise convergence. The dimension of the compact metric space gP is one less than the height of

P, which happens (for simplicial complexes) to be one less than the order of P. There is a distinguished continuous map spt : gP $\to$ P taking each $\mu \in$ gP to its support. Given a simplicial complex P and a subset G of P, the (geometric) <u>body</u> of G is $\mathrm{spt}^{-1}(G)$.

g is not functorial on simplicial complexes and continuous maps; more precisely, since I have not said what to do with maps and there is a functor, the traditional (linear) geometric realization is not functorial, though it is so on simplicial maps.

The <u>(first)</u> <u>barycentric subdivision</u> $P_{(1)}$ or bP of a finite partially ordered set P is the set of non-empty chains in P ordered by inclusion. This is a simplicial complex, evidently of the same height as P. There is a distinguished continuous map max : bP $\to$ P taking each chain to its greatest element. b is a functor (continuous = order-preserving $\Rightarrow$ chain-preserving) and max is a natural transformation.

(The natural geometric realization is gotten by iterating b. The maps max : $b^{n+1}P \to b^nP$ give an inverse system whose limit is a coherent space $b^\omega P$. Examining the successive barycentric subdivisions of a geometric simplex, one sees that the closed points of $b^\omega P$ form a subspace min $b^\omega P$ homeomorphic with $gP_{(1)}$. The functor min b^ω will not be developed in this paper.)

1.5. Lemma. A finite open cover C of a normal locale can be realized by a morphism into gN(C) that goes into the body of the grade G(C).

Proof. Observe that any partition of unity subordinated to the cover $\{U_i\}$, $\{f_i : X \to R\}$, will give a morphism f into the nerve realizing $\{U_i\}$, by $f(x)(U_i) = f_i(x)$. If we have merely non-negative morphisms $g_i : X \to R$ supported by U_i, and Σg_i is nowhere zero, this will give a partition of unity by $f_i = g_i/\Sigma g_i$. To stay in the body of the grade we must avoid each (open) simplex of the nerve that is not in the grade; that is (simplifying the notation), when $U_1 \wedge \ldots \wedge U_k$ is covered by the other U's we want the support of $\min(g_1,\ldots,g_k)$ covered by the supports of the other g's.

Urysohn's Lemma (8) gives us a partition of unity $\{g_i^0\}$ subordinated to $\{U_i\}$. Our program is now to add functions g_i^j, non-negative but of absolute value at most 2^{-j}, so that at the j-th step the j-th simplex not in the grade is corrected. Those simplexes must be enumerated so that each occurs infinitely many times. If the j-th simplex is $\{U_i | i \in \sigma\}$, the correction is to add functions g_i^j, $i \notin \sigma$, supported each by U_i, so that their sum does not vanish on $\wedge[V_i^j : i \in \sigma]$, where V_i^j is the support of $\Sigma[g_i^k : k < j]$. That intersection is an F_σ, hence normal, and it is covered by the U_i, $i \notin \sigma$, so the correction can be made. Then put $g_i = \Sigma g_i^j$, and we are done.

1.6. Lemma. For every grade $e : X \to Q$ of a completely normal locale there is a grade $e' : X \to bQ$ such that $(\max)e' = e$.

Proof. Let $\{U_i\}$ be the grade cover $\{e^*(\uparrow(i)) | i \in Q\}$. For any incomparable pair U_i, U_j, the sublocales $U_i - U_j$ and $U_j - U_i$ are contained in $U_i \vee U_j$ and are disjoint closed there; so they have disjoint neighborhoods V_{ij}, V_{ji} respectively. Those are open in an open part, thus open, and $V_{ij} \subset U_i$ since it is in $U_i \vee U_j$ and disjoint from $U_j - U_i$; similarly $V_{ji} \subset U_j$.

Let V_i be the intersection of all V_{ij} (U_j incomparable with U_i). Each V_i is open, contained in U_i, disjoint from V_j if U_i, U_j are incomparable. To see that they cover we may imagine points as usual. A point x is generic in some U_i, so it is certainly not in any incomparable U_j; for all such j, $x \in V_{ij}$, and $x \in V_i$. Moreover $e'(x) = \{i \in Q | x \in V_i\}$ is a chain (for incomparable i,j give incomparable U's and disjoint V's), non-empty since the V_i cover. The inverse image of a principal upper set in bQ is an open intersection of V_i's, so e' is a morphism. $\max(e'(x))$ is the i for which x is generic in U_i; that is $e(x)$, as required.

Note, this lifting property characterizes completely normal locales. For normality of open parts suffices (8), and separating H and K, disjoint closed in open $U \subset X$, is effected by lifting from the grade P of $\{X, U - H, U - K\}$ to bP.

Proof of 1.4. In normal X a cover C of height n can be shrunk by 1.5 to a cover $\{f^*(U_i)\}$ given by $f : X \to B$, B separable metric, and $\{U_i\} = C'$ covering B with the same grade. 1.6 gives $e' : B \to bG(C')$ realizing a cover of B that subdivides C' and has nerve a subcomplex of bG(C'), of order at most n; and f^* takes this to a refinement of C of no greater order.

Katětov's (1950) dimension is defined, for completely regular spaces only (except that it might as well be completely regular locales), by replacing general open sets in the definition of dim by cozero sets, which are sets $\{x | f(x) > 0\}$ for real-valued continuous functions f. A finite cover $\{U_i\}$ by such sets has, so to speak, a partition of unity not only subordinated to it but "ordinated" to it; the fit is exact, the map to the nerve goes into the grade, and the adjustments of 1.5 are unnecessary. Katětov dimension via height = Katětov dimension via order.

Remark. Since Katětov's dimension for normal T_1 spaces coincides with dim, one is tempted to take the simpler proof just indicated as proving 1.4. However, it is not a proof. Part of our subject matter is words, and the effect of a verbal change in a definition depends on its wording. I do not know whether mindim changes if one puts "height" for "order". What is more important is the broader question how height behaves for uniform covers.

2 GRADUATIONS AND GDIM

We define a _graduation_ G of a locale X as a set of finite open covers (briefly : covers) of X such that

(1) every two covers in G have a common subdivision in G;

(2) every cover having a subdivision in G belongs to G;

(3) if C is in G then $G^*(C)$ is in G ;

(4) the union of G is a basis of X.

Of course, (2) and (3) are bookkeeping axioms serving principally to make some correspondences between graduations and other attachments bijective. Here is one of them.

2.1. Every graduation consists of all finite covers contained in its union B . B is a basis and a sublattice of the frame T(X); conversely, for every sublattice basis B of T(X), the set of all finite covers contained in B is a graduation.

Proof. For any finite cover $C \subset B$ there are finitely many $C_i \in G$ whose union contains C. The C_i have a common subdivision $D \in G$, and $G^*(D)$ subdivides C by 1.3; so $C \in G$. The rest is evident.

For T_0 spaces X, everything (definition of graduation, 2.1, and what follows) goes the same as for locales. We may just note "(or space)" from time to time.

A _graduated locale_ is a locale X with a graduation G. The covers in G are _basic covers_ of (X,G), and their elements _basic parts_. A (G-) _decisive filter_ in X is a proper filter F of sublocales of X such that (1) F contains a least element of each basic grade cover C, $\mu(F,C)$, and (2) if D is a subdivision of C, $\mu(F,D) \leq \mu(F,C)$. (N.B. (2) does not follow from (1) in three-point spaces.) Two decisive filters are G-_equivalent_ if they contain the same elements of each basic grade cover.

2.2. A filter F in a graduated locale is decisive if and only if $F \cap B$ is prime, and two decisive filters F, F' are G-equivalent if and only if $F \cap B = F' \cap B$.

Proof. If $F \cap B$ is prime then in any finite sublattice of B the least element of F is join-irreducible, so this is the least element of F in the corresponding grade cover; for (2), enlarging the sublattice can only diminish its least element in F. If a proper filter F has $F \cap B$ non-prime then it contains some $x \vee y$ in B but neither x nor y; in the grade covers $\{1, x \vee y\}$ and $\{1,x,y\}$ or $\{x,y\}$ (exactly one of the latter two is a grade cover) F has respectively least element $x \vee y$, and no element $\leq x \vee y$. Finally, equivalence of decisive filters with the same trace on B is obvious, and if $x \in B$ is in just one of them then they treat $\{1,x\}$ differently.

Having the prime filters of a general distributive lattice, we have the general coherent space. Call the graduation of a coherent space given by the lattice of compact open sets its _coherent graduation_; call the resulting graduated spaces _coherent_. Define a _morphism_ of graduated locales, (X,G) to (Y,H), as a locale morphism $f : X \to Y$ such that f^* takes basic parts to basic parts; equivalently, f^* takes basic covers to basic covers. Without difficulty one sees:

2.3. A coherent space has a smallest graduation, the coherent one. These form a reflective subcategory of graduated locales. They are characterized by: each minimal decisive filter is the filter of neighborhoods of a point. Any graduated locale (X,G) and its coherent reflection rX have isomorphic bases of basic parts, by the morphism r^* induced by the reflection map $r : X \to rX$. Every locale embedding s of X in a coherent space such that s^* is injective on compact open sets is the coherent reflection map for a unique graduation of X.

The morphisms of coherent graduated spaces (the continuous maps f such that f^* preserves compact open sets) are called _coherent maps_. The category CohSp of coherent graduated spaces is dual to the category DLat of distributive lattices (with 0 and 1) (Johnstone 1982). As every distributive lattice is the direct limit of its finite sublattices, so every coherent space is the inverse limit of its finite epi-quotients by coherent maps, i.e., of the grades of its basic covers. Duality gives this only relatively in CohSp, but CohSp is reflective, hence limit-closed, in the biggest category in sight, the graduated locales. Locales form a full subcategory of graduated locales by taking the finest graduation, by all finite covers. T_0 spaces are not a subcategory, but sober spaces are limit-closed in spaces, so CohSp is limit-closed in everything. In particular, every limit of finite T_0 spaces is coherent.

2.4. A graduation of a locale induces graduations on its sublocales. Coherent reflection preserves embeddings of graduated locales; thus the coherent reflection of $(X,\mathcal{G})$ is the smallest coherent graduated locale containing it.

Proof. If $i : X \to Y$ is a locale embedding, i^* is a surjective frame morphism and takes lattice bases to lattice bases. Preservation of (graduated) embeddings i holds because i^* is surjective for lattices of basic parts, making its dual, the coherent reflection of i, a coherent embedding.

The Krull dimension of a graduated locale or space, $\mathrm{Ldim}(X,\mathcal{G})$, is the least n such that every basic cover has a basic subdivision of height at most $n + 1$; ∞, if there are no such n. (L can stand for lattice, locale, or limit. Perhaps lattice is best since the lattice $\mathcal{B}$ determines it: $T(X)$ is an extraneous ornament.)

Recalling that if a cover C refines D, C is said to be finer than (the coarser) D, let us say if C subdivides D that C is tighter, D is looser. Define a boundary part of $(X,\mathcal{G})$ as a sublocale of X which is a finite join of boundaries of basic parts. To every basic cover $C = \{C_i\}$ is associated a boundary part ∂C, the join of all the boundaries ∂C_i. The boundary of a finite join or meet of open parts is contained in the join of their boundaries; hence ∂C is determined by the lattice generated by C, and $\partial D \geq \partial C$ if D is tighter than C.

2.5. If $\mathrm{Ldim}(X,\mathcal{G}) = n$ then every boundary part of X has Krull dimension at most $n - 1$ in the induced graduation.

Proof. A boundary part B is ∂C for some basic cover C, and $B \leq \partial E$ for every subdivision E of C. If B has Krull dimension k there is a basic cover D subdividing C such that for each tighter basic cover E, $G^*(E)$ contains a chain of $k + 1$ elements all different on B. Then if E_0 is the least element of such a chain, $E_0 \wedge B$ has non-zero intersection with the boundary of some $E_i \in G^*(E)$; $E_0 \wedge E_i \neq 0$, so it is a join of one or more elements of $G^*(E)$, and $\mathrm{Ldim}(X,\mathcal{G}) \geq k + 1$.

A few words on the possible converse of 2.5. If every boundary part is empty, $\mathcal{B}$ is a lattice of open-closed parts. Ldim zero requires a field (Boolean algebra) $\mathcal{B}'$, which exists, of course, changing the graduation. One step higher, there are 0-dimensional boundaries to attend to. Basic parts split boundary parts (i.e., the meets are relatively open-closed). But boundary parts need not split each other. There is a coherent example Y. Let X be the compact metric space $\omega + 1$ plus a dense point; for Y, wedge together two copies of X at ω.

The graduated dimension gdim X of a locale X is the minimum of Ldim(X,$\mathcal{G}$) over all graduations $\mathcal{G}$ of X. The Krull dimension Kdim X of a topological space X is one less than the maximum length of a chain of irreducible closed sets -- thus, in a sober space, of a chain of points. Note, Kdim and ind are clearly monotone for subspaces (and ind for sublocales). Ldim is clearly monotone for graduated sublocales; by 2.4, gdim is monotone.

2.6. For every locale or T_0 space X, gdim X $\geq$ ind X. For compact Y, gdim Y $\geq$ Ind Y. For coherent Z, gdim Z = Kdim Z = ind Z.

Proof. From 2.5, Ldim(X,$\mathcal{G}$) $\geq$ ind X by an obvious induction. As for Ind Y, Y compact, it is -1 if Ldim(Y,$\mathcal{G}$) is. Assuming Ldim(Y,$\mathcal{G}$) $\leq$ n - 1 implies Ind Y $\leq$ n - 1, suppose Ldim(Y,$\mathcal{G}$) = n. For any closed part Y_0 of Y and any neighborhood N of Y, by compactness there is a finite join U of basic parts with $Y_0 \leq U \leq N$, and by 2.5, ∂U has Ldim (in the induced graduation) at most n - 1; so Ind(∂U) $\leq$ n - 1 as required. For coherent Z, we have gdim Z $\geq$ ind Z. Since a chain of r points has ind = r - 1 and Z is sober, ind Z $\geq$ Kdim Z. Finally, representing Z as inverse limit of its coherent grades G_α, the compact open sets are just the sets $p_\alpha^{-1}(U)$, U open in G_α, since such sets form a basis and compactness and directness prevail. Any finite cover by such sets has a subdivision given by one G_α, of height at most Kdim G_α. If Kdim Z = n one easily shows by König's Lemma that the G_α do not finally have greater Krull dimension; so gdim Z $\leq$ Kdim Z, completing the proof.

The equality of Kdim and ind for coherent spaces is known (Golan 1983).

2.7. Theorem. gdim X $\leq$ dim X for metrizable locales X.

Proof. If dim X = n then X has a completion Y with dim Y = n ((7) above). Now we need my extension (Isbell 1961) of Freudenthal's (1937) theorem for compact metric spaces: a completely metrizable space Y with dim Y $< \infty$ has a standard irreducible representation as inverse limit of some finite-dimensional geometric simplicial complexes K_i. The crux is "standard", meaning that the bonding maps $K_j \to K_i$ are simplicial into some barycentric subdivision $(K_i)_{(k)}$. Irreducibility implies that a subsequence of the K_i's (which suffice) have the same dimension as Y, by Theorem 2 (Isbell 1961. Irreducibility means that no projection $p_i : Y \to K_i$ can be pushed down within a closed maximal simplex σ, each point y of Y moving only within the closed carrier of $p_i(y)$, so as to uncover the interior of σ.) Then Y has a basis consisting of sets $p_i^{-1}(U)$, U running over stars of vertices in all barycentric subdivisions of all K_i, which would be what we need if the K_i were finite complexes. That is unobtainable, but we can modify the basis. First, we use only the $(K_i)_{(k)}$, not the K_i themselves. Then, note that the barycentric subdivision $K_{(1)}$ of an n-dimensional complex is (n + 1)-partitionable, i.e., its vertices fall into classes $V_0,\ldots,V_n$ such that no simplex of $K_{(1)}$ has two vertices in the same class. (V_i is the set of centroids of i-dimensional simplexes of K.) And note:

2.8. In an r-partitionable geometric complex let A be the set of unions of open stars of vertices. Every finite cover in A has a finite subdivision in A of height at most r.

Proof. A union of stars of vertices is the star-neighborhood S^* of a subset S of the set of vertices. $S \mapsto S^*$ is bijective and preserves inclusion and union. So given finitely many S_i covering $\cup V_j$, pass to the atoms of the field of sets generated by the S_i and the V_j. The resulting cover has order, hence height, at most r.

This finishes off 2.7 since the sets $B_{ik} = \{p_i^{-1}(U)$: U a union of stars of vertices in $(K_i)_{(k)}\}$ are directed by inclusion, so that a finite cover contained in their union is contained in one of them.

Some concluding remarks about spaces or locales of very low dimension. If ind X = 0, of course the basis of open-closed parts gives a graduation of Krull dimension 0, and gdim X = 0. This applies to Roy's (1968) example and shows gdim X < dim X can occur for metric spaces.

Almost nothing more. However, just below the level of Filippov's (1970) example Y, compact Hausdorff with dim Y = 1, ind Y = 2, Ind Y = 3, there is Vedenisov's (1939) theorem that ind Y = 1 implies Ind Y = 1 (Y compact Hausdorff). I do not know about gdim Y. The "lowest" example I know for gdim X > ind X is a normal space with ind X = 1, VI. 16 in (Isbell 1964). The proof there that every Hausdorff compactification Y has ind Y ≥ 2 extends to every compact topological space containing X (and the coherent ones suffice to show gdim X ≥ 2) when the lemma VI. 18 on the space W of all countable ordinals is extended as follows: Every compact space or locale containing W has a single point p each of whose neighborhoods contains all but countably many points of W. To prove that, note that the closed cofinal subsets of W form a filter base, which must be contained in an ultrafilter, which must have a limit point p.

The occasion may justify mentioning that the last example, a substitute for an example of Smirnov (1958) which seems to require regularity of $2^{\aleph_0}$, is in effect a tower of very strange bricks (of ambiguous dimension). The basic brick is Dowker's (1955).

References

Čech, E. [1933]. Contribution to the theory of dimension. Časopis Pěst. Mat. Fys. **62**, 277-91 (Czech).

Dowker, C. H. [1953]. Inductive dimension of completely normal spaces. Quart. J. Math. Oxford Ser. 2 **4**, 267-81.

[1955]. Local dimension of normal spaces. Quart J. Math. Oxford Ser. 2 **6**, 101-20.

Engelking, R. [1968]. Outline of General Topology. Elsevier, Amsterdam.

Filippov, V. V. [1970]. On bicompacta with non-coinciding inductive dimensions. Dokl. Akad. Nauk SSSR **192**, 289-93 (Russian). English translation: Soviet Math. Dokl. **11** (1970), 635-8.

Freudenthal, H. [1937]. Entwicklungen von Räumen und ihre Gruppen, Comp. Math. **4**, 145-234.

Golan, J. [1983]. Two dimensions defined on the spectrum of a noncommutative ring. Comm. Algebra **11**, 2299-2310.

Isbell, J. [1961]. Irreducible polyhedral expansions. Indag. Math. **23**, 242-8.

[1964]. Uniform Spaces. Amer. Math. Soc., Providence.

[1972]. Atomless parts of spaces. Math. Scand. **31**, 5-32.

Johnstone, P. T. [1982]. Stone Spaces. Cambridge University Press, Cambridge.

Katětov, M. [1950]. A theorem on the Lebesgue dimension. Časopis Pěst. Mat. Fys. **75**, 79-87.

Kelley, J. L. [1955]. General Topology. van Nostrand, Princeton.

Lebesgue, H. [1921]. Sur les correspondances entre les points de deux espaces. Fund. Math. **2**, 256-85.

Papert, D. and Papert S. [1958]. Sur les treillis des ouverts et les paratopologies. Sém. Ehresmann 1957-58, exposé 1.

Pasynkov, B. [1965]. On the spectral decomposition of topological spaces. Mat. Sb. **66**, 35-79 (Russian).

Roy, P. [1968]. Nonequality of dimensions for metric spaces. Trans. Amer. Math. Soc. **134**, 117-32.

Smirnov, Ju. M. [1958]. An example of a zero-dimensional normal space having infinite covering dimension. Dokl. Akad. Nauk SSSR **123**, 40-2.

Vedenisov, N. [1939]. Remarks on the dimension of topological spaces. Uč. Zap. Mosk. Gos. Univ. **30**, 131-46.

A GEOMETRICAL APPROACH TO DEGREE THEORY AND THE LERAY-SCHAUDER INDEX

J. Dugundji
University of Southern California, Los Angeles, CA 90089

This article is expository, aiming to develop the degree and the Leray-Schauder theory in an elementary manner that does not require extensive knowledge either of Analysis ([2][5][7][8][10]) or of Algebraic Topology ([3][4]). No applications will be given here; many standard ones, illustrating how the theory is used, are found in the references. Although this topic is not directly related to Dowker's research, his work helped popularize some of the basic concepts, so that their further development and use in this paper would appear to make it suitable for inclusion in the present volume.

A. DEGREE IN $\mathbb{R}^n$

Let U be a bounded open set in $\mathbb{R}^n$ and let $C(\overline{U},\mathbb{R}^n)$ be the space of all continuous maps $f : \overline{U} \to \mathbb{R}^n$ taken with the sup metric. In general terms, the (Brouwer) degree problem is to assign an integer $d(f,U)$ to each $f \in C(\overline{U},\mathbb{R}^n)$ that indicates the minimal number of zeros that f, and all functions sufficiently close to f, must have.

Clearly, local constancy of $f \mapsto d(f,U)$ is not possible if the functions are allowed to have zeros on the boundary ∂U of U, so attention must be restricted to the (open) subspace $C_0(\overline{U},\mathbb{R}^n) = \{f \mid f : (\overline{U},\partial U) \to (\mathbb{R}^n,\mathbb{R}^n - 0\}$. The development then starts by choosing a dense set $A \subset C_0(\overline{U},R^n)$ which is such that each $\varphi \in A$ has only finitely many zeros in U. Counting these zeros algebraically (examples such as $x \mapsto x^2$ on $]-1,1[$ show that a simple count need not be locally constant) gives a degree function that turns out to be locally constant on the subspace A. Defining the degree $d(f,U)$ for each $f \in C_0(\overline{U},\mathbb{R}^n)$ to be the degree of any sufficiently close approximation $\varphi \in A$ to f then yields the required degree function.

In the analytic approach, A is usually taken to be the C^∞ functions having 0 as regular value, and requires some knowledge about C^∞ approximation of continuous functions, and of Sard's theorem. In our approach we take A to consist of piecewise linear maps; the development then requires little more than fairly simple linear algebra, and generally known facts from PL topology.

1. GENERIC PL MAPS OF POLYHEDRA

Given an n-simplex $\sigma^n = (p_0,\dots,p_n) = \{ \sum_0^n \lambda_i p_i \mid \sum_0^n \lambda_i = 1, 0 \leq \lambda_i \leq 1, i = 0,\dots,n\}$ in some $\mathbb{R}^s$, a map $\varphi:\sigma^n \to \mathbb{R}^k$ is called affine if $\varphi(\sum_0^n \lambda_i p_i) = \sum_0^n \lambda_i \varphi(p_i)$.

Let $K \subset \mathbb{R}^s$ be a (finite) polyhedron. A continuous $\varphi: K \to \mathbb{R}^n$ is called PL if it is affine on each simplex, and PL^n if it is also a homeomorphism on each simplex of dimension $\leq n$. By slightly moving the images of the vertices, any PL map can be converted to a PL^n map; in fact

1.1 Lemma. Let $f:K \to \mathbb{R}^n$ be a PL map that is PL^n on some subpolyhedron Q. Then for each $\epsilon > 0$ there is a PL^n map $\varphi:K \to \mathbb{R}^n$ with $\varphi|Q = f|Q$ and $\| f - \varphi \| < \epsilon$.

Proof. For any finite set $S \subset \mathbb{R}^n$, let $F(S)$ be the union of the flats spanned by all subsets of $k \leq n$ elements of S; by Baire's theorem, $F(S)$ has empty interior. Let Q^o be the vertices of Q and order the vertices $p_1,\dots,p_s$ of K not in Q. Choose $y_1 \notin F[f(Q^o)]$ with $|y_1 - f(p_1)| < \epsilon$ and, proceeding recursively, choose $y_n \notin F[f(Q^o) \cup \{y_1,\dots,y_{n-1}\}]$ with $|y_n - f(p_n)| < \epsilon$, n=2,...s. Defining

$$\begin{aligned} \varphi(q_i) &= f(q_i) & q_i &\in Q^o \\ \varphi(p_i) &= y_i & i &= 1,\dots,s \end{aligned}$$

and extending affinely over each simplex gives the desired map φ : for any σ^n the images of the vertices do not lie in any $k < n$ flat, so $\varphi|\sigma^n$ is a homeomorphism; and if $x = \sum_0^n \lambda_i p_i \in K - Q$ we find $|f(x) - \varphi(x)| = | \sum_0^n \lambda_i [f(p_i) - \varphi(p_i)]| < \epsilon$. ∎

More important for our purposes is the PL^n approximation of arbitrary continuous maps:

1.2 Theorem. Let $f:K \to \mathbb{R}^n$ be a continuous map that is PL^n on some subpolyhedron $Q \subset K$. Then for each $\epsilon > 0$ there is a subdivision $\hat{K}$ of K and a PL^n map $\varphi:\hat{K} \to \mathbb{R}^n$ with $\varphi|Q = f|Q$ and $\|\varphi - f\| < \epsilon$.

Proof. By uniform continuity, there is a $\delta > 0$ with $|f(x) - f(y)| < \epsilon$ whenever $|x - y| < \delta$. Subdivide K barycentrically sufficiently often to get $\hat{K}$ with mesh $\hat{K} < \delta$, set $\varphi(\hat{p}) = f(\hat{p})$ for each vertex $\hat{p}$ of $\hat{K}$, and extend affinely over each simplex of $\hat{K}$ to get a PL map $\varphi: \hat{K} \to R^n$. If $\bar{\sigma}$ is a simplex of Q and $\hat{\sigma} \subset \bar{\sigma}$ a simplex of $\hat{K}$, the maps φ,f are both PL and coincide on the vertices of $\hat{\sigma}$, so $f|\hat{\sigma} = \varphi|\hat{\sigma}$, and the $\hat{\sigma}$ patch up to give $\varphi|Q = f|Q$. Finally if $x \in \hat{K}$ is in the interior of the simplex $(\hat{p}_0 \dots \hat{p}_n) \in \hat{K}$ then $|x - \hat{p}_i| < \delta$ for $i = 0,\dots,n$ so $|f(x) - \varphi(x)| = |f(x) - \sum_0^n \lambda_i f(\hat{p}_i)| < \epsilon$. Now apply the lemma to the PL map φ. ∎

Let K^s denote the s-skeleton of K, i.e. the subpolyhedron consisting of all the simplices of dimension $\leq s$. For any continuous $f:K \to \mathbb{R}^n$, we call $f(K^{n-1})$ its set of critical values, and all the points in $\mathbb{R}^n - f(K^{n-1})$ its regular values.

The regular values of any PL map $\varphi:K \to \mathbb{R}^n$ are an open dense set in $\mathbb{R}^n$: the critical values are contained in the union of the finitely many flats spanned by the $\varphi(\sigma^{n-1})$, $\sigma^{n-1} \in K$, each of which has no interior, so their union has no interior.

We will call a PL^n map $\varphi:K \to \mathbb{R}^n$ generic PL if 0 is a regular value. Thus, a PL map $\varphi:(K,K^{n-1}) \to (\mathbb{R}^n, \mathbb{R}^n) - 0)$ is generic PL if $\varphi|\sigma^n$ is a homeomorphism for each n-simplex $\sigma^n \in K$.

2. POLYHEDRAL DOMAINS IN $\mathbb{R}^n$; DEGREE.

An open set $U \subset \mathbb{R}^s$ is called a polyhedral domain if its closure $\bar{U}$ is a finite polyhedron; we shall assume each polyhedral domain given with a fixed simplicial subdivision. A continuous map $f:(\bar{U}, \partial U) \to (\mathbb{R}^n, \mathbb{R}^n - 0)$ is called generic if it is generic PL on some subdivision of $\bar{U}$. Denoting the set of generic maps by $A(\bar{U}, \mathbb{R}^n)$, we have

2.1 Theorem. If U is a polyhedral domain in $\mathbb{R}^n$ then $A(\overline{U},\mathbb{R}^n)$ is dense in $C_0(\overline{U},\mathbb{R}^n)$.

Proof. Starting with $f \in C_0(\overline{U},\mathbb{R}^n)$ and any $\epsilon > 0$ apply 1.2 to get a PL^n map with $\|\varphi - f\| < \epsilon/2$; since the regular points of φ are open dense, there is a regular value y_0 with $\|y_0\| < \epsilon/2$. Then $\varphi - y_0$ is generic and $\|(\varphi - y_0) - f\| < \epsilon$. ■

As stated in the introduction, we next consider the zeros of a $\varphi \in A(\overline{U},\mathbb{R}^n)$. Since $\dim \overline{U} = n$ and φ is PL^n on some subdivision, φ has only finitely many zeros, and at most one in the interior of each n-simplex. We now count these algebraically, by determining for each $p \in \varphi^{-1}(0)$ whether or not φ reverses the orientation of the simplex containing p in its interior.

To motivate the formula (cf. also [1][3][9]), recall that if $(u_1,\ldots,u_n)$ is a basis determining an orientation of $\mathbb{R}^n$, an ordered n-simplex $\sigma^n = (p_0,\ldots,p_n) \subset \mathbb{R}^n$ is, by definition, oriented as $\mathbb{R}^n$ or in the opposite way, according as the sign of the linear transformation $L: (u_1,\ldots,u_n) \to (p_1-p_0,\ldots,p_n-p_0)$ is positive or negative. Letting $((p_i)_1,\ldots,(p_i)_n) \in \mathbb{R}^n$ be the coordinates of the p_i, the determinant of L is

$$\det L = \begin{vmatrix} (p_1)_1 - (p_0)_1, \ldots, (p_n)_1 - (p_0)_1 \\ \vdots \qquad\qquad \vdots \\ (p_1)_n - (p_0)_n, \ldots, (p_n)_n - (p_0)_n \end{vmatrix} = \begin{vmatrix} 1 & \cdots & 1 \\ (p_0)_1 & \cdots & (p_n)_1 \\ \vdots & & \vdots \\ (p_0)_n & & (p_n)_n \end{vmatrix}.$$

We abbreviate $\det L / |\det L|$ by $[p_0,\ldots,p_n]$.

If $\varphi:\sigma^n \to R^n$ is an affine homeomorphism of the simplex σ^n then the product $[p_0,\ldots,p_n]\cdot[(p_0),\ldots,\varphi(p_n)]$ expresses whether or not σ^n and $\varphi(\sigma^n)$ have the same orientation. It is clearly independent of the particular orientation chosen for R^n, and of the order in which the vertices are written.

2.2 Definition. Let U be a polyhedral domain in $\mathbb{R}^n$ and let $\varphi:\overline{U} \to R^n$ be generic. If $p \in \varphi^{-1}(0)$ belongs to the interior of the n-simplex $(p_0,\ldots,p_n)$, then the index $J(\varphi,p)$ of φ at p is $J(\varphi,p) = [p_0,\ldots,p_n][\varphi(p_0,\ldots,\varphi(p_n)]$ and the (Brouwer) degree of φ on U is

$$d(\varphi,U) = \Sigma\{J(\varphi,p)|p \in \varphi^{-1}(0)\}.$$

As another description, it is trivial to establish.

2.3 Choosing any basis for $\mathbb{R}^n$ and letting L be the linear part of the affine homeomorphism $\varphi|\sigma^n$, then $J(\varphi,p) = (-1)^\lambda$ where λ is the number of negative eigenvalues of L, each counted with its multiplicity. ■

In particular, our $J(\varphi,p)$ depends only on the affine map $\varphi|\sigma^n$ and not on any special simplex $(q_0,\ldots,q_n) \subset (p_0,\ldots,p_n)$ containing the zero that is used to compute it. Moreover, the degree of φ is invariant under all sufficiently small translations of φ:

2.4 Let $\varphi:\overline{U} \to \mathbb{R}^n$ be generic. Then there is a neighborhood V(0) of 0 such that $d(\varphi - y, U) = d(\varphi,U)$ for all $y \in V$.

Proof. Let σ_i^n, $i = 1,\ldots, N$ be the set of n-simplices of $\overline{U}$ having a zero in their interior. Noting that $\epsilon_i = \text{dist}(0, \varphi(\partial\sigma_i^n)) > 0$ for $i = 1,\ldots,N$ and that $\eta = \text{dist}[0, \varphi(\overline{U} - \bigcup_1^N \sigma_i^n)] > 0$ the V(0) can be taken as a ball of radius $\epsilon < \min\{\epsilon_1,\ldots,\epsilon_N,\eta\}$. ■

3. LOCAL CONSTANCY.

Our next objective is to show $d(\varphi,U)$ locally constant on $A(\overline{U},\mathbb{R}^n)$. This will follow easily after we show that $d(\varphi,U)$ is invariant under homotopies $\phi:(\overline{U},\partial U)\times I \to (\mathbb{R}^n, \mathbb{R}^n - 0)$. The discussion of homotopy invariance is somewhat lengthy, and requires some attention to detail; but the effort gives in fact a more general version of homotopy invariance.

We start with the simple

3.1 Let $W \subset \mathbb{R}^{n+1}$ be a polyhedral domain, and $\varphi:W \to R^n$ generic. Then each component of $\varphi^{-1}(0)$ is homeomorphic either to the unit interval I or to a circle; and in the first case, its endpoints must both be on ∂W.

Proof. Assume $\sigma^{n+1} \cap \varphi^{-1}(0) \neq \emptyset$; because $\varphi|\sigma^{n+1}$ is affine and dim Im $\varphi = n$, we find that $(\varphi|\sigma^{n+1})^{-1}(0)$ is a line segment. It cannot be contained in any n-face σ^n of σ^{n+1} because $\varphi|\sigma^n$ is a homeomorphism, so it must join points in the interiors of two distinct n-faces of σ^{n+1}. If any of these n-faces is the face of some other $\hat{\sigma}^{n+1}$ the segment continues into $\hat{\sigma}^{n+1}$. Thus, the component is a finite string of segments, each meeting the next at a point in the interior of some σ^n. Since $\varphi^{-1}(0) \cap \sigma^{n+1}$ consists of at most one segment, the string can never cross itself. Therefore, if the string returns to a previous point, the component is homeomorphic to a circle, otherwise it is homeomorphic to I. In the latter case, the component cannot have an endpoint c in the interior of W: for then c would be in the interior of some n-simplex σ^n and also the center of an (n+1)-ball B contained in W; if B^+, B^- are the intersections of B with the two open half-spaces determined by the linear span of σ^n both are non-empty (n+1)-dimensional subsets of W, which must therefore be contained in two distinct simplices having σ^n as face; and as we have seen, the component will continue across σ^n. ∎

Let W be a bounded polyhedral domain in $\mathbb{R}^{n+1} = \mathbb{R}^n \times \mathbb{R}$ with $\overline{W} \subset \mathbb{R}^n \times I$. We will assume that

$$\overline{W}_0 = \overline{W} \cap (\mathbb{R}^n \times 0)$$
$$\overline{W}_1 = \overline{W} \cap (\mathbb{R}^n \times 1)$$

are both non-empty, and call the closure of $(\partial W) - (\overline{W}_0 \quad \overline{W}_1)$ the vertical boundary $\hat{\partial} W$ of W. With this notation, the basic result is

3.2 <u>Theorem</u>. Let $\Phi:(\overline{W},\hat{\partial}W) \to (\mathbb{R}^n,\mathbb{R}^n - 0)$ be continuous and generic on $\overline{W}_0 \cup \overline{W}_1$. Then $d[\Phi|\overline{W}_0, W_0] = d[\Phi|\overline{W}_1, W_1]$.

<u>Proof</u>. We can assume Φ generic. For, let $\epsilon = \mathrm{dist}[0, \Phi(\hat{\partial}W)] > 0$. Using 1.2, there is a PL^n generic map $H:\overline{W} \to \mathbb{R}^n$ with $\|H - \Phi\| < \epsilon/4$ and $H|\overline{W}_0 = \Phi|\overline{W}_0 \equiv \varphi_0$, $H|\overline{W}_1 = \Phi|\overline{W}_1 \equiv \varphi_1$. Now, by 2.4 there is a neighborhood V(0) such that $d[\varphi_i - y, W_i] = d(\varphi_i, W_i), i = 0,1$; for all $y \in V(0)$ and, since the regular values of H are dense, the V(0) contains a regular value $\|\hat{y}\| < \epsilon/4$ of H. Then $H - \hat{y}$ is generic, $d(\varphi_i - \hat{y}, W_i) = d(\varphi_i, W_i)$ i = 0,1 and, because $\|H - \hat{y} - \Phi\| < \epsilon/2$, the $H - \hat{y}$ has no zeros on $\hat{\partial}W$. Thus, to prove the theorem, we can assume Φ is generic.

To count zeros, we need to consider only the components of $\Phi^{-1}(0)$ that meet $\overline{W}_0 \cup \overline{W}_1$: any component that does not meet these sets will not give points affecting the calculation of $d(\varphi_0, W_0)$ or $d(\varphi_1, W_1)$.

Let then L be a component of $\Phi^{-1}(0)$ containing $a_0 \in \text{Int}\, \sigma_0^n \subset \overline{W}_0$. This component cannot be homeomorphic to a circle since the (n+1)-simplex having σ_0^n as face would then contain two distinct line segments of L. Thus, L must be homeomorphic to I, having one endpoint at a_0 and the other on ∂W; since Φ has no zero on $\hat{\partial} W$, the other endpoint must therefore be either in $\overline{W}_0$ or in $\overline{W}_1$.

Suppose L meets successively the n-simplexes $\sigma_0^n, \sigma_1^n, \ldots, \sigma_k^n$ at points $a_i \in \text{Int}\sigma_i^n$, where $\sigma_i^n, \sigma_{i+1}^n$ are faces of some $\sigma_{i,i+1}^{n+1}$. Letting $\sigma_i^n = (p_0^i, \ldots, p_n^i)$ define (cf. [1][3][9])

$$\Delta(\sigma_i^n, \Phi, a_{i+1}) = [p_0^i, \ldots, p_n^i, a_{i+1}] \cdot [\Phi(p_0^i), \ldots, \Phi(p_n^i)]$$

Observing that $[p_0, \ldots, p_n, \xi] = \pm[p_0, \ldots, p_n, \eta]$ according as ξ, η are or are not on the same side of the flat spanned by the n-simplex $(p_0, \ldots, p_n)$, it follows easily from properties of determinants that

$$\Delta(\sigma_i^n, \Phi, \alpha_{i+1}) = -\Delta(\sigma_{i+1}^n, \Phi, a_i) = \Delta(\sigma_{i+1}^n, \Phi, a_{i+2})$$

so that $\Delta[\sigma_i^n, \Phi, a_{i+1}]$ is constant along L.

A direct calculation shows $\Delta[\sigma_0^n, \Phi, a_1] = J(\varphi_0, a_0)$. If L ends at $a_k \in \overline{W}_0$ then

$$J(\varphi_0, a_0) = \Delta[\alpha_{k-1}^n, \Phi, a_k] = -\Delta[\alpha_k^n, \Phi, a_{k-1}] = -J(\varphi_0, a_k)$$

and the two zeros together give no contribution to $d(\varphi_0, W_0)$. If L ends at $a_k \in \overline{W}_1$ then

$$J(\varphi_0, a_0) = \Delta[\sigma_{k-1}^n, \Phi, a_k] = -\Delta[\sigma_k^n, \Phi, a_{k-1}] = +J(\varphi_1, a_k).$$

The argument is clearly reversible, starting from $\overline{W}_1$. Thus, only the zeros on components L running from $\overline{W}_0$ to $\overline{W}_1$ will contribute to the degrees of φ_0, φ_1, and each gives the same amount, therefore $d(\varphi_0, W_0) = d(\varphi_1, W_1)$ and the proof is complete. ∎

Remark. Notice that this result is more general than homotopy invariance: it allows simultaneous deformation of the domain $W \cap [\mathbb{R}^n \times t]$ and the map $\Phi | W \cap (\mathbb{R}^n \times t)$.

With this, we are now ready to define the degree $d(f,U)$ for each $f \in C_0(\overline{U},R^n)$ by approximating with elements of $A(\overline{U},\mathbb{R}^n)$. However, before proceeding, it is convenient to state explicitly a trivial general result that is frequently used;

3.3 Let $f:X \to \mathbb{R}^n$ be continuous and $A \subset X$ such that $\text{dist}(0,f(A)) = \alpha > 0$. If $g:X \to \mathbb{R}^n$ satisfies $\|g-f\| < \alpha/2$, then $g(A)$ does not contain 0. ∎

3.4 Definition. Let U be a polyhedral domain in $\mathbb{R}^n$ and $f:(\overline{U}, \partial U) \to (\mathbb{R}^n, \mathbb{R}^n - 0)$ continuous. Choose a generic $\varphi \in A(\overline{U},\mathbb{R}^n)$ such that $\|f - \varphi\| < \frac{1}{2}\text{dist}\ [0,f(\partial U)]$. The degree of f on U is $d(f,U) = d(\varphi,U)$.

This is independent of the particular φ selected. For, let $\psi \in A(\overline{U},\mathbb{R}^n)$ also satisfy the condition. Letting K, K' be the subdivisions of $\overline{U}$ on which φ, ψ, are generic PL, it is known ([1]) that there is a simplicial subdivision of the standard polyhedron $\overline{U} \times I$ that coincides with K on $\overline{U} \times 0$ and with K' on $\overline{U} \times 1$: working with the simplices of $\overline{U} \times I$ that do not lie in $\overline{U} \times 0$ or in $\overline{U} \times 1$, proceed recursively by dimension, subdividing each such σ^{k+1} into the simplices $(b,y_0,\ldots,y_k)$ where b is the barycenter of σ^{k+1} and $(y_0\ldots y_k)$ runs through all the k-simplices found on the boundary of σ^{k+1}. Taking $\overline{U} \times I$ with this subdivision, and setting $\Phi(x,t) = (1 - t)\ \varphi(x) + t\psi(x)$ we find that $\Phi | U \times 0 \cup U \times 1$ is generic; since

$$\|f(x) - \Phi(x,t)\| \leqslant (1-t)\|f(x) - \varphi(x)\| + t\|f(x) - \psi(x)\| < \frac{1}{2}\text{dist}\ (0,f(\partial U))$$

it follows from 3.3 that Φ has no zeros on $\hat{\partial}(U \times I)$ and therefore, by 3.2, that $d(\varphi,U) = d(\psi,U)$ as required.

Noting next that if $f,g \in C_0(\overline{U},\mathbb{R}^n)$ are sufficiently close, then the same $\varphi \in A(\overline{U},\mathbb{R}^n)$ can be used to approximate both of them, leads to the main theorem on (Brouwer) degree:

3.5 <u>Theorem</u>. Let U be a polyhedral domain in $\mathbb{R}^n$. Then $f \mapsto d(f,U)$ defines a locally constant (therefore continuous) function $d: C_0(\overline{U},\mathbb{R}^n) \to Z$ satisfying

1. (Normalization) If $T:\overline{U} \to \mathbb{R}$ is an affine heomeomorphism and $T(u) = 0$ for some $u \in U$, then $d(T,U) = (-1)^\lambda$ where λ is the number of negative eigenvalues in the linear part of T, each counted with its multiplicity.
2. (Additivity) For any pair of disjoint polyhedral domains $U_1, U_2 \subset U$ and for any $f \in C_0(\overline{U},\mathbb{R}^n)$ having all its zeros in $U_1 \cup U_2$ then
$$d(f,U) = d(f,U_1) + d(f,U_2)$$
3. (Homotopy) If $H:(\overline{U},\partial U) \times I \to (\mathbb{R}^n,\mathbb{R}^n - 0)$ is a homotopy, then $d(H|U \times 0, U) = d(H|U \times 1, U)$

All these are obvious from the definitions; the homotopy property comes by taking an appropriate generic approximation to the given homotopy and using 3.2; more generally, our remark shows that d(f,U) is invariant under suitable simultaneous continuous deformations of both f and U.

The properties of 3.5 characterize the degree function:

3.6 <u>Theorem</u>. Let U be a polyhedral domain in $\mathbb{R}^n$. If $\hat{d}: C_0(\overline{U},\mathbb{R}^n) \to Z$ is any function with the normalization, additivity, and homotopy properties, then $\hat{d}(f,U)$ coincides with the Brouwer degree function $d(f,U)$

<u>Proof</u>. We use the conditions to calculate $\hat{d}(f,U)$. By 2.1 and 3.3 the homotopy property shows $\hat{d}(f,U) = \hat{d}(\varphi,U)$ for some generic $\varphi \in A$. Then $\varphi^{-1}(0) = \{u_1,\ldots,u_k\}$ and choosing a set of disjoint simplices with $u_i \in \operatorname{Int} \sigma_i^n$, $i = 1,\ldots,k$ gives, by additivity, that $\hat{d}(\varphi,U) = \sum_1^k \hat{d}(\varphi,\operatorname{Int} \sigma_i^n)$; the normalization property then shows $\hat{d}(\varphi,\operatorname{Int} \sigma_i^n) = d(\varphi,\operatorname{Int} \sigma_i^n)$. ∎

As a consequence, our d(f,U) is independent of the given simplicial subdivision of $\overline{U}$.

Notice that by taking $U = U_1 = U_2 = \emptyset$, the additivity property implies, in a purely formal way, first that $d(f,\emptyset) = 0$ and then the useful

3.7 Excision). If $f(\overline{U}, \partial U) \to (\mathbb{R}^n, \mathbb{R}^n - 0)$ has all its zeros in a polyhedral domain $V \subset U$, then $d(f,U) = d(f,V)$. □

It is an immediate consequence that, if $f(\overline{U}) \subset \mathbb{R}^n - 0$, then $d(f,U) = 0$, and this, in turn leads to the fundamental

3.8 Theorem (Solution property). If $d(f,U) \neq 0$, then $f(U)$ contains a neighborhood of 0.

Proof. If $d(f,U) \neq 0$, then $f(u) = 0$ has a solution in U. By local constancy, $d(g,U) \neq 0$ for all g sufficiently close to f, so for some $\epsilon > 0$ all $g = f - y_0$ with $\|y_0\| < \epsilon$ have zeros in U, and this proves the theorem. ■

Another important property is that the degree of an $f \in C_0(\overline{U}, \mathbb{R}^n)$ depends only on the boundary values:

3.9 If $f,g \in C_0(\overline{U}, \mathbb{R}^n)$ satisfy $f|\partial U = g|\partial U$ then $d(f,U) = d(g,U)$

since $H(x,t) = (1-t)f(x) + tg(x)$ is a homotopy $H:(\overline{U}, \partial U) \times I \to (\mathbb{R}^n, \mathbb{R}^n - 0)$; somewhat more generally,

3.10 (Rouché) If $f,g \in C_0(\overline{U}, R^n)$ satisfy $|f(x) - g(x)| < f(x)$ for each $x \in \partial U$, then $d(f,U) = d(g,U)$

since in this case, too, the homotopy H cannot have a zero on $\partial U \times I$. We use this to calculate the index of a C^1 function at a regular zero:

3.11 Let $f:(\overline{U}, \partial U) \to (\mathbb{R}^n, \mathbb{R}^n - 0)$ be C^1 on U and let ξ be a regular zero for f (i.e. $f'(\xi)$ is non-singular). Then in a sufficiently small neighborhood $V(\xi)$ the map $f|V(\xi)$ is homotopic to the affine $x \mapsto f'(\xi)(x - \xi)$ by a homotopy $H:(V, \partial V) \times I \to (\mathbb{R}^n, \mathbb{R}^n - 0)$ and therefore $J(f,\xi) = \det f'(\xi)/ |\det f'(\xi)|$.

Proof. Since $f'(\xi)$ is invertible, we have $\|f'(\xi)(x-\xi)\| \geq m\|x-\xi\|$ for all $(x-\xi)$ and suitable $m > 0$. Choose $V(\xi)$ so small that $\|f(x) - f'(\xi)(x-\xi)\| < \frac{m}{2} |x-\xi|$ on $V(\xi)$; then

$f'(\xi)(x-\xi) \neq 0$ on ∂V and, by 3.3, we therefore have $f(V,\partial V) \to (R^n, R^n - 0)$; moreover $|f(x) - f'(\xi)(x-\xi)| < |f'(\xi)(x-\xi)|$ on ∂V so 13.10 applies. ■

4. EXTENSION TO ARBITRARY OPEN SETS.

The condition $f:(\overline{U},\partial U) \to (\mathbb{R}^n, \mathbb{R}^n - 0)$ that the functions in $C_o(\overline{U},\mathbb{R}^n)$ satisfy can be expressed by saying that the zero-set Z(f) of f is contained in U, a statement that makes no explicit reference to ∂U. Using this formulation, we can remove the requirement that U be polyhedral, or even bounded, by requiring that the zero-sets be compact:

4.1 <u>Definition.</u> Let $W \subset \mathbb{R}^n$ be an arbitrary open set and $f = \overline{W} \to \mathbb{R}^n$ a continuous function having a compact $Z(f) \subset W$. The degree d(f,W) is defined by choosing any polyhedral domain U with $Z(f) \subset U \subset W$ and setting $d(f,W) = d(f,U)$.

This definition is independent of the particular polyhedral U selected: if V were another, then since the intersection of polyhedral domains is polyhedral, excision gives $d(f,U) = d(f,U \cap V) = d(f,V)$. This definition evidently has the properties 1 and 2 of Theorem 3.5. The homotopy property 3 requires a compact homotopy $H:\overline{W} \times I \to \mathbb{R}^n$ (i.e. $\cup\{Z(H_t) | 0 \leq t \leq 1\}$ must be a compact set in W) to assure that 3.2 can be used; and if W is unbounded (so that the sup metric cannot be used) the local constancy of d(f,W) is valid if $C_0(\overline{W},\mathbb{R}^n)$ is taken with the topology of uniform convergence.

The entire degree theory can be extended to any n-dimensional real vector space E; If $U \subset E$ is open and $f:U \to E$ has a compact $Z(f) \subset U$, choose any linear isomorphism $h:\mathbb{R}^n \to E$ and define

$$d(f,U) = d[h^{-1}fh,\ h^{-1}(U)].$$

Because

$$[h^{-1}(p_0),\dots,h^{-1}(p_n)]\cdot[h^{-1}fhh^{-1}(p_0),\dots,h^{-1}(p_0)] =$$

$$(\det h^{-1})^2[p_0,\dots,p_n][f(p_0),\dots,f(p_n)]$$

this definition is independent of the particular isomorphism is used. It can also be extended to finite-dimensional complex vector spaces.

5. FIXED POINT INDEX IN $\mathbb{R}^n$

Let $U \subset \mathbb{R}^n$ be open, $f:\overline{U} \to \mathbb{R}^n$, and let Fix(f) denote the fixed-point set of f. Observing that

$$\text{Fix}(f) = Z(\text{id}-f)$$

we will use 4.1 to develop an index that indicates the minimal number of fixed points that f must have.

5.1 Definition. Let $U \subset \mathbb{R}^n$ be open. (a) An $f:\overline{U} \to \mathbb{R}^n$ is called admissible if $U \cap \text{Fix } f$ is compact. (b) The fixed-point index of an admissible $f:\overline{U} \to \mathbb{R}^n$ is $L(f,U) = d[(\text{id} - f), U]$.

For admissible maps on an open $U \subset \mathbb{R}^n$ the properties of degree immediately translate to properties of the index; for example

5.2 (Additivity) For an pair of disjoint open $V_1, V_2 \subset U$ if $\text{Fix }(f) \subset V_1 \cup V_2$ then $L[f,U] = L[f,V_1] + L[f,V_2]$

5.3 (Homotopy) Let $H_t:\overline{U} \times I \to \mathbb{R}^n$ be a homotopy with $\cup\{\text{fix } H_t \mid 0 \le t \le 1\}$ compact. Then $L(H|\overline{U} \times 0, U) = L(H|\overline{U} \times 1, U)$.

5.4 If $c = \overline{U} \to \mathbb{R}^n$ is the constant map $u \mapsto c_0$ then $L(c,U) = 1$ if $c_0 \in U$, and $L(c,U) = 0$ otherwise.

5.5 (Existence) If $L(f,U) \neq 0$, then f has at least one fixed point in U.

The crucial property of the index, which is useful in extending the concept to more general spaces, is

5.6 (Commutativity) Let $U \subset \mathbb{R}^n$, $V \subset \mathbb{R}^k$ be open, and let $f:U \to \mathbb{R}^k$, $g:V \to \mathbb{R}^n$ be continuous. If $g \circ f$ is admissible, so also is $f \circ g$ and $L[g \circ f, U] = L[f \circ g, g^{-1}(U)]$.

A proof can be found in [4][6]. As an important consequence,

5.7 Let $U \subset \mathbb{R}^n$ be open, and let f be an admissible map with $f(U) \subset L$, where L is some flat in $\mathbb{R}^n$. Then $L(f,U) = L[f|U \cap L, U \cap L]$

Proof. Letting $i: U \cap L \hookrightarrow U$ be the inclusion, 5.6 gives

$$L[f,U] = L[if,U] = L[fi, i^{-1}(U)] = L[f|U \cap L, U \cap L] \quad \blacksquare$$

Questions of unicity of the fixed-point index reduce to those for the degree, since each fixed-point index $L(f,U)$ determines a degree d_L by the formula $d_L(f,U) = L[id-f, U]$.

6. EXTENSION TO INFINITE-DIMENSIONAL NORMED SPACES; THE LERAY-SCHAUDER INDEX.

In an infinite-dimensional normed space E, there can be no degree theory applicable to all continuous maps: for if $\overline{U}$ is the unit ball in E, it is well-known that there is a retraction $r:\overline{U} \to \partial U$; since $r|\partial U = id|\partial U$ the 3.9 and 5.4 would imply that $d(r,U) = d(id,U) = 1$, giving the contradiction that $r(x) = 0$ has a solution in U. The class of maps for which a degree, or index, theory can be defined must therefore be restriced.

For any space X, a map $f:X \to E$ is called compact if $\overline{f(X)}$ is compact; it is finite-dimensional if f(X) is contained in a finite-dimensional subspace. The Leray-Schauder fixed-point index will be defined for compact admissible maps on an open set, so we start by recalling some properties of such maps.

Let $U \subset E$ be a bounded open set and $f:\overline{U} \to E$ an admissible compact map; then $\eta(f) = \inf\{\|x-f(x)\| \mid x \in \partial U\}$ is positive. Using Schauder projections, for each $\epsilon < \frac{1}{2}\eta(f)$ there are admissible finite-dimensional maps f_ϵ with $\|f(x) - f_\epsilon(x)\| < \epsilon$ for all $x \in U$; and for any two such approximations, $H(x,t) = (1-t)f_\epsilon(x) + tf_{\epsilon'}(x)$ is a compact admissible homotopy. This leads to the

6.1 Definition. Let $f:\overline{U} \to E$ be an admissible compact map on a bounded open set. Choose any finite-dimensional approximation $g:\overline{U} \to L$ with $|f(x) - g(x)| < \frac{1}{2}\eta(f)$ for $x \in U$. The Leray-Schauder fixed-point index $LS(f,U)$ of f on U is $LS(f,U) = L[g|U \cap L, U \cap L]$

This definition is independent of the finite-dimensional approximation chosen. For if $g_i : U \to L_i, i=1,2$ are any two approximations, choose a finite-dimensional subspace $L \supset L_1 \cup L_2$ so we have $g_i : U \cap L \to L$ for $i=1,2$. Since $g_i(U \cap L) \subset L_i$ we find first from 5.7 that

$$L[g_i|U \cap L, U \cap L] = L[g_i|U \cap L_i, U \cap L_i] \quad i=1,2$$

and then, because of the homotopy of g_1 and g_2 that they have the same index on $U \cap L$:

$$L[g_1|U \cap L, U \cap L] = L[g_2|U \cap L, U \cap L].$$

The properties of the index in $\mathbb{R}^n$ yield, in a straightforward manner the corresponding properties 5.2 - 5.6 for the Leray-Schauder index for compact admissible maps; and as before, the excision property allows us to relax the requirement that U be bounded.

Still more generally, the Leray-Schauder index can be extended to compact admissible maps $f:\overline{U} \to X$ on open subsets U of arbitrary metric ANR spaces X ([6]). For, by the Arens-Eells theorem, X be can be embedded as a closed subset of a normed space E, and is therefore a retract of a neighborhood $V \supset X$ in E. Let $r:V \to X$ be a retraction. Given an open $U \subset X$ and compact admissible $f:\overline{U} \to X$ consider the diagram

$$\overline{r^{-1}(U)} \xrightarrow{r} \overline{U} \xrightarrow{f} X \xhookrightarrow{j} V$$

where $j:X \hookrightarrow V$ is inclusion, and define $L(f,U) = LS[jfr, r^{-1}(U)]$. This is independent of the particular V and r that have been selected: if $\rho:W \to X$ is a retraction of some other neighborhood onto X, and $\eta:X \longrightarrow W$ is the inclusion, then by the commutativity

$$L[jfr,r^{-1}(U)] = L[j\rho\eta fr,r^{-1}(U)] = L[\eta frj\rho\ (j\rho)^{-1}r^{-1}(U)] = L[\eta f\rho,\rho^{-1}(U)]$$

because $\rho\eta = ri = \text{id}$. The properties 5.2 - 5.6 for this index follow from those of LS.

BIBLIOGRAPHY

1. Alexandroff, P. and Hopf, H. (1935). Topologie. Springer-verlag, Berlin.
2. Amann, H. (1974). Lectures on some fixed-point theorems. Monografias de Matematica # 17, IMPA, Rio de Janeiro.
3. Cronin, J. (1964). Fixed points and topological degree in non-linear analysis. Math. Surveys #11, AMS, Providence.
4. Dold, A. (1972). Lectures on algebraic topology. Springer-Verlag Berlin-Heidelberg.
5. Eisenack, G. and Fenske, C. (1978). Fixpunkttheorie. Bibliographisches Institut, Zurich.
6. Granas, A. (1972). Leray-Schauder index and fixed-point theory for arbitrary ANR's. Bull. Soc. Math. France 100, 209-228.
7. Krasnoselskii, M. and Zabreiko, P. (1975). Geometric methods of non-linear analysis. NAUKA, Moscow.
8. Lloyd, N., (1978). Degree theory. Cambridge Tract # 73, Cambridge University Press.
9. Peitgen H. and Siegburg, H. (1981). An ε-perturbation of Brouwer's definition of degree. Lecture Notes in Mathematics # 886, 331-368, Springer-Verlag.
10. Schwartz, J.T. (1969). Nonlinear functional analysis. Gordon & Breach, New York.

ON DIMENSION THEORY

B.A. Pasynkov
Moscow State University, Moscow, USSR

This article consists of two parts. In the first, the author's findings concerning the monotinicity of the dimension dim, and the dimension dim of topological products, are published. The second contains recent discoveries of Soviet topologists (chiefly my students' results) in dimension theory.

Throughout, a space is understood to be a topological space, a mapping is a continuous mapping, and the dimension (if not otherwise stated) is the dimension dim, defined by refining finite functionally open covers by covers of the same type and suitable multiplicity [3].

I

This part contains proofs of results published in [7] concerning the most general conditions under which the inequalities

$$\dim A \leq \dim X , \qquad A \subseteq X \tag{*}$$

$$\dim X \times Y \leq \dim X + \dim Y \tag{**}$$

hold. Also coincidence of the local dimension with the global dimension is proved for all topological groups.

For any set $\mathfrak{a}$ we denote by $\mathfrak{a}_f$ the collection of all non-empty finite subsets of $\mathfrak{a}$.

1 DEFINITION 1. Let X be a space and A be a subspace. We shall say that a set $U \subseteq X$ is a piecewise extension (from A to X) of a set $O \subseteq A$ if O is clopen in $U \cap A$. We shall say also that the collection $\eta = \{U_\alpha : \alpha \in \mathfrak{a}\}$ of subsets of the space X piecewise extends (from A to X) the collection $\omega = \{O_\alpha : \alpha \in \mathfrak{a}\}$ of subsets of the set A if U_α is a piecewise extension of O_α from A to X for each $\alpha \in \mathfrak{a}$.

DEFINITION 2. The subset A of the space X is called a d-right (resp. d-posed) subset if, for each functionally open set O in A, there exists a cover $\omega = \{O_\alpha : \alpha \in \mathfrak{A}\}$ of the set O which is σ-locally finite in A and which can be piecewise extended (resp. extended) to a collection $\eta = \{U_\alpha : \alpha \in \mathfrak{A}\}$ of sets functionally open in X. If moreover the collection η can always be chosen in such a way that, for each $\alpha \in \mathfrak{A}$, a subset D_α of X can be found such that $D_\alpha \supseteq U_\alpha$ and $\dim D_\alpha \leq n$, then the subset A is called (n,d)-right (resp. (n,d)-posed), for $n = -1, 0, 1, \ldots$.

Evidently, a d-right (d-posed) subset of the space X is $\dim X$-right ($\dim X$-posed).

The notion of d-posedness was introduced in a somewhat different way in [1]. Evidently d- or (n,d)-rightness follows from d- or (n,d)-posedness, and d-posedness, in its turn, follows from C^*- and even from cozero-embeddedness [5] (≡ Z-embeddedness [4]).

LEMMA 1. For a subset A of the space X the following statements are equivalent:

(a) A is d-, (n,d)-right;

(b) for any finite functionally open cover Ω of the set A there exists a functionally open and σ-locally finite refinement $\omega = \{O_\alpha : \alpha \in \mathfrak{A}\}$ of Ω which can be piecewise extended to a collection $\eta = \{U_\alpha : \alpha \in \mathfrak{A}\}$ of sets functionally open in X (and, in the case of (n,d)-rightness, for each $\alpha \in \mathfrak{A}$ a subset D_α of X can be found such that $D_\alpha \supseteq U_\alpha$ and $\dim D_\alpha \leq n$).

<u>Proof</u>. The proof is given for d-rightness only, the case of (n,d)-rightness being analogous.

Let condition (a) hold and let $\Omega = \{A_i : i = 1, \ldots, k\}$ be a functionally open cover of the set A. For each i we can find a cover ω_i of the set A_i which is σ-locally finite in A and can be piecewise extended (from A to X) to a collection η_i of sets functionally open in X. Then the collection $\omega = \cup\{\omega_i : i = 1, \ldots, k\}$ will be the required σ-locally finite refinement of Ω.

Now let condition (b) hold. Consider a set O, functionally open in A. It can be represented as the union of a countable collection of sets F_i, $i = 1,2,\ldots$, functionally closed in A. By hypothesis, each binary cover $\Omega_i = \{0, A \setminus F_i\}$ has a refinement ω_i, functionally open and σ-locally finite in A, which can be piecewise extended to a collection which is functionally open in X. Let us denote by ω_i' the subcollection of all those elements of the collection ω_i which are contained in 0. Then the collection $\omega = \cup\{\omega_i' : i = 1,2,\ldots\}$ will be a cover of the set 0 which is σ-locally finite in A and can be piecewise extended (from A to X) to a collection functionally open in X. □

THEOREM 1. The inequality $\dim A \le n$, $n = 0,1,2,\ldots$, follows from the (n,d)-rightness of the subset A of the space X.

First let us prove a generalization of Proposition 10 from [6], §3.

PROPOSITION 1. Let a space A, an inverse system $\{X_\lambda, p_{\mu\lambda}; \lambda \in \Lambda\}$ and mappings $\varphi_\lambda : A \to X_\lambda$, $\lambda \in \Lambda$, be given, such that

(a) $\varphi_\lambda = p_{\mu\lambda} \circ \varphi_\mu$, $\lambda < \mu$.

Let a σ-locally finite open cover $\omega = \{0_\alpha : \alpha \in \mathcal{A}\}$ of the space A be fixed. Suppose that for each $a \in \mathcal{A}_f$, an index $\lambda(a) \in \Lambda$, a subset D_a of $X_{\lambda(a)}$ and a functionally open set U_a in $X_{\lambda(a)}$ can be chosen such that

(b) $U_a \subseteq D_a$, $\dim D_a \le n$;

(c) $\lambda(a) \ge \lambda(b)$ if $a \supset b$;

(d) the set $0_a = \cap\{0_\alpha : \alpha \in a\}$ is included in $\varphi_{\lambda(a)}^{-1} U_a$ and, in the case of $|a| = 1$, it is also clopen in $\varphi_{\lambda(a)}^{-1} U_a$.

Then there exists an ω-mapping from the space A to a metrizable space R with $\dim R \le n$, $n = 0,1,2,\ldots$.

Proof. The proof is based on Proposition 9 from [6], §3.

Without loss of generality we can suppose that

(e) $p_{\lambda(a)\lambda(b)} U_a \subseteq U_b$, $b \subset a \in \mathcal{A}_f$;

otherwise the set $\cap\{p_{\lambda(a)\lambda(b)}{}^{-1}U_b : b \subseteq a\}$ must be taken instead of U_a. Then evidently

(f) O_a is a clopen subset of $\varphi_{\lambda(a)}{}^{-1}U_a$ for every $a \in \mathcal{A}_f$.

Let us take, for each $a \in \mathcal{A}_f$, a mapping θ_a from $X_{\lambda(a)}$ to $I_a = [0,1]$ such that $U_a = \theta_a{}^{-1}(0,1]$.

Fix $a \in \mathcal{A}_f$ and suppose that for each $b \subset a$ (if such exist) a metrizable space R_b, an open subset V_b of R_b and mappings

$$g_b : X_{\lambda(b)} \to R_b, \quad h_b : R_b \to I_b, \quad \pi_{bb'} : V_b \to V_{b'}, \quad b' \subset b,$$

have been constructed such that the following conditions are fulfilled:

(1) $\dim V_b \leq n$;

(2) $U_b = g_b{}^{-1}V_b$, $g_b U_b = V_b$;

(3) $g_{b'} \circ p_{\lambda(b)\lambda(b')} = \pi_{bb'} \circ g_b$ on U_b, $b' \subset b$;

(4) $\theta_b = h_b \circ g_b$;

(5) the set $R_b \setminus V_b$ consists only of one point $\bar{\bar{o}}_b$.

We shall denote the diagonal product of the mappings θ_a and $g_b \circ p_{\lambda(a)\lambda(b)}$, $b \subset a$, by ξ_a and the projections of the product $\prod_a = I_a \times \prod\{R_b : b \subset a\}$ onto the factors I_a and R_b by pr_a and pr_b respectively. According to the factorization theorem (Theorem 2 and Corollary 2 from [6] or Lemma 2.2 from [3]) there exist a metrizable space S_a and mappings $g_a' : D_a \to S_a$, $h_a' : S_a \to \prod_a$ such that $h_a' \circ g_a' = \xi_a$ on D_a and $\dim S_a \leq \dim D_a \leq n$. We may certainly suppose that $S_a = g_a' D_a$. Consider $W_a = (pr_a \circ h_a')^{-1}(0,1]$. It is clear that $U_a = (g_a')^{-1}W_a$ and $g_a' U_a = W_a$. The set of points of the space R_a is obtained from the set W_a by adding to it a point $\bar{\bar{o}}_a \notin W_a$. Denote the mapping which is equal to 0 at the point $\bar{\bar{o}}_a$ and coincides with $pr_a \circ h_a'$ on W_a by $h_a : R_a \to I_a$. We take the inverse images of all open sets in I_a and all open subsets of the set W_a as base for the topology of R_a. It is clear that the space R_a is metrizable and that the mapping h_a is continuous. Denote the set W_a, considered as a subset of the space R_a, by V_a. Evidently V_a is homeomorphic to W_a. Thus $\dim V_a \leq \dim S_a \leq n$. The mapping $g_a : X_{\lambda(a)} \to R_a$, which coincides with g_a' on U_a and maps $X_{\lambda(a)} \setminus U_a$ to the point $\bar{\bar{o}}_a$, is continuous and $g_a{}^{-1}V_a = U_a$, $g_a U_a = V_a$, $\theta_a = h_a \circ g_a$.

For the mapping

$$\pi_{ab} = pr_b \circ h_a' : (V_a \equiv W_a) \to R_b , \quad b \subset a ,$$

we have the relation on U_a :

$$\pi_{ab} \circ g_a = pr_b \circ h_a' \circ g_a' = pr_b \circ \xi_a = g_b \circ p_{\lambda(a)\lambda(b)} ;$$

and (because of the relation (e))

$$\pi_{ab}(V_a \equiv W_a) = \pi_{ab} \circ g_a U_a = g_b \circ p_{\lambda(a)\lambda(b)} U_a \subseteq g_b U_b = V_b .$$

Thus the relations (1)-(5) are also fulfilled for $b=a$.

The constructions described allow us to use induction on the number of elements in the set $a = (\alpha_1, \ldots, \alpha_k) \in \mathcal{A}_f$ to construct, for any $a \in \mathcal{A}_f$, a metrizable space R_a , an open subset V_a of R_a and mappings $g_a : X_{\lambda(a)} \to R_a$, $h_a : R_a \to I_a$, $\pi_{ab} : V_a \to V_b$, $b \subset a$, which satisfy the statements (1)-(5) (where b and b' are changed into a and b respectively).

For each $a \in \mathcal{A}_f$, the mapping $f_a : A \to R_a$ is defined to be equal to $g_a \circ \varphi_{\lambda(a)}$ on O_a and to map $A \setminus O_a$ to the point $\bar{\bar{o}}_a$ (i.e. f_a coincides with $g_a \circ \varphi_{\lambda(a)}$ on $(A \setminus \varphi_{\lambda(a)}^{-1} U_a) \cup O_a$). This mapping is continuous, by condition (f).

It is clear that $f_a^{-1} V_a = O_a$ and $\pi_{ab} \circ f_a = f_b$ when $b \subset a$.

Thus, for the spaces R_a, their subsets V_a and the mappings f_a and π_{ab}, all the conditions of Proposition 9 from [6], §3 are fulfilled. From there the proof of the proposition follows. □

THEOREM 1'. Let a mapping $f: A \to X$ have the following property:

(#) any finite functionally open cover of the space A has a σ-locally finite refinement ω, such that for each $O \in \omega$, a set $D(O)$ and a functionally open set $U(O)$ can be found in X for which

(1) $U(O) \subseteq D(O)$, $\dim D(O) \leq n$,

(2) O is clopen in $f^{-1} U(O)$ (and accordingly is functionally open in A).

Then $\dim A \leq n$.

<u>Proof</u>. Consider a finite functionally open cover Ω of the space A and let $\omega = \{O_\alpha : \alpha \in \mathfrak{A}\}$ be its refinement given by condition (#). The set $\mathfrak{A}_f$ is directed by inclusion. It is easily seen that the space A, the inverse system $\{X_a = X, p_{ab} = id_X ; \mathfrak{A}_f\}$, the mappings $\varphi_a = f$, the cover ω, the sets $U_a = \cap\{U(O_\alpha) : \alpha \in a\}$ and sets D_a equal to any one of the D_α for $\alpha \in a$, satisfy the conditions of Proposition 1. Therefore there exists an ω-mapping (thus an Ω-mapping) of the space A onto a metrizable space R with $\dim R \leq n$. The relationship $\dim A \leq n$ follows from the arbitrariness of the cover Ω. □

Theorem 1 follows from Theorem 1′ and Lemma 1 (the inclusion map of A into X being taken as the mapping f of Theorem 1′). From Theorem 1 we have:

COROLLARY 1. If a subset A of a space X is represented as the union of a collection ω, σ-locally finite in A, of sets functionally open in X, and if $\dim O \leq n$, $O \in \omega$, then $\dim A \leq n$, $n = 0, 1, 2, \ldots$.

<u>Proof</u>. Any functionally open finite cover Ω of the set A is refined by the cover $\Omega \wedge \omega = \{W \cap O : W \in \Omega, O \in \omega\}$, which is σ-locally finite in A and functionally open in X ; and we have $W \cap O \subseteq O$, $\dim O \leq n$. □

Corollary 1 generalizes Theorems 2.4 and 2.5 of [5]; it can also be deduced from Theorem 7.3 of [4].

THEOREM 2′. Let the mapping of spaces $f: A \to X$ have the following property:

(##) any finite functionally open cover Ω of A has a σ-locally finite refinement ω such that, for each $O \in \omega$, a functionally open subset $U(O)$ of X can be found with O clopen in the inverse image $f^{-1}U(O)$.

Then $\dim A \leq \min\{\dim X' : fA \subseteq X'\} \leq \dim X$.

This result follows from Theorem 1′ by changing $U(O)$ into $U(O) \cap X'$ and putting $D(O) = X'$, where $X' \subseteq fA$.

DEFINITION 3. The mapping of spaces $f: A \to X$ is called strongly decomposing if the condition (##) is fulfilled.

For more about decomposing mappings, see [11].

From Theorem 2′ (if f is the inclusion map of A into X) we have:

THEOREM 2. If a subset A of a space X is d-right, then $\dim A \leq \min\{\dim X' : A \subseteq X' \subseteq X\} \leq \dim X$.

COROLLARY 2 ([1]). The inequality (*) holds if A is d-posed in X. In particular, (*) holds if A is cozero- (Z-) embedded (Theorem 5.16 of [4] and Theorem 1.3 of [5]) or if A is the union of a collection, σ-locally finite in A, of functionally open sets of X. (This is a generalization of Corollary 2.6 of [5].)

A space X is called completely regular at the point $x \in X$ if the set $\{x\}$ is functionally separated from the complement of each neighbourhood of the point x.

PROPOSITION 2. If a space X is completely regular at all points of a completely (in particular, strongly) paracompact subspace A (for example, if the space X is completely regular), then A is d-right in X.

COROLLARY 3. If a space X is completely regular at all points of its completely paracompact subspace A, then the inequality (*) holds.

This corollary generalizes Proposition 5.18 of [4] and the theorem of [9].

QUESTION. Will a completely (strongly) paracompact subset of a Tychonoff (normal, paracompact) space be d-posed?

We note also:

PROPOSITION 3. If $\dim X = 0$ and $A \subseteq X$, then the inequality $\dim A \leq 0 = \dim X$ is equivalent to the d-rightness of A in X.

Proof. Half the proposition follows from Theorem 2. Now suppose we have $\dim A \leq 0$. Then any set functionally open in A is the union of a countable collection of sets clopen in A and X is a piecewise extension (from A to X) of any set clopen in A. □

2 Let us apply the above findings to topological groups.

THEOREM 3. For any topological group G, $\operatorname{loc}\dim G = \dim G$.

Proof. Take a neighbourhood V of the unit of the group G such that $\dim [V] = \operatorname{loc}\dim G = n$. We can refine the uniform cover $\{V.g : g \in G\}$ by a locally finite and functionally open cover ω. Then for any $O \in \omega$ there exists $g = g(O) \in G$ with $O \subseteq V.g$ and (according to Corollary 2) $\dim O \leq \dim [V.g] = n$. According to Corollary 1, we get the inequality $\dim G = \dim \cup\omega \leq n$. □

COROLLARY 4. If a separated group G is locally completely paracompact, then $\dim G \leq \operatorname{ind} G$.

Proof. If the closure $[V]$ of a neighbourhood V of the unit of the group is completely paracompact, then $\dim G = \operatorname{loc}\dim G \leq \dim [V] \leq \operatorname{ind} [V] = \operatorname{ind} G$. □

3 In this section, X is the Tychonoff product of spaces $X_i \neq \emptyset$, $i \in I$, and

$$n = \sup\{\dim X_{i_1} + \ldots + \dim X_{i_k} : i_1, \ldots, i_k \in I, \quad k \in \mathbf{N}\}.$$

A set of the kind

$$\{x = (x_i) \in X : x_{i_j} \in O_{i_j}, \ j = 1, \ldots, k\},$$

where O_{i_j} is functionally open in X_{i_j}, $j = 1, \ldots, k$, we shall term a functionally open rectangle of the product X. A clopen subset of a functionally open rectangle is called a functionally open rectangular piece. A cover of the product X by functionally open rectangular pieces (rectangles) may then be called functionally open piecewise rectangular (rectangular).

DEFINITION 4. The product X is called piecewise rectangular (rectangular) if each of its finite functionally open covers has a

σ-locally finite functionally open piecewise rectangular (rectangular) refinement.

It is clear that piecewise rectangularity of X follows from its rectangularity. For finite I, the definition of rectangularity is given in [6].

It is known (see, for instance, [3]) that for any space Y it is possible to define a unique (in a natural sense) Tychonoff space τY and surjective mapping $\tau_Y : Y \to \tau Y$ such that for every mapping $f: Y \to Z$ into a Tychonoff space Z there is a mapping $g: \tau Y \to Z$ for which $f = g \circ \tau_Y$.

It is easy to verify that

$$\tau X \equiv \prod\{\tau X_i : i \in I\} \quad \text{and} \quad \tau_X \equiv \prod\{\tau_{X_i} : i \in I\}$$

if and only if each functionally open set in X is a union of functionally open rectangles of the product X. In particular, it is so if the product X is rectangular. (For finite I this statement was proved in [2].)

In general the statement is not true for a piecewise rectangular product X. But the connection between τX and the product $\prod = \prod\{\tau X_i : i \in I\}$ can be decribed quite simply.

LEMMA 2. Let the product X be piecewise rectangular, let $p = \prod\{\tau_{X_i} : i \in I\}$, let $\prod' = \prod\{\beta(\tau X_i) : i \in I\}$ and let q be the natural inclusion of $\prod$ into $\prod'$. Then the mapping $p' = q \circ p : X \to \prod'$ is strongly decomposing and $\dim X \leq \dim \prod'$.

Proof. Consider a finite functionally open cover Ω of the product X. There exists a functionally open piecewise rectangular and σ-locally finite refinement ω of Ω. Fix, for each $O \in \omega$, a functionally open rectangle $V(O)$ such that O is clopen in $V(O)$. The set $W(O) = p(V(O))$ is a functionally open rectangle of the product $\prod$ and $p^{-1}(W(O)) = V(O)$. Let

$$W(O) = \{y = (y_i) \in \prod : y_{i_j} \in W_{i_j},\ j = 1, \ldots, k\},$$

where the W_{i_j} are functionally open in τX_{i_j}. There are functionally open sets U_{i_j} in $\beta(\tau X_{i_j})$ such that $W_{i_j} = U_{i_j} \cap \tau X_{i_j}$. Then the set

$$U(O) = \{z = (z_i) \in \prod' : z_{i_j} \in U_{i_j},\ j = 1, \ldots, k\},$$

is a functionally open rectangle of $\prod'$ and $q^{-1}U(0) = W(0)$. Hence $(p')^{-1}U(0) = V(0)$. We have proved p' to be a strongly decomposing mapping. The inequality $\dim X \le \dim \prod'$ follows from Theorem 2′. □

LEMMA 3. We have $\dim \prod' \le n$ for any product X (where $\prod'$ is defined as in Lemma 2).

Proof. The inequality is evident if there is an infinite collection of non-zero-dimensional spaces among the X_i. Suppose now we have a finite subset $F = \{i_1, \dots, i_m\}$ of I such that $\dim X_i = 0$ for $i \in I \setminus F$. It is clear that in this case $n = \dim X_{i_1} + \dots + \dim X_{i_m}$. Let $\prod_F = \prod\{\beta(\tau X_i) : i \in F\}$ and $\prod'' = \prod\{\beta(\tau X_i) : i \in I \setminus F\}$.

Since the inequality (**) is satisfied for bicompacta, since the product of any collection of zero-dimensional bicompacta is zero-dimensional and since $\dim \beta(\tau Y) = \dim(\tau Y) = \dim Y$ for any space Y (see, for example, [3]), it follows that $\dim \prod' \le \dim \prod_F + \dim \prod'' = \dim \prod_F \le n$. □

From Lemmas 2 and 3 we have:

THEOREM 4. If the product X is piecewise rectangular, then $\dim X \le n$.

COROLLARY 5. If $\dim X_i = 0$, $i \in I$, then $\dim X = 0$ if and only if the product X is piecewise rectangular.

Proof. If $\dim X = 0$, then the product is piecewise rectangular, as any finite functionally open cover of the space X has a finite open and disjoint refinement consisting of sets clopen in the functionally open rectangle X. The second half of the statement follows from Theorem 4. □

EXAMPLE. The square (and any other power) of the Sorgenfrey line is zero-dimensional [8] and so is piecewise rectangular; but, as has been shown by Zolotarev, it is not rectangular.

Let us characterize the piecewise rectangularity of products of Tychonoff spaces, using the notion of d-rightness.

PROPOSITION 4. If the X_i, $i \in I$, are Tychonoff spaces, then the product X is piecewise rectangular (rectangular) if and only if (in the notation of Lemma 2) $X \equiv \prod \equiv p'X$ is a d-right (d-posed) subset of the product $\prod'$.

<u>Proof</u>. Under our conditions, $\tau X \equiv X$, $p \equiv \tau_X \equiv id_X$, $p' \equiv q$ and $\beta(\tau X_i) \equiv \beta X_i$, $i \in I$.

Let the product X be piecewise rectangular. Then according to Lemma 2 the inclusion $p' \equiv q$ is strongly decomposing. Thus the d-rightness of the set $X \equiv p'X$ in $\prod'$ follows from Lemma 1.

Now let the set $X = p'X$ be d-right in $\prod'$. Consider a finite functionally open cover Ω of the space X. According to Lemma 1, it has a σ-locally finite functionally open refinement ω such that, for each $O \in \omega$, a functionally open subset $U(O)$ can be found in $\prod'$ for which O is clopen in $X \cap U(O)$. The set $U(O)$ is σ-bicompact. Hence there exists a countable collection of functionally open rectangles $U_j(O)$, $j = 1,2,\ldots$ in $\prod'$ such that $U(O) = \cup\{U_j(O) : j = 1,2,\ldots\}$. It is clear that the collection $\{O \cap U_j(O) : O \in \omega,\ j = 1,2,\ldots\}$ is a σ-locally finite functionally open piecewise rectangular refinement of Ω. Piecewise rectangularity of the product X is proved.

The case of rectangularity of X and its d-posedness in $\prod'$ is treated analogously. □

We note that, for the case $|I| < \aleph_0$, the d-posedness of X in $\prod'$ is deduced from the rectangularity of X in [1].

From Propositions 2 and 4 we have:

PROPOSITION 5. If all the X_i, $i \in I$, are Tychonoff spaces and the space X is completely (in particular, strongly) paracompact, then the product X is piecewise rectangular.

QUESTION. Will the product X be rectangular under the conditions of Proposition 5 (and when $|I| = 2$) ?

4 Let $S = \{X_\lambda, \pi_{\mu\lambda} : \lambda \in L\}$ be an inverse system of spaces and let X be the limit of S.

DEFINITION 4. The inverse system S is said to be piecewise rectangular (rectangular) if, for each finite functionally open cover Ω

of the limit X, there exists a σ-locally finite open refinement ω of Ω such that for every $O \in \omega$ an index $\lambda = \lambda(O)$ and a functionally open set $U(O)$ in X_λ can be found such that O is a clopen subset of $\pi_\lambda^{-1} U(O)$ ($O = \pi_\lambda^{-1} U(O)$).

THEOREM 5. Let the inverse system S be piecewise rectangular (or, in particular, rectangular) and $\dim X_\lambda \le n$ for each $\lambda \in L$. Then $\dim X \le n$, $n = 0,1,2,\ldots$.

<u>Proof</u>. We shall write τ_λ instead of τ_{X_λ}. For any pair of indices $\lambda, \mu \in L$, $\mu > \lambda$, there exists a unique mapping $\tau\pi_{\mu\lambda}$: $\tau X_\mu \to \tau X_\lambda$ such that

$$\tau_\lambda \circ \pi_{\mu\lambda} = \tau\pi_{\mu\lambda} \circ \tau_\mu .$$

Thus we have an inverse system $\tau S = \{\tau X_\lambda, \tau\pi_{\mu\lambda}; \lambda \in L\}$ with limit X' and projections $\pi'_\lambda : X' \to \tau X_\lambda$. The mappings $\tau_\lambda \circ \pi_\lambda : X \to \tau X_\lambda$ define a mapping $\tau_S : X \to X'$ such that

$$\tau_\lambda \circ \pi_\lambda = \pi'_\lambda \circ \tau_S , \quad \lambda \in L .$$

The spaces τX_λ are Tychonoff spaces. Therefore we have an inverse system $\beta\tau S = \{Y_\lambda, \rho_{\mu\lambda}; \lambda \in L\}$ consisting of the Stone-Čech bicompactifications Y_λ of τX_λ and the extensions $\rho_{\mu\lambda}$ of the mappings $\tau\pi_{\mu\lambda}$ to these bicompactifications. Let Y be the limit of $\beta\tau S$. Evidently there exists a natural embedding h of X into Y such that, for the natural embeddings $h_\lambda : \tau X_\lambda \to Y_\lambda$ and projections $\rho_\lambda : Y \to Y_\lambda$ we have

$$h_\lambda \circ \pi'_\lambda = \rho_\lambda \circ h , \qquad \lambda \in L ;$$

so

$$h_\lambda \circ \tau_\lambda \circ \pi_\lambda = \rho_\lambda \circ h \circ \tau_S , \quad \lambda \in L .$$

The relations $\dim Y_\lambda = \dim \tau X_\lambda = \dim X_\lambda \le n$ and the bicompactness of the Y_λ, $\lambda \in L$, give us ([26], [32]) the inequality $\dim Y \le n$. The theorem is proved if we show that $h \circ \tau_S$ is a strongly decomposing mapping (see Theorem 2$'$).

Let us consider any finite functionally open cover Ω of X. Let ω be a σ-locally finite open refinement of Ω such that for every $O \in \omega$ there exist an index $\lambda = \lambda(O)$ and a set $U(O)$ functionally open in X_λ such that O is clopen in $\pi_\lambda^{-1} U(O)$.

The image $U'(0) = \tau_\lambda U(0)$ is functionally open in τX_λ and $\tau_\lambda^{-1} U'(0) = U(0)$. So

$$\pi_\lambda^{-1} U(0) = (\tau_\lambda \circ \pi_\lambda)^{-1} U'(0) = (\pi'_\lambda \circ \tau_S)^{-1} U'(0) .$$

Consequently we can find a set $V(0)$ functionally open in Y_λ such that $h_\lambda^{-1} V(0) = U'(0)$. Then 0 is clopen in

$$(h_\lambda \circ \tau_\lambda \circ \pi_\lambda)^{-1} V(0) = (h \circ \tau_S)^{-1} \rho_\lambda^{-1} V(0) ,$$

and $\rho_\lambda^{-1} V(0)$ is functionally open in Y. □

COROLLARY 6. Let $\dim X_\lambda \le 0$ for every $\lambda \in L$. Then $\dim X \le 0$ if and only if the system S is piecewise rectangular.

To complete our considerations we add the following (easily proved) result:

PROPOSITION 6. Let all X_λ, $\lambda \in L$, be Tychonoff spaces. Then S is piecewise rectangular (rectangular) if the set $hX \equiv h(\tau X)$ is d-right (d-posed) in the limit Y of $\beta\tau S \equiv \beta S$ (defined as in the proof of Theorem 5).

II

1 The question of the relations between the various dimensional invariants - and, in particular, between the covering dimension dim, the small inductive dimension ind and the large inductive dimension Ind - is a major problem in dimension theory. Many topologists (in particular A.L. Lunc, O.V. Lokucievskiǐ, P. Vopěnka, S.Mardešić, B.A. Pasynkov, V.V. Fedorchuk, V.V. Filippov, etc.) have studied this question in the class of bicompacta and constructed examples of bicompacta which differ in the dimensions dim and ind.

V.A. Chatyrko has made further advances concerning this question. Responding positively to Mardešić's conjecture, Chatyrko constructed, for any $n = 2, 3, \ldots$, a separable chainable bicompactum S_n of countable character with $\operatorname{ind}_x S_n = n$ for every point $x \in S_n$ (so in particular $\operatorname{ind} S_n = n$). The chainable bicompacta are in a sense the simplest among all bicompacta X with $\dim X = 1$ (topologically they differ arbitrarily little from a segment; or, more precisely, for every cover ω they possess an ω-mapping onto a segment). For this reason, Chatyrko's bicompacta occupy an extremal position among bicompacta with non-coinciding dimensions dim and ind.

It should be noted that, before this, only chainable bicompacta with $\mathrm{ind}\, X = 2$ had been constructed ([27], [33], [15]).

Chatyrko did not limit himself to constructing inductively finite-dimensional chainable bicompacta and constructed, first, chainable bicompacta S' and S'' with the dimensions $\mathrm{ind}\, S' = \omega_0$ and $\mathrm{ind}\, S'' = \omega_0 + 1$ and, secondly, a separable chainable bicompactum S of countable character without any (even transfinite) inductive dimension. The bicompactum S is also remarkable for the fact that each infinite connected subbicompactum is homeomorphic to it.

QUESTION 1. Does a chainable bicompactum S_α with $\mathrm{ind}\, S_\alpha = \alpha$ exist for every ordinal number α ?

QUESTION 2. Does a chainable bicompactum C with $\mathrm{ind}\, C < \mathrm{Ind}\, C$ exist?

QUESTION 3. Is it possible to construct, for each $n = 2, 3, \ldots$, chainable bicompacta C_n and D_n with $\mathrm{ind}\, C_n = \mathrm{ind}\, D_n = n$, which satisfy the following requirements:

a) C_n is the union of chainable bicompacta C_n' and C_n'' with $\max\{\mathrm{ind}\, C_n', \mathrm{ind}\, C_n''\} < n$;

b) D_n is the union of chainable bicompacta D_n^i with $\mathrm{ind}\, D_n^i = 1$, $i = 0, 1, \ldots, n$?

A bicompactum $C_2 = D_2$ was constructed in [33]. A bicompactum C_3 was constructed by Chatyrko, and apparently it is possible to construct bicompacta C_n, $n \geq 3$, using his method.

Chatyrko's method for constructing the bicompacta S_n is a specialization and development of Fedorchuk's method of completely closed mappings ([19], [20]). Chatyrko's method allowed him to clarify the relation between the dimensions dim and ind, not only in the class of chainable bicompacta, but also in the class of homogeneous bicompacta. Recall that a space X is called homogeneous if, for any two of its points, x and x', there exists a self-homeomorphism h such that $hx = x'$. Before Chatyrko's findings, only one homogeneous bicompactum F with $\dim F < \operatorname{ind} F$ had been known: namely the homogeneous bicompactum F, with $\dim F = 1$, $\operatorname{ind} F = 2$, constructed by Fedorchuk [20]. Chatyrko constructed, for each $n = 2, 3, \ldots$, a homogeneous separable bicompactum T^n of countable character, with $\dim T^n = 1$, $\operatorname{ind} T^n = n$. He also constructed a homogeneous separable bicompactum T of countable character with $\dim T = 1$ and without any (even transfinite) dimension ind.

QUESTION 4. Does a homogeneous bicompactum X with $\operatorname{ind} X < \operatorname{Ind} X$ ($< \omega_0$) exist?

QUESTION 5. Does a homogeneous bicompactum T_α with $\dim T_\alpha < \infty$ and $\operatorname{ind} T_\alpha = \alpha$ exist for each ordinal number $\alpha \geq \omega_0$?

In Question 5 the homogeneity of T_α can be replaced by a weaker requirement: $\operatorname{ind}_x T_\alpha = \alpha$ for all points $x \in T_\alpha$. Note that, put on a dense set of points x in T_α, this condition is a tractable one [16].

QUESTION 6. Do there exist homogeneous infinite-dimensional and weakly infinite-dimensional (countable-dimensional, weakly countable-dimensional) bicompacta?

In connection with the article by Luxemburg [25] we can ask:

QUESTION 7. Does there exist a homogeneous metrizable bicompactum X with $\omega_0 < \operatorname{ind} X < \operatorname{Ind} X$?

If we omit the requirement that the homogeneous space be bicompact, then, as was proved by V.K. Bel'nov, its homogeneity will not impose any relations between the dimensions dim, ind and Ind. Indeed, the homogeneous space $B(X)$ constructed by Bel'nov in [14], associated with any (hereditarily) normal space X, is again (hereditarily) normal and has the same dim, ind and Ind as X. If, in addition, X is a paracompactum, then $B(X)$ is also a paracompactum.

Among the homogeneous spaces there are spaces that are algebraically homogeneous. A space X is called algebraically homogeneous if there exists a topological group G with a closed subgroup H such that $X = G/H$. If the group G is 'good' (for example, almost metrizable [36]; in particular, locally bicompact or even Čech-complete), then [37]

$$\dim X = \operatorname{Ind} X . \tag{1}$$

If, in addition, the group G is locally bicompact, then we even have [34]

$$\dim X = \operatorname{ind} X = \operatorname{Ind} X . \tag{2}$$

The following questions still remain open:

QUESTION 8. Is the second equality of (2) valid for metrizable groups (and metrizable algebraically homogeneous spaces)?

QUESTION 9. Is at least one of the equalities (2) valid for every algebraically homogeneous bicompactum?

It is worth mentioning that the properties of algebraically homogeneous bicompacta can differ essentially from the properties of bicompact groups. For example, the bicompactum called the 'two arrows' space of Alexandroff is algebraically homogeneous and: (a) it is zero-dimensional but is not homeomorphic to a Cantor cube or even dyadic, (b) it has countable character but is not metrizable. D.B. Motorov has constructed an algebraically homogeneous bicompactum M which is hereditarily normal, has countable character, is locally connected and finite-dimensional (more precisely, $\operatorname{Ind} M = 1$), but which is non-metrizable, non-dyadic and has Souslin number equal to the cardinality of the continuum. Note also that the bicompactum M is 3-homogeneous: i.e. any three distinct points may be mapped by a self-homeomorphism into any other three distinct points.

Questions 5 - 7 make sense also for algebraically homogeneous bicompacta.

2 In the last section we spoke about the relations between different dimensional invariants on the same topological space. In this section we shall consider the behaviour of the same dimensional invarant on different, yet in some sense equivalent, spaces.

The article [21] by M.I. Graev showed that two metrizable bicompacta which are M-equivalent (i.e. whose free topological groups are isomorphic) have the same dimension dim . D.S. Pavlovskiĭ extended Graev's result to the case of ℓ-equivalent metrizable bicompacta. (We recall that spaces X and Y are called ℓ-equivalent if the linear topological spaces $C_p(X)$ and $C_p(Y)$ of all continuous functions on X and Y, when considered in the topology of pointwise convergence, are linearly homeomorphic.) Moreover, as Pavlovskiĭ found [31], for compact non-zero-dimensional polyhedra ℓ-equivalence is equivalent to the equality of their dimensions. A.V. Arhangelskiĭ [13] and L.G. Zambakhidze [43] extended Pavloskiĭ's result to all (non-metrizable) bicompacta; and V.G. Pestov got the final result by stating the equality $\dim X = \dim Y$ for any (Tychonoff) ℓ-equivalent spaces X and Y [39].

Ju. A. Burov has extended the results mentioned above in two directions. First, he proved the equality $\dim_G X = \dim_G Y$ for any two finite-dimensional (in the sense of dim) ℓ-equivalent spaces X and Y and a finitely-generated commutative group G (where $\dim_G X = c\dim_G \beta X$ and $c\dim_G X \le n$ if every mapping from a closed subset of the space X to the Eilenberg-MacLane complex $K(G,n)$ is extendable over X). Pestov's theorem proves this in the case $G = \mathbf{Z}$. Secondly, Burov showed that if, in Pavlovskiĭ's theorem on metrizable compacta, we change the dimension dim to ind or Ind (both equal to dim in the finite-dimensional case), then the result does not hold in the infinite-dimensional case. Let us describe this finding of Burov in detail.

In [40] Ju. M. Smirnov constructed metrizable bicompacta S_α with $\operatorname{Ind} S_\alpha = \alpha$ for all numbers $\alpha < \omega_1$. Evidently $\operatorname{ind} S_{\omega_0+1} = \omega_0 + 1$. Burov showed that: (a) for any limit number $\alpha < \omega_1$, the bicompactum $S_{\alpha+n}$ is ℓ-equivalent to a metrizable bicompactum $S'_{\alpha+n}$ with $\operatorname{Ind} S'_{\alpha+n} = \alpha$; (b) for $\alpha = k.\omega_0 + n$, $k \ge 1$, $n \ge 0$, S_α is ℓ-equivalent to a metrizable bicompactum S'_α with $\operatorname{Ind} S'_\alpha = \omega_0$; (c) for $\omega_0 \le \alpha \le 2.\omega_0$, the bicompactum S_α is ℓ-equivalent to S_{ω_0} .

QUESTION 10. Is the equality Ind X = Ind Y true for normal ℓ- (or even M-) equivalent spaces X and Y (where $\max\{\mathrm{Ind}\, X, \mathrm{Ind}\, Y\} < \omega_0$)?

According to Burov, the answer is positive for totally normal and M-equivalent X and Y. This is a consequence of Burov's statement that, if a space X is M-equivalent to a space Y, then X is a 'good' countable union of sets locally closed in X which are homeomorphic to sets locally closed in Y.

An analogous but weaker result was obtained by Burov for ℓ-equivalent X and Y. Using this, he showed that for ℓ-equivalent perfectly normal subparacompacta X and Y there exists a representation of the space X as a σ-discrete union of closed sets, homeomorphic to closed subsets of the space Y, and Ind X = Ind Y.

QUESTION 11. Do the equalities Ind X = Ind Y and ind X = ind Y hold for M-equivalent metrizable infinite-dimensional bicompacta X and Y?

3 We have already spoken of infinite-dimensional spaces. Let us now consider two more topics in this area.

First consider local finite-dimensionality. A space X is considered to be locally finite-dimensional if, for each $x \in X$, there exists $n = n(x)$ and a neighbourhood Ox such that $\dim [Ox] \le n$.

In [42] Wenner described a locally bicompact and σ-bicompact space W, which is universal in the class of all locally finite-dimensional metrizable spaces of countable weight. In [24] Luxemburg proved the existence of a universal space in the class of locally finite-dimensional metrizable spaces of weight $\le \tau$, for any τ, but without presenting a concrete space of this kind.

L.Ju. Bobkov [17] showed that: (1) the product $W \times B(\tau)$, where $B(\tau)$ is the generalized Baire space of weight τ (i.e. $B(\tau)$ is the product of a countable collection of spaces N_τ which consist of τ isolated points), is universal in the class of all locally finite-dimensional strongly metrizable (in particular, strongly paracompact metrizable) spaces of weight $\le \tau$; (2) the partial product [35] $P(W, \{\sigma_i : i \in \mathbf{N}\}, N_\tau)$, where $\{\sigma_i : i \in \mathbf{N}\}$ is a countable base of W, is universal in the class of all locally finite-dimensional metrizable spaces of weight $\leq \tau$.

Bobkov also extended Wenner's theorem to the non-metrizable case.

Let M be the class of all normal and countably paracompact T_2-spaces X which satisfy the condition $\operatorname{loc}\dim F = \dim F$ for every closed set F of X. (This holds, for example, in the class of all paracompacta.)

It turns out [17] that, in the class of all locally finite-dimensional spaces $X \in M$ of weight $\leq \tau$, there exists a universal locally bicompact and σ-bicompact space L_τ. (We cannot require the space L_τ to be bicompact, as a locally finite-dimensional bicompactum is finite-dimensional.)

It follows from this statement that any locally finite-dimensional space $X \in M$ of weight $\leq \tau$ has a locally finite-dimensional, locally bicompact and σ-bicompact extension of weight $\leq \tau$.

Bobkov also stated [17] that there exists a universal (Čech-complete and paracompact) space in the class of locally finite-dimensional spaces $X \in M$ of metrizable weight [6] $\mu w X \leq \theta$ and of bicompact weight [6] $bw X \leq \tau$.

Bobkov obtained the theorems about universal spaces using his factorization theorems [17].

QUESTION 12. Is it possible in Bobkov's theorems to eliminate the requirement that $X \in M$ (taking X to be completely regular)?

Now let us consider the transfinite small inductive dimension ind.

As shown by Smirnov [41], $\operatorname{Ind} X < \omega_1$ for any metrizable space X; and $\operatorname{Ind} S_\alpha = \alpha$ for Smirnov's bicompactum S_α, $\alpha < \omega_1$. (Smirnov's bicompacta were mentioned in section 2.) B.T. Levshenko showed [23] that, for any number $\alpha < \omega_1$, there exists a subbicompactum $L(\alpha)$ of Smirnov's bicompactum with $\operatorname{ind} L(\alpha) = \alpha$. It is clear that $\operatorname{ind} D = \omega_1$ for the discrete union D of all Smirnov's bicompacta.

The question of the possible values of the dimension ind in the class of all metrizable spaces was open (see, for example, [18]) until recently.

It turns out [38] that for any ordinal number α there exists a strongly paracompact metrizable space R_α with $\text{ind}\, R_\alpha = \alpha$.

The proof of this fact is based on consideration of the following dimensional invariant. We define $\text{Ind}^+ X = -1$ if and only if $X = \emptyset$; $\text{Ind}^+ X \le \alpha$, in the case of a non-limit ordinal number α , if between any two disjoint sets A and B , closed in X , there exists a partition C with $\text{Ind}^+ C < \alpha$; $\text{Ind}^+ X \le \alpha$, in the case of a limit ordinal α , if for each $x \in X$ there exists a neighbourhood Ox with $\text{Ind}^+ [Ox] < \alpha$.

For a strongly paracompact T_2-space X the dimensions $\text{ind}\, X$ and $\text{Ind}^+ X$ are defined (or not) simultaneously, $\text{ind}\, X \le \text{Ind}^+ X$, and for each ordinal number α there exists a number $\xi(\alpha)$ such that $\text{Ind}^+ X \le \xi(\text{ind}\, X)$.

The strongly paracompact metrizable spaces T_α described below, for any α , are analogous to Smirnov's compacta and have the property that $\text{Ind}^+ T_\alpha = \alpha$.

Take the n-cube Q^n as T_n for $n = 0, 1, 2, \ldots$. If, for a number $\beta \ge \omega_0$, all T_α with $\alpha < \beta$ have already been constructed, then we take T_β equal to: 1) the discrete union of all T_α with $\alpha < \beta$ if β is a limit number; 2) the single-point extension of the space $T_{\beta-1}$, obtained by adding the point $\bar{\bar{o}}_\beta$ if $\beta = (\beta - 1) + 1$ and $\beta - 1$ is a limit number, where we require that the set $\{\bar{\bar{o}}_\beta\}$ be closed and that the collection of sets $O_{n\beta} = \{\bar{\bar{o}}_\beta\} \cup \bigcup \{T_{\alpha+i} : \alpha$ is 0 or a limit number less than β, and $i = n, n+1, \ldots \}$, $n = 0, 1, 2, \ldots$, be a base at the point $\bar{\bar{o}}_\beta$; 3) $T_{\beta-1} \times Q^1$ if both β and $\beta - 1$ are non-limit numbers.

The invariant Ind^+ can be considered to be an extension of the dimension Ind over the wider class of metrizable spaces, as (1) $\text{Ind}^+ R = \text{Ind}\, R + 1$ if the dimension Ind of a metrizable space R is defined and transfinite, and (2) for a strongly paracompact complete metric space R the dimension $\text{Ind}^+ R$ is defined if and only if R is countable-dimensional.

The natural behaviour of the invariant Ind^+ is illustrated, for example, by the fact that on the discrete union I of all n-cubes, $n \in \mathbf{N}$, the dimension Ind is not defined, but $\text{Ind}^+ I = \omega_0$ (which conforms with our intuitive notions).

4 In the preceding section Bobkov's theorem about embedding any locally finite-dimensional strongly metrizable space of weight $\leq \tau$ in the product $W \times B(\tau)$ was stated. It is an analogue to Morita's and Nagata's theorems about the embedding of (n-dimensional) strongly metrizable spaces of weight $\leq \tau$ in the product $I^{\infty} \times B(\tau)$, where I^{∞} is the Hilbert cube (resp. in the products $Q^{2n+1} \times B(\tau)$ and $U^n \times B(\tau)$, where U^n is the universal n-dimensional metrizable bicompactum).

We can strengthen some of these results.

As is known, the components of any star-finite open cover of a space are finite or countable (see for example [12], ch.1, §6.3). A star-finite open cover of a space is called finite-component if all the components of the cover are finite. A space is called superparacompact if each open cover has a finite-component refinement.

D.K. Musaev has shown [28] that properties of superparacompacta (= superparacompact T_2-spaces) are close to those of bicompacta. In particular, the space X/C corresponding to the decomposition of a superparacompactum into its components is a paracompactum with $\dim X/C = 0$, and the quotient mapping of X onto X/C is perfect. Musaev deduces that (n-dimensional in the sense of dim) superparacompacta of weight $\leq \tau$ coincide topologically with closed subsets of products of zero-dimensional paracompacta of weight $\leq \tau$ and the Tychonoff cube I^{τ} (resp. the n-dimensional universal bicompactum of weight τ).

Musaev also showed that the metrizable (n-dimensional) superparacompacta of weight $\leq \tau$ coincide topologically with closed subsets of products of zero-dimensional metrizable spaces of weight $\leq \tau$ and I^{∞} (resp. U^n). If, in addition, an (n-dimensional) superparacompactum is Čech-complete, then it can be embedded in $I^{\infty} \times B(\tau)$ (resp. in $U^n \times B(\tau)$ and $Q^{2n+1} \times B(\tau)$) as a closed set.

5 In this, the last section, we shall once more consider homogeneity.

According to the Alexandroff theorem, every metrizable compactum is an image of the Cantor discontinuum, which is zero-dimensional and homogeneous. A.V. Ivanov [22] showed (in CH) that every bicompactum of countable character is an image of a zero-dimensional bicompactum of countable character. D.B. Motorov proved that every zero-dimensional bicompactum of countable character can be considered as a retract of a homogeneous bicompactum of the same type. Thus we have the following analogue of the Alexandroff theorem: (CH) every bicompactum of countable character is an image of a zero-dimensional and homogeneous bicompactum of the same type.

In this connection we recall the result of N.G. Okromeshko [30]: every Tychonoff space is an open image of a stratifiable space which is homogeneous and zero-dimensional in the sense of the dimension dim .

Remark

After this article had been written, Filippov's paper [44] appeared, containing proofs of the statements in [1].

REFERENCES

1 V.V. Filippov, On normally posed subspaces, Abstracts of Moscow Int. Conference (Moscow, 1980) 90.
2 T. Hoshina & K. Morita, On rectangular products of topological spaces, Topology Appl. 11 (1980) 47-57.
3 K. Morita, Čech cohomology and covering dimension for topological spaces, Fund.Math. 87 (1975) no.1, 31-52.
4 K. Morita, Dimension of general topological spaces, Surveys in General Topology (New York, 1980) 297-336.
5 K. Nagami, Dimension of non-normal spaces, Fund.Math. 109 (1980) 113-121.
6 B.A. Pasynkov, Factorization theorems in dimension theory, Uspekhi Mat.Nauk 36 (1981) no.3, 147-175.
7 B.A. Pasynkov, On the monotonicity of dimension, Dokl.Akad.Nauk SSSR 267 (1982) no.3, 548-552.
8 J. Terasawa, On the dimension of some non-normal product spaces, Sci.Tokyo Kyoiku Daygaku 11 (1972) 167-174.
9 V.M. Zolotarjev, On the dimension of subspaces, Vestnik Moskov.Univ. Mat.Meh. (1975) no.5, 10-12.
10 A.V. Zareluva, On a theorem of Hurevicz, Matem.Sbornik 60 (1963) no.1, 18-28.
11 A.Z. Zareluva, On the equality of dimensions, Matem.Sbornik 62 (1963) no.3, 295-319.

12 P.S. Alexandroff & B.A. Pasynkov, Introduction to Dimension Theory (Moscow, 1973).
13 A.V. Arhangel'skiǐ, Principle of τ-approximation and a condition of equality of the dimensions of bicompacta, Dokl.Akad.Nauk SSSR 252 (1980) no.4, 777-780.
14 V.K. Bel'nov, Dimension of topologically homogeneous spaces and free homogeneous spaces, Dokl.Akad.Nauk SSSR 238 (1978) no.4 781-784.
15 L.Ju. Bobkov, A snakelike bicompactum of countable character and with non-coinciding dimensions, Geometry of Immersed Manifolds (MGPI, Moscow, 1979) 110-113.
16 L.Ju. Bobkov, Partial products and dimension, Soobsc.Akad.Nauk Gruzin.SSR 102 (1981) no.2, 277-279.
17 L.Ju. Bobkov, On universal locally finite-dimensional spaces, Uspekhi Mat.Nauk 37 (1982) no.2, 191-193.
18 R. Engelking, Transfinite dimension, Surveys in Topology (ed. G.M. Reed, New York, 1980) 131-161.
19 V.V. Fedorchuk, On bicompacta with non-coinciding dimensions, Dokl. Akad.Nauk SSSR 182 (1968) 275-277.
20 V.V. Fedorchuk, An example of a homogeneous bicompactum with non-coinciding dimensions, Dokl.Akad.Nauk SSSR 198 (1971) no.6, 1283-1286.
21 M.I. Graev, Theory of topological groups, Uspekhi Mat.Nauk 5 (1950) no.2, 3-56.
22 A.V. Ivanov, On zero-dimensional inverse images of bicompacta of countable character, Uspekhi Mat.Nauk 35 (1980) no.6, 161-162.
23 B.T. Levshenko, Spaces of transfinite dimension, Matem.Sbornik (1965) no.2, 255-266.
24 L. Luxemburg, On universal metric locally finite-dimensional spaces, Gen.Topol.Appl. 10 (1979) no.3, 283-290.
25 L. Luxemburg, On compact metric spaces with non-coinciding transfinite dimensions, Pacific J.Math. 93 (1981) no.2, 339-386.
26 S. Mardešić, On covering dimension and inverse limits of compact spaces, Illinois J.Math. 4 (1960) no.2, 278-291.
27 S. Mardešić, Chainable continua and inverse limits, Glasnik Mat.Fiz. Astr. 14 (1959) no.3, 219-231.
28 D.K. Musaev, On superparacompact spaces, Dokl.Akad.Nauk Uzbek SSR (1983) no.2, 5-6.
29 D.K. Musaev, On dimension of superparacompact spaces, Dokl.Akad. Nauk Uzbek SSR (1983) no.9, 8-9.
30 N.G. Okromeshko, On retractions of homogeneous spaces, Dokl.Akad. Nauk SSSR 268 (1983) no.3, 548-551.
31 D.S. Pavlovskiǐ, On spaces of continuous functions, Dokl.Akad.Nauk SSSR 253 (1980) no.1, 38-41.
32 B.A. Pasynkov, On polyhedral spectra and dimension of bicompacta, in particular bicompact topological groups, Dokl.Akad.Nauk SSSR 121 (1958) no.1, 45-48.
33 B.A. Pasynkov, On snakelike bicompacta, Czechoslovak Math.J. 13 (1963) no.88, 473-476.
34 B.A. Pasynkov, On coincidence of different definitions of dimension for factor-spaces of locally bicompact groups, Uspekhi Mat. Nauk 17 (1962) no.5, 129-135.

35 B.A. Pasynkov, Partial topological products, Trudy Moscov.Mat.Obsc. 13 (1965) 136-245.
36 B.A. Pasynkov, Almost metrizable groups, Dokl.Akad.Nauk SSSR 161 (1965) no.2, 281-284.
37 B.A. Pasynkov, On the dimension of spaces with a compact group of transformations, Uspekhi Mat.Nauk 31 (1976) no.5, 112-120.
38 B.A. Pasynkov, On transfinite dimensions, Abstracts of Leningrad Int.Topol.Conf. ('Nauka', Leningrad, 1982) 123.
39 B.G. Pestov, Coincidence of the dimensions dim of ℓ-equivalent spaces, Doklady Akad.Nauk SSSR 266 (1982) no.3, 553-556.
40 Ju.M. Smirnov, On universal spaces for some classes of spaces, Izv.Akad.Nauk SSSR 23 (1959) 185-196.
41 Ju.M. Smirnov, On transfinite dimension, Matem.Sbornik 58 (1962) no.4, 415-422.
42 B.R. Wenner, A universal separable metric locally finite-dimensional space, Fund.Math. 80 (1973) no.3, 283-286.
43. L.G. Zambakhidze, On relations between dimensional and cardinal functions of spaces embedded in spaces of a special type, Soobsc.Akad.Nauk Gruzin.SSR 100 (1980) no.3, 557-560.

44. V.V. Filippov, On normally posed subspaces, Trudy Mat.Inst.Akad.Nauk SSSR 154 (1983) 239-251.

AN EQUIVARIANT THEORY OF RETRACTS

Sergey Antonian (USSR, Yerevan)

Introduction.

The purpose of this paper is to give a survey of that part of the theory of topological transformation groups (or G - spaces) which is related to the notion of absolute (neigbourhood) retracts and associated notions. The viewpoint we adopt is to regard the usual theory of retracts as the theory of equivariant retracts with a trivial group acting and try to generalize this theory to the case when the acting group is a not necessarily trivial topological group. It will be observed that the same idea is used in many works on equivariant algebraic topology. In this way G.Bredon [1], [2] and Th.Bröcker [3] have constructed, for discrete acting groups, the equivariant homology and cohomology. S.Illman [4] has studied the equivariant singular homology and cohomology when the acting group is a compact Lie group. I.M.James & G.B.Segal [5], [6], T.Matumoto [7] developed the theory of equivariant homotopy. But here we are not concerned with these other works on equivariant algebraic topology.

§ 1. Notational conventions and definitions

A topological transformation group (ttg) is a triple $\langle G, X, \pi \rangle$, where G is a topological group with the unit element e, X a topological space and $\pi : G \times X \to X$ a continuous mapping such that $\pi(e,x)=x$ and $\pi(g,\pi(h,x))=\pi(gh, x)$ for all $x \in X$, $g, h \in G$ (by gh we denote the group operation (multiplication) in G).

The mapping π is called the action of G on X and the group G is called the acting group. If $\langle G, X, \pi \rangle$ is a ttg, we shall call X a G - space. We shall in general not explicitly name the action π and simply write gx for $\pi(g, x)$. An invariant subset or a G - subset of a G - space $\langle G, X, \pi \rangle$ is a subset A of X such that $\pi(G \times A) \subset A$. It is obvious that the same group G acts (according to the same "rule") on every invariant subset A of X.

A linear ttg, or a linear G - space, is a ttg $\langle G, X, \pi \rangle$ in which X is a topological vector space and $\pi(g, \lambda x + \mu y) = \lambda \pi(g, x) + \mu \pi(g, y)$, for all $g \in G$, $x, y \in X$ and for all scalars λ and μ.

Let X be a G - space and $x \in X$. We define the isotropy group at x denoted by G_x by $G_x = \{ g \in G, gx=x \}$. The subspace $G(x) = \{ gx; g \in G \}$ is called the orbit of x. Clearly the orbits of any two points in X are either equal or disjoint, in other words X is partitioned by its orbits. We denote the set of all orbits of X by X/G. Let $p: X \to X/G$ denote the natural map taking X into its orbit $G(x)$. Then X/G endowed with the quotient topology from the projection p is called the orbit space of X. If H is a subgroup of G we denote by (H) the conjugacy class of H in G, i.e. $(H) = \{ gHg^{-1}; g \in G \}$. Since clearly $G_{gx} = g G_x g^{-1}$ it follows that $\{ G_y; y \in G(x) \} = (G_x)$, which is called the orbit type of $G(x)$. The union of orbits of type (H) will be denoted by $X_{(H)}$ and we write $\overline{X}_{(H)} = p(X_{(H)})$. By $X[H] = \{x \in X ; hx = x$ for every $h \in H\}$ we denote the subspace of fixed points of H on X. A continuous map $f: X \to Y$, X and Y are G - spaces, is called a G - map or an equivariant map if $f(gx) = gf(x)$ for all $g \in G$, $x \in X$. Correspondingly we have the notion of a G - homeomorphism, a G - retraction, etc.

For any class T of topological spaces we denote by T^G the class of all G - spaces X such that X as a topological space belongs to the class T. A G - space Y is called a neighbourhood (resp., absolute) extensor of T^G if for every $X \in T^G$ and every invariant closed subset A of X, every equivariant map $f: A \to Y$ can be extended to an equivariant map $f': U \to Y$ on some invariant neighbourhood U of A in X (resp., on all of X). In this case we write $Y \in G - ANE(T)$ (resp. $Y \in G - AE(T)$). A G - space Y is called a neighbourhood (resp. absolute) retract of T^G if $Y \in T^G$ and for every closed equivariant embedding $i: Y \to Z$ into an object $Z \in T^G$ the image $i(Y)$ is an equivariant retract of some open invariant neighbourhood in Z (resp. of all of Z). In this case we write $Y \in G - ANR(T)$ (resp. $Y \in G - AR(T)$). Of course, every absolute (neighbourhood) extensor of the class T^G belonging to T^G is an absolute (neighbourhood) retract.

Let X and Y be G - spaces, and f, $f': X \to Y$ two G - maps. A G - homotopy from f to f' is a G - map $H: I \times X \to Y$ (G acts on $I \times X$ by acting trivially on I, that is $g(t, x) = (t, gx)$, $g \in G$, $(t, x) \in I \times X$) such that $H(o, x) = f(x)$, $H(1, x) = f'(x)$ for all $x \in X$. A G - space X is said to be G - contractible if the identity map of X is G - homotopic to a constant map $f: X \to X$, where the image $f(X) = x_0$ is a G - fixed point, i.e. $x_0 \in X[G]$. Correspondingly X is said to be locally G - contractible (at $x \in X$) if every G_x - neighbourhood U of x contains a G_x - neighbourhood V of x such that V is G_x - contractible in U.

Unless stated otherwise the letter G always denotes an arbitrary Hausdorff topological group. All spaces considered are completely regular and Hausdorff. Throughout this paper the letter K denotes any one of the following important classes of topological spaces

C - all compact Hausdorff spaces

M - all metrizable spaces

CM - all compact metrizable spaces

P - all p - paracompact[1/] spaces

N - all normal spaces

§ 2. Extensions of G - maps

In the ordinary theory of retracts the Tietze-Urysohn and Dugundji extension theorems play very important roles. The first analogous result for G - spaces has been established by A.Gleason [8] in 1950:

[1/] A space is said to be p - paracompact if it can be mapped in a continuous and perfect manner onto some metrizable space.

Theorem 1 ([8]). Let G be a compact Lie group acting on a euclidean space E by means of orthogonal transformations. Then the G - space E is a G - $AE(N)$ space.

In 1973 Jan Jaworowski [9] proved the following generalization of the well-known Dugundji extension theorem [10] :

Theorem 2 ([9]). Let G be a cyclic group of prime order. Let L be a locally convex linear G - space and V be an invariant convex subspace of L. Then $V \in G$ - $AE(M)$.

In 1980 the author [11] , [12] established the following general equivariant extension theorem, which implies the above theorems 1 and 2 and is an equivariant analogue of the Dugundji extension theorem for an arbitrary compact acting group.

Theorem 3 ([11] , [12]). Suppose that G is a compact group. Let V be a complete[1/] convex invariant subset of a locally convex linear G - space Z and let A be an invariant closed subset of a G - space X. Then each equivariant map $f:A \to V$, which has a continuous extension over X can be extended to an equivariant map $f':X \to V$. If either Z is finite-dimensional, or G is finite, then the completeness of V can be omitted.

As an immediate consequence of this theorem we obtain the following equivariant version of the Dugundji extension theorem.

Theorem 4 ([11] , [12]). Let G, Z and V be as in

1/ i.e. V is a complete uniform subspace of the topological vector space Z endowed with the natural group uniformity.

the above theorem 3. Then $V \in G - AE(M)$. If either Z is finitedimensional or G is finite, then the completeness of V can be omitted.

Now the above theorem 2 of J.Jaworowski is an immediate consequence of theorem 4 when $G = Z_p$ the cyclic group of prime order p. Analogously, Gleason's theorem 1 follows from theorem 3 if we take a compact Lie group for G and a euclidean orthogonal G - space for V.

Observe that the assumption about the completeness of V in theorems 3 and 4 is essential (in contradistinction to the ordinary Dugundji extension theorem). This follows from the following:

<u>Example 1</u> ([11] , [12]). Let G be the circle group, and let Z be the Banach space of all complex-valued continuous functions $f:G \to C$ where G acts on Z by the rule $(g, f) \to gf$, $(gf)(z) = f(zg)$, for all $g, z \in G$, $f \in Z$. Obviously Z is a linear G - space. Let $f_0(z) = e^z$, $z \in G$ and let V be the convex hull of the orbit $G(f\ )$ in Z. It is easy to show that V is not complete and the fixed point set $V[G]$ is empty. Then the next simple proposition implies that V is not a G - AE(M) space.

<u>Proposition 1</u> ([11] , [12]). Let G be an arbitrary group and let T^G be a class of G - spaces such that for each element $X \in T^G$ the discrete sum $X' = X \cup \{a\}$ belongs to T^G, where $\{a\}$ is a one-point trivial G - space.
Then for every $Y \in G - AR(T)$ the fixed point set $Y[G]$ is non-empty.

The assumption about compactness of the group G in theorems 3 and 4 is essential too. This is implied by the following.

Example 2 ([11] , [12]). Let G be the additive group of reals and let Z be the space of all continuous functions $f: G \to G$ endowed with the compact-open topology and pointwise defined algebraic operations. It is clear that Z becames a linear G - space if we define the action by the rule: $(g, f) \to gf$, $(gf)(x) = f(x + g)$, for all $g, x \in G$, $f \in Z$. It is not hard to show that Z is locally convex, complete and metrizable. Let $\bar{f}: G \to G$ be the identity map and V be the orbit $G(\bar{f})$. It is easy to check that V is convex and complete in Z and that the fixed point set $V[G]$ is empty. Then proposition 1 implies that V is not a G - AE(M) (and hence G - AR(M)) space.

In connection with theorem 4 and example 1 we raise the following:

Question 1. Let G be an infinite compact group and let Z be a noncomplete normed linear G - space. Is it true that $Z \in G$ - AE(M)?

The following more general question was formulated by Yu.M.Smirnov [13] .

Question 2. Let G be a compact metrizable group and $X \in K^G$. Suppose that for every subgroup H of G the fixed point set $X[H]$ is an ANR(K) (resp., an AR(K)) space. Is it true that $X \in G - ANR(K)$ (resp., $X \in G - AR(K)$)?

The next theorem asserts that these conditions for $X[H]$ are necessary in order that $X \in G - ANR(K)$ (resp. $X \in G - AR(K)$).

Theorem 5 ([14]). Let G be a locally compact σ - compact group. Then for every $G - ANR(K)$ (resp., $G - AR(K)$) space X and for every subgroup H of G, the space $X[H]$ is an $ANR(K)$ (resp., an $AR(K)$) space.

In the case when the group G is either compact or metrizable, this theorem has been obtained by Yu.M. Smirnov [13]. Later on J. de Vries [15] presented the following strengthened version of theorem 5.

Theorem 5′ ([15]). If G is a metrizable group, H a compact subgroup of G, and a ttg $\langle G, X, \pi \rangle \in G - AE(M)$, then $\langle H, X, \pi \rangle \in H - AE(M)$. If G is an arbitrary group, H a locally compact subgroup of G such that $G = B \cdot H$ for some compact set $B \subset G$, and if $\langle G, X, \pi \rangle \in G - AE(C)$, then $\langle H, X, \pi \rangle \in H - AE(C)$.

In several particular important cases positive answers to question 2 are given by J.Jaworowski [9], [16 - 18] and R.Lashof [19].

Theorem 6 ([16]). Let G be a compact Lie group, X be a locally compact separable metric finite-dimensional G-space with a finite number of orbit types, A a closed G - subset of X and $f: A \to Y$ a G - map to a locally compact separable metric G - space Y. If for every orbit type (H) in $X \setminus A$, $Y[H] \in ANR(M)$ (resp., $Y[H] \in AR(M)$), then f has a neighbourhood G - extension (resp., a G - extension over X).

In [19] R.Lashof removed the local-compactness assumption in theorem 6. Then J.Jaworowski [18] modifying Lashof's proof showed that the finite-dimensionality assumption can also be weakened to that of having a finite structure, as follows.

Definition ([18]). A G - space X is said to have a finite structure if it has a finite number of orbit types and if for each orbit type (H), the orbit bundle $X_{(H)} \to \overline{X}_{(H)}$ (equivalently, its principal bundle $X_H \to \overline{X}_H$) has a finite trivializing cover; i.e., there exists a finite open cover $U_1, \ldots, U_m$ of $\overline{X}_{(H)}$ such that each $p^{-1}(U_i)$, i = 1, 2, ..., m, is G - equivalent to $G/H \times U_i$.

Theorem 7 ([18]). Let G be a compact Lie group, let X be a separable metric G - space and let A be a closed G - subset of X such that $X \setminus A$ is of finite structure. Let $f: A \to Y$ be an equivariant map of A into a metrizable G - space Y. If for each orbit type (H) in $X \setminus A$, $Y[H]$ is an ANR(M) (resp. an AR(M)) then f has an equivariant extension over an invariant neighbourhood of A (resp. over X).

This theorem was then used to give the following characterization of G - ANR's and G - AR's.

Theorem 8 ([18]). Let G be a compact Lie group and let X be a separable metric G - space with a finite structure. Then $X \in G$ - ANR(M) (resp. $X \in G$ - AR(M)) iff for each closed subgroup H of G, $X[H] \in$ ANR(M) (resp. $X[H] \in$ AR(M)).

Recently M.Madirimov [20] with the aid of modifying Jaworowski-Lashof's method generalized the above theorem 6 of Jaworowski.

Theorem 9 ([20]). Let G be a compact Lie group, X a separable metric G - space and A a closed G - subset of X such that $X \setminus A$ has a finite number of orbit types and $\dim (X \setminus A) \leq n + 1$. Let $f: A \to Y$ be an equivariant map of A into a metrizable G - space. If for each orbit type (H) in $X \setminus A$, $Y[H]$ is locally connected (resp., locally connected and connected) in all dimensions $\kappa = 0, 1, \ldots, n$, then f has a neighbourhood G - extension (resp., a G - extension over X).

As in the topological case, finite dimensional G - ANR's can be characterized by local G - contractibility:

Proposition 2 ([17]). Let G be a compact Lie group and X a metric G - space. If $X \in G$ - ANR(M) then X is locally G -contractible. Conversely, if X is finite-dimensional and locally G - contractible then X is a G - ANR(M).

Proposition 3 ([11], [12]). Let G be a compact group. If X is a G - AR(K) space then X is G - contractible. Conversely, if X is G - contractible and $X \in G$ - ANR(K), then $X \in G$ - AR(K).

From propositions 2 and 3 above we immediately obtain the following characterization of finite-dimensional G - AR's

Proposition 4. Let G be a compact Lie group and let X be a finite dimensional metric G - space. Then X is a G - AR(M) space if and only if X is G - contractible and locally

G - contractible.

3. Equivariant embeddings and linearizations.

In the ordinary theory of retracts and in shape theory theorems about the closed embeddings of spaces of a given class T in an absolute retract of the same class play a very important role. Therefore in the equivariant case the following question naturally arises.

Question 3. Can every object of the class K^G be equivariantly embedded into some G - AR(K) space?

Historically the first equivariant embedding theorem was obtained by G.Mostow [21] .

Theorem 10 ([21]). Let G be a compact Lie group. If X is a finite-dimensional separable metric G - space with a finite number of orbit types, then there is an equivariant embedding of X as an invariant subspace of some euclidean space with an orthogonal G - action. If X is also locally compact then the embedding may be taken to be as a closed subspace.

Since by Gleason's theorem 1 every euclidean G - space is a G - AE(N) and hence a G - AR(M) space, then by Theorem 10 we have a partial affirmative answer to question (3). In the infinite-dimensional case for different classes of G - spaces the question (3) has been affirmatively answered by Yu.M.Smirnov [22] and the present author [14] , [25] . To formulate these results, let us recall that by the linearization of an action of a given group G on a space X we mean the equivariant embedding of X into some linear G - space.

For finite-dimensional G - spaces the linearization problem is positively solved by the above theorem 10 of G.Mostow and in the infinite-dimensional case by Yu.M. Smirnov [22] and J. de Vries [23] .

Theorem 11 ([22] , [23]). Let G be a locally compact group and X be a G - space. Then every continuous map $f:X \to Y$ into a space Y induces an equivariant map $f:X \to C(G, Y)$ by the formula $f(x)(g) = f(gx)$, for every $x \in X$, $g \in G$. If in addition f is a topological embedding (resp., with closed image) then f is an equivariant embedding (resp., with closed image). Here $C(G, Y)$ denotes the G - space of all continuous mappings $\varphi : G \to Y$ endowed with the compact-open topology and with action of G defined by the rule $(g, f) \to gf$; $(gf)(x) = f(xg)$ for all $g, x \in G$, $f \in C(G, Y)$.

Since every G - space X can be topologically embedded in some power $\mathbb{R}^{\tau}$ of the real line, by theorem 11 X can be equivariantly embedded in the locally convex linear G - space $C(G, \mathbb{R}^{\tau})$. Thus the linearization problem is solved. Returning to question (3), let us observe, that if X is a metrizable G - space, then by the Wojdyslawski embedding theorem [24] X admits a closed embedding into a convex subset V of a Banach space. Consequence, by theorem 11, X can be embedded as a closed invariant subset into the G - space $C(G, V)$. If in addition G is σ - compact, the space $C(G, V)$ is metrizable. Finally, note that by the next simple proposition 5, $C(G, V)$ is a G - AR(M) space.

Proposition 5. Let G be a compact (resp. locally compact)

group and T a certain class of spaces. If $Y \in ANE(T)$ (resp. $Y \in AE(T)$), then $C(G, Y) \in G - ANE(T)$ (resp. $C(G, Y) \in G - AE(T)$).

Thus we have obtained the following partial answer to question 3 originally established by Yu.M. Smirnov (see [22],[12]).

Theorem 12. Let G be a locally compact σ- compact group. Then every metrizable G - space admits a closed equivariant embedding into a $G - AR(M)$ space.

For the classes C and CM an affirmative answer to question 3 follows from the following equivariant version of the classical Tychonov embedding theorem due to present author [14], [25] .

Theorem 13 ([14] ,[25]). Let τ be an infinite cardinal and let G be a compact group of weight $\leq \tau$. Then on the topological power $\mathbb{R}^{\tau}$ of the real line there exists an action π of G such that

(1) the cube I^{τ} , $I = [0, 1]$ is an invariant subset of $\mathbb{R}^{\tau}$ under this action π ,

(2) every G - space X of weight $\leq \tau$ can equivariantly be embedded in I^{τ} .

Moreover, there exists a linear structure on $\mathbb{R}^{\tau}$ such that $\mathbb{R}^{\tau}$ is a locally convex topological vector space such that

(3) I^{τ} is a convex subset of $\mathbb{R}^{\tau}$,

(4) the action π is linear.

By the above theorem 3 of author, the G - space I^{τ} , which we call the Tychonov G - cube, will be a $G - AE(N)$ space. Therefore, every object of the class C (resp., CM) can be equivariantly embedded in a $G - AR(C)$ (resp., in a $G - AR(CM)$ space (namely in a Tychonov G - cube of the same weight).

Observe that recently J. de Vries extended theorem 13 to the case of an arbitrary locally compact σ - compact group (to be published). In connection with theorem 13 let us formulate two recent unpublished results of the present author.

Theorem 14. Let G be a compact group and τ be an infinite cardinal number. Then on $\mathbb{R}^{\tau}$ with its ordinary product topology and linear structure there exist a linear action of G and an invariant convex compactum B_{τ} of weight = τ, such that

(1) B_{τ} is an invariant subset of the linear G - space $\mathbb{R}^{\tau}$,

(2) every G - space of weight $\leq \tau$ can equivariantly be embedded in B_{τ},

(3) B_{τ} is a G - AE(N) space.

Theorem 15. Let G be a compact group. For every p - paracompact G - space X of infinite weight τ there exist a normed linear G - space L, a linear action of G on $\mathbb{R}^{\tau}$ and an invariant compactum $B_{\tau} \subset \mathbb{R}^{\tau}$ of weight = τ such that

(1) X can be equivariantly embedded in the G - space $L \times B_{\tau}$ as a closed subspace,

(2) the G - space $L \times B_{\tau}$ is a G - AR(P) space.

Thus the question (3) is affirmatively solved for each of the classes CM, C, M and P, when the acting group G is compact. In the case of an arbitrary locally compact group G the question (3) still remains open.

Corollary. Let G be a compact (resp., locally compact and σ - compact) group. Then for each of the classes CM^G, C^G and P^G (resp., for M^G) the notions of absolute (neighbourhood) retract and absolute (neighbourhood) extensor coincide.

Now, when the question (3) is solved affirmatively, many important results of the ordinary theory of retracts can be easily extended to the equivariant case, and the proofs are completely analogous to the proofs in the ordinary case. For example, the Borsuk homotopy extension theorem, the additional theorem for AR(M) and ANR(M) spaces, the theorems about ANR(M) and AR(M) - mapping spaces, etc. are also valid in the equivariant case (see [11] , [12] , [26]).

4. Orbit spaces of G - ANR's and G - AR's

Finally let us formulate the following question closely connected with many problems of the equivariant theory of retracts.

Question 4. Let G be a compact group (even a compact Lie group). Is it true that if $X \in G$ - ANR(M) (resp., X G-AR(M)) then $X/G \in$ ANR(M) (resp., $X/G \in$ AR(M)).

In some particular cases E.Floyd [27] and P.Conner [28] , [29] established the following stronger results.

Theorem 16 ([27] , [28]). Let G be either finite or circle group and X a compact finitedimensional metric G - space with a finite number of orbit types. If X is an ANR(M) (resp., an AR(M)) space then X/G is also an ANR(M) (resp. an AR(M)) space.

Theorem 17 ([29]). Let G be a compact Lie group with commutative component of identity and X be a compact connected finite dimensional metric G - space with a finite number of orbit types. If X is an ANR(M) space then X/G is also an ANR(M) space.

Note the following particular case of our question 4.

Let X be a space and X^n denote the n - fold cartesian product of X. Let G be the permutation group on n letters. Then if $g \in G$ and $x = \{x_k\} \in X^n$, define $gx = \{ x_{g^{-1}(k)} \}$. Thus X^n becomes a G - space whose orbit space denoted by $SP^n X$ is called the n - fold symmetric product of X associated with G. By proposition 5 above $X^n \in G$ - ANR(M) (resp., $X^n \in G$ - AR(M)) whenever $X \in$ ANR(M) (resp., $X \in$ AR(M)). Therefore the following question is a very special case of question 4.

Question 5. Is it true that $SP^n X$ is an ANR(M) (resp. an AR(M)) space whenever X is?

By theorem 16 above, if X is a finite-dimensional compact ANR(M) (resp. AR(M)) space, then so is $SP^n X$. In [30] J.Jaworowski proved this result without the finite-dimensional restriction. But in the general (non-compact) case question 5 still remains open.

References

[1] Bredon G., Introduction to compact transformasion groups, Academic Press, 1972.

[2] Bredon G., Equivariant cohomology theories, Lecture Notes in Math., v. 34 (1967).

[3] Bröcker Th. - Singuläre definition der äquivarianten Bredon Homologie, Manuscripta Math., 5 (1971), 91-102.

[4] Illman S. - Equivariant singular homology and cohomology MEMO, AMS, vol.1, Issue 2, N 156 (1975).

[5] James I.M., and Segal G.B. On equivariant homotopy type, Topology. vol. 17(1978), 267-272.

[6] James I.M., and Segal G.B. On equivariant homotopy theory. Lecture Notes in Math., 788(1980) 316-330.

[7] Matumoto N. On G - CW complexes and a theorem of J.H.C. Whitehead, J.Fac.Sci.Univ. Tokyo, 18(1971), 363-374.

[8] Gleason A., Spaces with a compact Lie group of transformations Proc. AMS, vol1, no.1, 1950, 35-43.

[9] Jaworowski J. Equivariant extensions of maps, Pacific J. Math., 45(1973), 229-244.

[10] Dugundji J. An extension of Tietze's theorem, Pacific J. Math., 1(1951), 353-367.

[11] Antonian S.A., Retracts in the category of G - spaces. Bull. Acad. Polon. Sci., Ser. Math. v.28, no.11-12 (1980), 613-618 (Russian).

[12] Antonian S.A., Retracts in the categories of G - spaces. Izvestia Acad. Nauk. Arm.SSR, Ser. Math., v.15, no.5(1980), 365-378 (Russian).

[13] Smirnov Yu.M., Sets of H - fixed points are absolute extensors, Math. USSR Sbornik, vol. 27(1975), no.1, 85-92.

[14] Antonian S.A., Tychonov's theorem in the category of topological transformation groups, Doklady Akad. Nauk Arm.SSR, 71(1980), 212-216 (Russian).

[15] Vries J. de, Topics in the theory of topological transformation groups, Topological Structure II, Math. Centre.Tracts 116 (1979), 291-304 (Proc. Symp. in Amsterdam, 1978).

[16] Jaworowski J. - Extensions of G - maps and Euclidean G - retracts, Math. Zeitschrift, 146(1976), 143-148.

[17] Jaworowski J. - Extension properties of G - maps, Proc. Inter. Conf. Geometric Top., Warszawa 1980, 209-213.

[18] Jaworowski J. - An equivariant extension theorem and G - retracts with a finite structure, Manuscr. Math., 35(1981), 323-329.

[19] Lashof R., The equivariant extension theorem, Proc. Amer. Math. Soc., v. 83, N 1, 1981, 138-140.

[20] Madirimov M., On the J.Jaworowski equivariant extension theorem (Russian), Uspehi Mat.Nauk (to appear).

[21] Mostow G., Equivariant embeddings in Euclidean space, Annals of Math., 65, no.3, 1957, 432-446.

[22] Smirnov Yu.M., On equivariant embeddings of G - spaces (Russian), Uspehi Mat. Nauk, 31:5(1976), 137-147.

[23] Vries J. de, Universal topological transformation groups, Gen. Top. Appl., v. 5, no. 2 (1975), 107-122.

[24] Wojdislawski M., Retracts absolus et hyperspace des continus, Fund. Math. 32 (1939), 184-192.

[25] Antonian S.A., Smirnov Yu.M., Universal objects and compact

extensions of topological transformation groups (Russian), Dokl. Akad. Nauk SSSR, v. 257, no. 3(1981), 521-525.

[26] Antonian S.A., Mapping spaces are equivariant absolute extensors (Russian), Vestnik Moskow Univ. Ser. Math.-Mech., 6 (1981), 22-25.

[27] Floyd E., Orbit spaces of finite transformation groups II, Duke. Math. J., 20 (1953), 563-567.

[28] Conner P., Floyd E., Orbit spaces of circle groups of transformations. Ann. Math. v. 67, no. 1 (1958), 90-98.

[29] Conner P., Retraction properties of the orbit space of a compact topological transformation group, Duke. Math. J., 27 (1960), 341-357.

[30] Jaworowski J., Symmetric products of ANR's associated with a permutation group. Bull. Akad. Polon. Sci., Ser. Math., v. 20, no.8 (1972), 649-651.

Department of Mathematics
State University of Yerevan
375049 Yerevan-49
U.S.S.R.

P-EMBEDDING, LC^n SPACES AND THE HOMOTOPY EXTENSION PROPERTY

Kiiti Morita
Department of Mathematics, Sophia University,
Kioicho, Chiyoda-ku, Tokyo 102, Japan

1 *Introduction*

Throughout this paper, let X be a topological space and A a subset of X, and let λ be an infinite cardinal number. By an AR (resp. ANR) we always mean an AR (resp. ANR) for metric spaces.

A subspace A of X is said to be P^λ-embedded in X if for every locally finite cozero-set cover $\mathcal{U}$ of A with cardinality $\leq \lambda$ there exists a locally finite cozero-set cover $\mathcal{V}$ of X such that $\mathcal{V} \cap A$ refines $\mathcal{U}$, where $\mathcal{V} \cap A = \{V \cap A \mid V \in \mathcal{V}\}$; if A is P^λ-embedded in X for every λ, then A is said to be P-embedded in X (cf. Shapiro (1966)). It was proved by Dowker (1952) that X is collectionwise normal if and only if every closed subspace of X is P-embedded in X.

In Morita (1975 a) we proved

Theorem 1.1. A subspace A of X is P^λ-embedded in X if and only if any continuous map from A into any AR Y of weight $\leq \lambda$ can be extended over X, where Y is assumed, in addition, to be Čech complete unless A is a zero-set.

As is well-known, a pair (X, A) is said to have the homotopy extension property (abbreviated to HEP) with respect to a topological space Y if every partial homotopy

$$h_t : A \to Y \quad (0 \leq t \leq 1)$$

of an arbitrary continuous map $f : X \to Y$ has an extension

$$f_t : X \to Y \quad (0 \leq t \leq 1)$$

such that $f_0 = f$ —— that is, if every continuous map from the subspace $X \times \{0\} \cup A \times I$ of $X \times I$ into Y can be extended over $X \times I$, where I is the closed unit interval $[0, 1]$ in the real line.

The original homotopy extension theorem of K. Borsuk was generalized by Dowker (1956) as follows.

I. If X is countably paracompact and normal and if A is closed in X, then (X, A) has the HEP with respect to an ANR Y which is separable and Čech complete.

II. If X is countably paracompact and collectionwise normal and if A is closed in X, then (X, A) has the HEP with respect to an ANR Y, provided either Y is Čech complete or A is a G_δ set.

These results of Dowker were generalized in Morita (1975 a) as follows.

Theorem 1.2. A pair (X, A) has the HEP with respect to any ANR Y of weight $\leq \lambda$ if A is P^λ-embedded in X and either Y is Čech complete or A is a zero-set of X.

In a joint paper Morita & Hoshina (1975) the following theorem was proved.

Theorem 1.3. A pair (X, A) has the HEP with respect to any Čech complete ANR Y of weight $\leq \lambda$ if and only if A is P^λ-embedded in X.

Recently Sennott (1978 a) has defined a new notion of M^λ-embedding as follows.

Definition 1.1'. A subspace A of X is said to be M^λ-embedded in X if any continuous map from A into any AR of weight $\leq \lambda$ can be extended over X. If A is M^λ-embedded in X for every cardinal λ, then A is said to be M-embedded in X.

In the same paper Sennott proved

Theorem 1.3'. (X, A) has the HEP with respect to any ANR Y of weight $\leq \lambda$ if and only if A is M^λ-embedded in X.

On the other hand, another homotopy extension theorem was proved by Kuratowski (1950):

III. If X is a separable metric space and A is closed in X and if $\dim(X - A) \leq n$, then (X, A) has the HEP with respect to any separable metric space which is LC^n.

Here, for an integer $n \geq 0$ a topological space Y is called LC^n if for each point y of Y and each open neighborhood U of y there is an open neighborhood V of y contained in U such that every continuous map $f : S^k \to V$ is null homotopic in U for all integers k with $0 \leq k \leq n$, where S^k is a k-sphere. Y is called C^n if every continuous map $f : S^k \to Y$ is null homotopic for all integers k with $0 \leq k \leq n$. As usual, we make the convention that an arbitrary space is LC^{-1} as well as C^{-1}. Basic properties of LC^n and C^n separable metric spaces (cf. Kuratowski (1950)) were extended to the case of arbitrary LC^n and C^n metric spaces by Dugundji (1958).

The purpose of this paper is to establish the following two theorems which are analogous to the theorems stated above and contain III above; these results were announced in Morita (1978) and referred to in Sennott (1978 b).

Theorem 1.4. If $\dim X/A \leq n+1$ for an integer $n \geq -1$, then A is M^λ-embedded (resp. P^λ-embedded) in X if and only if any continuous map from A into any metric (resp. complete metric) space Y of weight $\leq \lambda$ which is LC^n and C^n can be extended over X.

Theorem 1.5. If $\dim X/A \leq n$ for a non-negative integer n, then (X, A) has the HEP with respect to any metric (resp. complete metric) space Y of weight $\leq \lambda$ which is LC^n if and only if A is M^λ-embedded (resp. P^λ-embedded) in X.

By the convention, Theorem 1.4 for the case $n = -1$ states:

Theorem 1.6. If $\dim X/A \leq 0$, then A is M^λ-embedded (resp. P^λ-embedded) in X if and only if any continuous map from A into any metric (resp. complete metric) space Y of weight $\leq \lambda$ can be extended over X.

As a corollary to Theorem 1.6 we obtain the following theorem proved by Hoshina (1977).

Theorem 1.7. If $\dim X = 0$ and A is metrizable, then A is M-embedded in X if and only if A is a retract of X.

Indeed, we have only to put $Y = A$ and to apply Theorem 1.6 to the identity map : $A \to A$.

As is known, X is normal (resp. collectionwise normal) if and only if every closed subset of X is $P^{\aleph_0}$-embedded (resp. P-embedded) in

X; and, if X is normal, dim $X \le n+1$ if and only if every continuous map from any closed subset of X into an (n+1)-shere S^{n+1} can be extended over X. Since S^{n+1} is LC^n and C^n, we have as a corollary to Theorem 1.4:

Theorem 1.8. Every closed subset of X is M^λ-embedded (resp. P^λ-embedded) in X and dim $X \le n+1$ if and only if every continuous map from any closed subset of X into any metric (resp. complete metric) space Y of weight $\le \lambda$ which is LC^n and C^n can be extended over X.

Theorem 1.8 contains a result which was reported in an abstract by Michael (1955) and, when applied to the case $n = -1$, yields also Propositions 5.1, 5.2 and 5.3 in Sennott (1978 b).

2 *Preliminalies*

By X/A we mean the quotient space obtained from X by contracting A to a point q_A. The covering dimension of a space X, written dim X, is defined to be the least integer n such that any finite normal cover of X is refined by a finite normal cover of order $\le n+1$ (cf. Morita (1980)). Then the following holds (cf. Theorem 5.8 of Morita (1980)).

Lemma 2.1. If $A \subset X$, then dim $X/A \le$ dim X.

Since $(X \times I)/(A \times I) \cong ((X/A) \times I)/(\{q_A\} \times I)$, we have by Theorems 5.9 and 8.3 of Morita (1980)

Lemma 2.2. dim $(X \times I)/(A \times I) = 1 +$ dim X/A.

It is to be noted that if X is metric and $A \subset X$ closed, then dim $X/A \le n$ is equivalent to $\dim(X - A) \le n$ (cf. Proposition 5.10 of Morita (1980)).

Concerning LC^n and C^n spaces the following theorems were established by Theorems 3.2 and 9.1 of Dugundji (1958) for the case $n \ge 0$.

Theorem 2.3. Let Y be a metric space and n an integer ≥ -1. Then the following properties are equivalent.

(a) Y is LC^n.

(b) If T is a metric space, $B \subset T$ closed, and $\dim(T - B) \le n+1$, then every continuous map $f : B \to Y$ can be extended over an open neighborhood W of B in T.

Theorem 2.4. Let Y be a metric space which is LC^n for some n with $n \geq -1$. Then the following properties are equivalent.

(a) Y is C^n.

(b) If T is a metric space, $B \subset T$ closed, and $\dim(T - B) \leq n+1$, then every continuous map $f : B \to Y$ can be extended over the whole space T.

If T is a metric space, $B \subset T$ closed and $\dim(T - B) \leq 0$, then there exists a retraction $r : T \to B$. This fact was proved essentially by Stone (1962) (cf. the proof of Theorem 3' in §2 of his paper. Cf. also Engelking (1969)). Therefore Theorems 2.3 and 2.4 are true also for the case $n = -1$.

Concerning M^λ-embedding, Hoshina (1977) obtained the following characterization.

Theorem 2.5. A is M^λ-embedded in X if and only if A is P^λ-embedded in X with the property (*):

(*) For any metric space T of weight $\leq \lambda$ and for any continuous map $f : X \to T$ there exist a metric space S of weight $\leq \lambda$, a continuous map $g : X \to S$ and a uniformly continuous map $h : S \to T$ such that $f = h \circ g$ and $g(A)$ is closed in S.

(Although the words "of weight $\leq \lambda$" following "a metric space S" and the adverb "uniformly" preceding "continuous map h" are lacking in the original statement of Theorem 2.5 in Hoshina (1977), the proof there actually shows the validity of Theorem 2.5 as stated here.)

Concerning P^λ-embedding, our argument in Morita (1975 a) shows the validity of Lemma 2.6 below.

Lemma 2.6. If A is P^λ-embedded in X and f is a continuous map from A into a metric space Y of weight $\leq \lambda$, then there exist a metric space T of weight $\leq \lambda$, a continuous surjective map $g : X \to T$ and a uniformly continuous map $\psi_o : g(A) \to Y$ such that $f = \psi_o \circ (g|A) : A \to Y$, where $g|A$ means the restriction of g to A with $g(A)$ as its range.

On the other hand, the arguments in §2 of Morita (1975 b) yield the following lemma.

Lemma 2.7. If $\dim X/A \leq n$ and $f : X \to T$ is a continuous map from X into a metric space T of weight $\leq \lambda$, then there exist a metric space S of weight $\leq \lambda$, a continuous surjective map $g : X \to S$

and a uniformly continuous map $h : S \to T$ such that $f = h \circ g$ and $\dim(S - Cl_S g(A)) \leq n$.

3 *Proof of Theorem 1.4*

Since an AR is LC^n and C^n for every $n \geq 0$ by Theorem 12.3 in Dugundji (1958) and for $n = -1$ by definition, the "if" part of Theorem 1.4 is a direct consequence of Theorem 1.1 and Definition 1.1'.

To prove the "only if" part of Theorem 1.4 let us assume first that

(i) A is P^λ-embedded in X and $\dim X/A \leq n+1$,

(ii) Y is metric, is LC^n and C^n, and has weight $\leq \lambda$,

(iii) $f : A \to Y$ is a continuous map.

Then by Lemma 2.6 there exist a metric space T of weight $\leq \lambda$, a continuous map $g : X \to T$ and a uniformly continuous map $\psi_o : g(A) \to Y$ such that

$$f = \psi_o \circ (g|A) : A \to Y. \tag{1}$$

Then it follows from Lemma 2.7 that there exist a metric space S of weight $\leq \lambda$, a continuous surjective map $h : X \to S$ and a uniformly continuous map $k : S \to T$ such that

$$g = k \circ h, \tag{2}$$

$$\dim(S - Cl_S h(A)) \leq n+1. \tag{3}$$

Let us now distinguish two cases.

Case (a): Y is a complete metric space. Since $\psi_o \circ (k|h(A)) : h(A) \to Y$, where $k|h(A) : h(A) \to k(h(A)) = g(A)$ is the restriction of k, is uniformly continuous, this map extends to a continuous map $\phi_o : Cl_S h(A) \to Y$. In view of (3) we can apply Theorem 2.4 to the metric space S and the map $\phi_o : Cl_S h(A) \to Y$. Hence ϕ_o extends to a continuous map $\phi : S \to Y$. Then the map $\phi \circ h : X \to Y$ is clearly an extension of $f : A \to Y$.

Case (b): A is M^λ-embedded in X. In this case, by applying Theorem 2.5 to the map $g : X \to T$, we can find a metric space T_1 of weight $\leq \lambda$ and two continuous maps $g_1 : X \to T_1$, $g_2 : T_1 \to T$ such that $g = g_2 \circ g_1$ and $g_1(A)$ is closed in T_1. Hence we may, and shall, assume that $g(A)$ is closed in T. Then we have $k(Cl_S h(A)) = g(A)$, and hence we can define a map $\phi_o : Cl_S h(A) \to Y$ by

$$\phi_o(s) = \psi_o(k(s)), \quad \text{for } s \in Cl_S h(A).$$

Now, in view of (3) we can apply Theorem 2.4 to the metric space S and the map ϕ_o. Therefore there is a continuous map $\phi : S \to Y$ which is an extension of ϕ_o. Since $\phi \circ h$ is an extension of $\psi_o \circ (g|A) : A \to Y$, $\phi \circ h$ is an extension of f over X.

Thus the proof of Theorem 1.4 is completed.

A modification of the above proof of Theorem 1.4 yields a generalization of the implication (a) $\Rightarrow$ (b) in Theorem 2.3.

Theorem 3.1. If A is M^λ-embedded (resp. P^λ-embedded) in X, if $\dim X/A \le n+1$, and if Y is a metric (resp. complete metric) space of weight $\le \lambda$ which is LC^n, then every continuous map $f : A \to Y$ can be extended over a cozero-set neighborhood U of A in X.

Indeed, the above proof shows that there exist a metric space S of weight $\le \lambda$ and two continuous maps $h : X \to S$, $\phi_o : Cl_S h(A) \to Y$ such that $f(x) = \phi_o(h(x))$ for $x \in A$, and $\dim(S - Cl_S h(A)) \le n+1$. By Theorem 2.3 there exist an open neighborhood V of $Cl_S h(A)$ in S and a continuous map $\phi : V \to Y$ which is an extension of ϕ_o. Since $U = h^{-1}(V)$ is a cozero-set of X, the map $\phi \circ h_o : U \to Y$, where $h_o : U \to V$ is defined by $h_o(u) = h(u)$ for $u \in U$, is a desired extension of f.

4 *Proof of Theorem 1.5*

To prove the "if" part, let us assume that A is M^λ-embedded (resp. P^λ-embedded) in X, that Y is a metric (resp. complete metric) space of weight $\le \lambda$, $\dim X/A \le n$, and that $f : C \to Y$ is a continuous map, where $C = X \times \{0\} \cup A \times I$. Then we have

$$\dim(X \times I)/C \le \dim(X \times I)/(A \times I) \le 1 + \dim X/A \le n+1$$

by Lemmas 2.1 and 2.2.

On the other hand, it follows from Theorems 1.1, 1.3, 1.3' and Definition 1.1' that C is M^λ-embedded (resp. P^λ-embedded) in $X \times I$.

Therefore by Theorem 3.1 there exist a cozero-set neighborhood U of C in $X \times I$ and a continuous map $g : U \to Y$ such that $g(x, t) = f(x, t)$ for $(x, t) \in C$.

Since, as is shown on p.80 of Morita & Hoshina (1975), there is a cozero-set B of X such that

$$A \times I \subset B \times I \subset U,$$

we can proceed similarly as in the argument on p.80 of Morita & Hoshina

(1975). Namely, since A is C-embedded in X and X - B is a zero-set of X disjoint from A, we construct a continuous function $\phi : X \to I$ such that

$$\phi(x) = \begin{cases} 1 & \text{for } x \in A, \\ 0 & \text{for } x \in X - B \end{cases}$$

and define a continuous map $\psi : X \times I \to U$ by $\psi(x, t) = (x, \phi(x)t)$, and finally we construct the composite $h = g\circ\psi : X \times I \to Y$. Then h is a desired extension of f. This proves the "if" part of Theorem 1.5.

Since any ANR is LC^n by Theorem 12.3 of Dugundji (1958), the "only if" part of Theorem 1.5 is a direct consequence of Theorems 1.3 and 1.3'. Thus the proof of Theorem 1.5 is completed.

5 *Remark*

Finally, as another corollary to Theorem 1.4 we have the following theorem, which was proved by Kuratowski (1950) for the case where X and Y are separable metric spaces, and by Morita (1980) for the case $Y = S^{n+1}$.

Theorem 5.1. Let $f, g : X \to Y$ be continuous maps such that $f|A \simeq g|A$. If A is M^λ-embedded (resp. P^λ-embedded) in X, if $\dim X/A \le n$, and if Y is a metric (resp. complete metric) space of weight $\le \lambda$ which is LC^n and C^n, then $f \simeq g$. Here, as usual, by "$\simeq$" we mean "is homotopic to".

Proof. Let us put

$$K = X \times \{0, 1\} \cup A \times I.$$

Then K is P^λ-embedded or M^λ-embedded in $X \times I$ according as A is P^λ-embedded or M^λ-embedded in X; this is seen by Theorem 3.4 of Morita & Hoshina (1975) and by Theorem 4.2 of Sennott (1978 a). By Lemmas 2.1 and 2.2 we have

$$\dim(X \times I)/K \le \dim(X \times I)/(A \times I) = 1 + \dim X/A \le n+1.$$

Therefore, Theorem 5.1 follows readily from Theorem 1.4.

References

Dowker, C. H. (1952). On a theorem of Hanner. Ark. f. Mat., 2, 307-13.
Dowker, C. H. (1956). Homotopy extension theorems. Proc. London Math. Soc., 6 (3), 100-16.

Dugundji, J. (1958). Absolute neighborhood retracts and local connectedness in arbitrary metric spaces. Compositio Math., 13, 229-46.

Engelking, R. (1969). On closed images of the space of irrationals. Proc. Amer. Math. Soc., 21, 583-6.

Hoshina, T. (1977). Remarks on Sennott's M-embedding. Sci. Rep. Tokyo Kyoiku Daigaku, Sect. A, 13, no.380, 284-9.

Kuratowski, C. (1950). Topologie II. Warszawa, Polish Scientific Publishers.

Michael, E. (1955). On a theorem of Kuratowski. Bull. Amer. Math. Soc., 61, (Abstract 631t), p.444.

Morita, K. (1975 a). On generalizations of Borsuk's homotopy extension theorem. Fund. Math., 88, 1-6.

Morita, K. (1975 b). The Hopf extension theorem for topological spaces. Houston J. Math., 1, 121-9.

Morita, K. (1978). Review of Hoshina's paper (1977) referred to above. Zentralblatt für Math., 357, #54010.

Morita, K. (1980). Dimension of general topological spaces. In Surveys in General Topology, ed. G. M. Reed, pp. 297-336. New York: Academic Press.

Morita, K. & Hoshina, T. (1975). C-embedding and the homotopy extension property. Gen. Topology and Appl., 5, 69-81.

Sennott, L. I. (1978 a). On extending continuous functions into a metrizable AE. Gen. Topology and Appl., 8, 219-28.

Sennott, L. I. (1978 b). Some remarks on M-embedding. Topology Proceedings 3, 507-20.

Shapiro, H. L. (1966). Extensions of pseudometrics. Can. J. Math., 18, 981-98.

Stone, A. H. (1962). Non-separable Borel sets. Rozp. Mat., 28.

Special Group Automorphisms and Special Self-Homotopy Equivalences

Peter Hilton

0. Introduction

In 1951 Serre published his famous paper [1] on the homology of fibre spaces, in which he applied Leray's theory of spectral sequences to tremendous effect to obtain powerful new results in homotopy theory. It was Hugh Dowker who drew my attention to this work, whose significance he had quickly appreciated. We were both at that time in Manchester and I learned a great deal from Hugh. Of his own original work, I recall most vividly his example of the topological product of two CW-complexes which failed to have the CW-topology; his beautiful proof of the equivalence of Čech and Vietoris cohomology in his study of the homology groups of relations; and his example of a pathological space whose dimension decreased when a point was judiciously added. I recall, too, that it was Dowker who first noticed Postnikov's work on the analysis of homotopy types and suggested to me that I translate it into English.

In this survey, we look at self-maps of fibre spaces of a special kind and the associated algebraic theory. What we are concerned with is the elucidation of a particular property of finitely generated abelian groups - and hence of simply-connected spaces of finite type - and a generalization to arbitrary abelian groups. This study was initiated by J. M. Cohen [2,3] in 1968, but his definitions contained an omission which invalidated his results; however, it is a pleasure to acknowledge the originality of his idea. Links between abelian group theory and homotopy theory in the study of self-maps are made through the Serre spectral sequence in homology and through the Postnikov and Cartan-Serre-Whitehead decompositions of homotopy type, so that the ideas shared with Hugh Dowker in the early 1950's here find application.

However, we carry the algebraic aspects of this work beyond the domain of abelian groups, since the basic idea is relevant to arbitrary groups. Indeed, this idea is relevant to any algebraic structure in which we may speak of a set of generators. We describe an automorphism ϕ of such a structure G as special or, more precisely, finitary if, for every x in G, there exists a finitely generated substructure K_x, containing x, such that $\phi | K_x$ is an automorphism of K_x. We will not deal in this generality in this survey, but we will discuss finitary automorphisms of arbitrary groups. By generalizing beyond the category of abelian groups, we are able to study spaces which are not simply-connected. In particular, we can deal with homologically nilpotent fibre spaces, that is, fibrations in which the fundamental group of the base acts nilpotently on the homology groups of the fibre. However, we do also pursue the group-theoretical aspects in directions not immediately motivated by applications to homotopy theory.

At the end of the paper we introduce the idea of a special, or finitary, self-homotopy equivalence of a (nilpotent) space, of the homotopy type of a CW-complex, by requiring that it induce special automorphisms in homology or homotopy, and we speculate on possible intrinsic characterizations of such self-equivalences.

Much of the work described in this survey is joint work with Manuel Castellet or Joseph Roitberg.

1. The Basic Definitions

Let $\phi: G \to G$ be a group-endomorphism. We call ϕ a finitary automorphism (fa) if, $\forall x \in G$, $\exists$ fg $K_x(\phi) \subseteq G$, such that $x \in K_x$ and $\phi | K_x : K_x \cong K_x$. Obviously such an endomorphism ϕ is an automorphism; but not all group-automorphisms are finitary.

Counterexamples

1. The automorphism $Q \xrightarrow{2} Q$ of the additive group of rationals is not finitary. Indeed, the only finitary automorphisms of Q are ± 1. This shows that finitary automorphisms tend to be rigid.

2. Let A be any abelian group and let G be the direct sum of countably many copies of A, indexed by the integers. Let $\phi: G \to G$ move each component one place to the right,

$$\begin{array}{ccccccccc} & & A & & A & & A & & \\ \cdots & \to & \bullet & \to & \bullet & \to & \bullet & \to & \cdots \end{array}$$

Plainly, if $A \neq 0$, ϕ is not finitary.

On the positive side, it is plain that all periodic automorphisms are finitary. More generally, we have

Theorem 1.1 Let $\phi: G \to G$ be an endomorphism and let n be a positive integer. Then ϕ is fa if and only if ϕ^n is fa.

Proof Let ϕ be fa. Then $K_x(\phi)$ serves as $K_x(\phi^n)$. Conversely let ϕ^n be fa. Then ϕ is certainly an automorphism and we set $K_x(\phi) = \langle K_x(\phi^n), \phi K_x(\phi^n), \ldots, \phi^{n-1} K_x(\phi^n) \rangle$. It is plain that $K_x(\phi)$ is fg, $x \in K_x(\phi)$, and $\phi | K_x(\phi)$ is an automorphism of $K_x(\phi)$.

Notice, too, that if $\phi: G \to G$ is an automorphism, then ϕ is finitary if and only if ϕ^{-1} is finitary; and that, for any isomorphism $\theta: G \to \bar{G}$, ϕ is finitary if and only if $\theta\phi\theta^{-1}$ is finitary.

We now invoke a stronger notion. Let $x \in G$ and let $0_x = 0_x(\phi)$ be the orbit of x under $\phi : G \to G$, that is, the set $\{\phi^n(x), n \geq 0\}$. Then we define $H_x = H_x(\phi)$ to be the group generated by 0_x; and we further define ϕ to be strongly finitary (sfa) if, for all $x \in G$, H_x is fg and $x \in \phi H_x$. It is plain that sfa $\to$ fa; but it is easy to show by counterexample that fa $\not\to$ sfa. On the other hand, we have

Theorem 1.2 (U. Stammbach) *Let* G *be locally noetherian. Then if* $\phi : G \to G$ *is finitary it is strongly finitary.*

This theorem has for us two principal interests: first, every locally nilpotent group is locally noetherian and, second, if G is locally noetherian, then we can establish that ϕ is sfa by exhibiting a suitable K_x. This is very helpful, since, in fact, our main results concern those ϕ which are sfa.

We point out that the notions of finitary and strongly finitary automorphisms may be applied to other algebraic structures. In particular they may be applied to P-local groups [4,6] and to modules [7]. However, the notion of finite generation must then, of course, be appropriate to the algebraic structure in question.

2. Hereditary Properties

Theorem 2.1 *Let* $\phi : G \to G$ *be* sfa *and let* $G' \subseteq G$ *be a subgroup such that* $\phi G' \subseteq G'$. *If* $\phi' = \phi | G' : G' \to G'$, *then* ϕ' *is* sfa.

Theorem 2.1 admits a converse statement.

Theorem 2.2 Let $\phi : G \to G$ *be an automorphism such that* (a) *for any* $G' \subseteq G$ *with* $\phi G' \subseteq G'$, $\phi | G'$ *is an automorphism of* G' *and* (b) *for any* $G' \subseteq G$ *with* $\phi^{-1} G' \subseteq G'$, $\phi^{-1} | G'$ *is an automorphism of* G', *then* ϕ *is* sfa.

<u>Theorem 2.3</u> <u>Let</u> $G' \to G \to G''$ <u>be exact at</u> G <u>and let</u>

$$\begin{array}{ccccc} G' & \to & G & \to & G'' \\ \downarrow\phi' & & \downarrow\phi & & \downarrow\phi'' \\ G' & \to & G & \to & G'' \end{array}$$

<u>be an endomorphism.</u> <u>If</u> ϕ', ϕ'' <u>are</u> sfa, <u>so is</u> ϕ .

We note that neither Theorem 2.1 nor Theorem 2.3 is true if we replace 'sfa' by 'fa'.

3. Homological Properties

We begin with a result on finitary automorphisms of locally noetherian groups which we will need in this section and which illustrates the usefulness of Theorem 1.2.

<u>Theorem 3.1</u> <u>Let</u> G <u>be locally noetherian and let</u> $\phi, \psi : G \to G$ <u>be commuting fa's.</u> <u>Then</u> $\phi\psi$ <u>is fa</u> .

<u>Proof</u> Since ψ is fa, it follows [6] that, for each $x \in G$, $\langle \psi^n x, n \in \mathbb{Z} \rangle$ is fg. Let $\langle \psi^n x \rangle = \langle y_1, y_2, \ldots, y_k \rangle$. For each i , $1 \le i \le k$, it now follows, since ϕ is fa, that $\langle \phi^m y_i , m \in \mathbb{Z} \rangle$ is fg. Thus

$$\langle \phi^m \psi^n x, m, n \in \mathbb{Z} \rangle = \langle \phi^{m_1} y_1, \ldots, \phi^{m_k} y_k, m_i \in \mathbb{Z} \rangle \qquad (3.1)$$

is fg. Moreover $\langle (\phi\psi)^n x, n \in \mathbb{Z} \rangle = \langle \phi^n \psi^n x, n \in \mathbb{Z} \rangle$, being a subgroup of (3.1), is fg, so $\phi\psi$ is fa.

<u>Remarks</u> (a) The composite of fa's (even if G is abelian) need not be fa. (b) A stronger theorem than Theorem 3.1, replacing the commuting condition by a nilpotency condition, is available [6], but we will not need it in this exposition.

We now consider Λ-modules, where Λ is an arbitrary unitary ring. We will only apply our results here to abelian groups, but the more general case is no more difficult to treat.

<u>Theorem 3.2</u> <u>Let</u> A <u>be a right</u> Λ<u>-module and</u> $\phi : A \to A$ <u>an automorphism</u>

which is finitary with respect to the underlying abelian group structure. Likewise let B *be a left* Λ*-module and* $\psi: B \to B$ *an automorphism which is finitary with respect to the underlying abelian group structure. Then* $\mathrm{Tor}_n^\Lambda(\phi,\psi)$ *is a finitary automorphism of* $\mathrm{Tor}_n^\Lambda(A,B)$, $n \geq 0$.

Proof We first handle the case $n = 0$. For each $a \in A$, $b \in B$, we have fg subgroups K_a, L_b of A,B, such that $a \in K_a$, $\phi|K_a : K_a \cong K_a$, and $b \in L_b$, $\psi|L_b : L_b \cong L_b$. Consider the diagram

$$\begin{array}{ccc} K_a \otimes L_b & \xrightarrow{i} & A \otimes_\Lambda B \\ \Big\downarrow{\scriptstyle \phi|K_a \otimes \psi|L_b} & & \Big\downarrow{\scriptstyle \phi \otimes_\Lambda \psi} \\ K_a \otimes L_b & \xrightarrow{i} & A \otimes_\Lambda B \end{array}$$

where i is induced by the obvious embeddings. Since the vertical arrows are automorphisms, the restriction of $\phi \otimes_\Lambda \psi$ is an automorphism of im i , which is fg. Write $K_{a\otimes b} = \mathrm{im}\ i$. Now consider an arbitrary element $x = \sum_{j=1}^{k} n_j(a_j \otimes b_j)$ of $A \otimes_\Lambda B$. Set

$$K_x = \langle K_{a_j \otimes b_j}\ ,\ j = 1,\dots,k \rangle .$$

Then K_x is fg, $x \in K_x$, and $(\phi \otimes_\Lambda \psi)|K_x$ is an automorphism of K_x . Thus $\phi \otimes_\Lambda \psi$ is fa.

We now suppose $n \geq 1$. Let $P \xrightarrow{\varepsilon} A$ be a projective resolution of A . Then $1 \otimes \psi : P \otimes_\Lambda B \to P \otimes_\Lambda B$ is finitary in each dimension and $\mathrm{Tor}_n^\Lambda(A,B) = H_n(P \otimes_\Lambda B)$. It follows from Theorem 2.1 that $\mathrm{Tor}_n^\Lambda(1,\psi): \mathrm{Tor}_n^\Lambda(A,B) \to \mathrm{Tor}_n^\Lambda(A,B)$ is finitary. Likewise let $Q \xrightarrow{\eta} B$ be a projective resolution of B ; then we infer that $\mathrm{Tor}_n^\Lambda(\phi,1): \mathrm{Tor}_n^\Lambda(A,B) \to \mathrm{Tor}_n^\Lambda(A,B)$ is finitary. Now $\mathrm{Tor}_n^\Lambda(1,\psi)$ and $\mathrm{Tor}_n^\Lambda(\phi,1)$ commute and their composite is $\mathrm{Tor}_n^\Lambda(\phi,\psi)$. Thus it follows from Theorem 3.1 that $\mathrm{Tor}_n^\Lambda(\phi,\psi)$ is finitary.

We include for completeness the following result from [5].

Theorem 3.3 *Let* $\phi:G\to G$ *be a finitary automorphism of the locally nilpotent group* G *. Then* $H_n\phi:H_nG\to H_nG$ *is finitary.*

We remark that, as proved in [8], we cannot take G arbitrary in Theorem 3.3. We do not know the situation if ϕ is strongly finitary.

4. An application in Group Theory

We will consider a group G which is given as the product of two of its subgroups, K and L , thus $G = \langle K,L\rangle$. We further suppose given an endomorphism $\phi:G,K,L \to G,K,L$. It is easy to construct an example when $\phi|K$, $\phi|L$ are automorphisms whereas ϕ is not an automorphism; indeed such an example can be constructed with G abelian. Thus the following result is significant.

Theorem 4.1 *Suppose that* $\phi|K$ *is an automorphism and* $\phi|H$ *is sfa. Then, provided that* H *or* K *is normal,* ϕ *is an automorphism which is sfa if and only if* $\phi|K$ *is sfa.*

Proof Suppose first that H is normal. We have

$$G/H \cong K/H\cap K$$

Since $\phi|H$ is sfa, it follows from Theorem 2.1 that $\phi|H\cap K$ is sfa. Since $\phi|K$ is an automorphism, so is the induced map on $K/H\cap K$ and hence on G/H. Since $\phi|H$ is an automorphism, so is ϕ . Of course, if ϕ is sfa, so is $\phi|K$. Conversely, if $\phi|K$ is sfa, then, by Theorem 2.1, ϕ induces an sfa on $K/H\cap K$ and hence on G/H . Since $\phi|H$ is sfa it follows from Theorem 2.3 that ϕ is sfa.

Now suppose that K is normal. We have

$$G/K \cong H/H\cap K$$

Since $\phi|H$ is sfa, ϕ induces an sfa on $H/H\cap K$ and hence on G/K . Since $\phi|K$ is an automorphism, so is ϕ , and, by Theorem 2.1 and 2.3, ϕ is sfa if and only if $\phi|K$ is sfa.

The question naturally arises as to whether it is necessary to assume H or K normal. As shown in [4], this condition can be dispensed with if G is locally nilpotent, provided that we assume both $\phi|K$ and $\phi|L$ to be finitary. Indeed, the following results are proved in [4]; especially relevant is Theorem 4.3.

Theorem 4.2 *Let* G *be nilpotent and* $\phi: G \to G$. *Then* ϕ *is* fa *if and only if* ϕ_{ab} *is* fa.

Theorem 4.3 *Let* G *be locally nilpotent,* $G = \langle K, L \rangle$, $\phi: G,K,L \to G,K,L$. *If* $\phi|K$ *and* $\phi|L$ *are* fa, *so is* ϕ.

Theorem 4.4 *Let* G *be locally nilpotent, generated by* S. *Let* $\phi: G \to G$ *be such that, for each* $s \in S$, $\exists$ fg $K_s \subseteq G$, *such that* $s \in K_s$ *and* $\phi|K_s: K_s \cong K_s$. *Then* ϕ *is a finitary automorphism.*

We conjecture that the conclusions of Theorem 4.1 would remain valid if we assumed G nilpotent, but no longer required H or K to be normal.

5. Applications in Homotopy Theory

We consider first a fibration $\mathfrak{F}$,

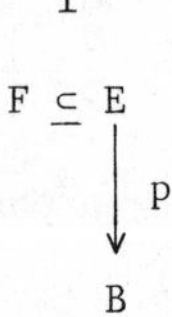

and a self-map f from $\mathfrak{F}$ to itself. All spaces are assumed path-connected, and the fibration is assumed to be *homologically nilpotent*, in the sense that $\pi_1 B$ acts nilpotently on $H_n F$, for all n. We proved in [5], using homological properties of fa's,

Theorem 5.1 (i) *If* $f_*: H_* B \to H_* B$ *and* $f_*: i_* H_* F \to i_* H_* F$ *are finitary automorphisms, then* $f_*: H_* F \to H_* F$ *and* $f_*: H_* E \to H_* E$ *are finitary automorphisms.*

(ii) *If* $f_*:H_*F \to H_*F$ *and* $f_*:p_*H_*E \to p_*H_*E$ *are finitary automorphisms, then* $f_*H_*B \to H_*B$ *and* $f_*:H_*E \to H_*E$ *are finitary automorphisms*.

We now describe a second application, to the fibration $\mathfrak{F}$, of the notion of finitary automorphisms. We describe the subgroup $p_{**}H_n(E,F)$ of H_nB as the subgroup of *transgressive elements* in H_nB, and then the boundary homomorphism in the homology sequence of the pair (E,G) induces a homomorphism t from the subgroup of transgressive elements of H_nB to a quotient group of $H_{n-1}F$. The homomorphism t is called the (homology) *transgression*, in dimension n, associated with the fibration $\mathfrak{F}$. Note that the range of t is $H_{n-1}F/i_*(\ker p_{**})$.

We now prove

Theorem 5.2 If $f_*:H_*E \to H_*E$ *is a finitary automorphism, and if* f_* *is also a finitary automorphism on the image of the transgression, then* $f_*:H_*F \to H_*F$ *and* $f_*:H_*B \to H_*B$ *are finitary automorphisms.*

Proof It follows from Theorems 4.1 and 4.2 of [5] that if, in the Serre spectral homology sequence for $\mathfrak{F}$, f_* is a finitary automorphism on E^2_{pq}, for $p < n$, $q = 0$, and for $p = 0$, $q < n$, then f_* is a finitary automorphism on E^2_{pq} for $p < n$, $q < n$. We will use this remark to prove, inductively with respect to n, that f_* is a finitary automorphism on E^2_{pq} for $p < n$, $q < n$. A second application of Theorem 4.2 of [5] will then establish the theorem.

Now we know that f_* is a finitary automorphism on the image of $d^n:E^n_{n0} \to E^n_{0,n-1}$. We also know that f_* is fa on each H_nE and hence on E^∞_{pq} for all pairs (p,q). Now the sequence

$$E^\infty_{n0} \rightarrowtail E^n_{n0} \xrightarrow{d^n} E^n_{0,n-1} \twoheadrightarrow E^\infty_{0,n-1}$$

is exact, so we infer, by heredity, that f_* is fa on E^n_{n0} and

$E^n_{0,n-1}$, for all n. Now $H_1B = E^2_{10} = E^\infty_{10}$ so f_* is fa on H_1B. Also f_* is fa on E^2_{01} and so it follows that f_* is fa on E^2_{pq} for $p<2$, $q<2$. Assume then that f_* is fa on E^2_{pq} for $p<n$, $q<n$ with $n \geq 2$. To achieve the inductive step it now suffices to show that f_* is fa on E^2_{n0} and E^2_{0n}. Consider the exact sequence

$$E^i_{n0} \rightarrowtail E^{i-1}_{n0} \xrightarrow{d^{i-1}} E^{i-1}_{n-i+1,i-2} \quad , \qquad 3 \leq i \leq n \quad .$$

We know that f_* is fa on $E^{i-1}_{n-i+1,i-2}$. Thus if we assume inductively (with respect to i) that f_* is fa on E^i_{n0}, it follows that f_* is fa on E^{i-1}_{n0}. Since f_* is fa on E^n_{n0}, we conclude that f_* is fa on E^2_{n0}.

Now consider the exact sequence

$$E^{i-1}_{i-1,n-i+2} \xrightarrow{d^{i-1}} E^{i-1}_{0,n} \twoheadrightarrow E^i_{0,n} \quad , \quad 3 \leq i \leq n+1$$

We know that f_* is fa on $E^{i-1}_{i-1,n-i+2}$; this follows, in particular, for $i=n+1$ from the already proven fact that f_* is fa on E^2_{n0}, hence on E^2_{n1} and E^n_{n1}. Thus if we assume inductively (with respect to i) that f_* is fa on E^i_{0n}, it follows that f_* is fa on E^{i-1}_{0n}. Since f_* is fa on E^{n+1}_{0n}, we conclude that f_* is fa on E^2_{0n}, so that the theorem is completely proved.

<u>Remarks</u> (i) The proof shows that it would have sufficed to assume, in addition to the transgression hypothesis, that f_* is fa on E^∞_{n0} and $E^\infty_{0,n-1}$, for all n. We would then conclude that f_* is also fa on H_*E.

(ii) There are obvious generalizations in which the assumptions on f_* are only made up to certain dimensions.

(iii) If we assume that E is homologically trivial, then we are left with the transgression hypothesis. Now, under this hypothesis on E, there is a <u>suspension</u> homomorphism $s: H_{n-1}F \to H_nB$ and the transgression is just the relation inverse to s . Thus the image of the transgression may be identified with the image of the suspension, and we conclude that, if E is homologically trivial, f_* is fa on H_*F and H_*B if (and only if) it is fa on the image of the suspension. This result is interesting if E is contractible. Then we conclude that, for a map $f:B \to B$, of a simply-connected space B , f_* is fa on H_*B (and on $H_* \Omega B$) if, and only if, f_* is fa on the image of the suspension homomorphism $s:H_{n-1} \Omega B \to H_n B$, for all n .

A further application to homotopy theory was made in [5], namely

<u>Theorem 5.3</u> <u>Let</u> $f:X \to X$ <u>be a self-map of the nilpotent space</u> X . <u>Then the following conditions are equivalent</u>:

(i) $f_*:\pi_* X \to \pi_* X$ <u>is</u> fa;

(ii) $f_*:H_* X \to H_* X$ <u>is</u> fa;

(iii) $f_*:\pi_1 X \to \pi_1 X$ <u>and</u> $f_*:H_* X \to H_* \tilde{X}$ <u>are</u> fa.

We note that, by Remark (iii) above, it suffices, at least when X is simply-connected, to require that f_* be fa on the image of the suspension homomorphism $H_{n-1} \Omega X \to H_n X$, for all n .

These results plainly suggest that we should describe a self-homotopy equivalence f of a nilpotent space X as <u>finitary</u> if it enjoys any of the equivalent properties of Theorem 5.3; there would be a similar notion for a self-fibre-homotopy equivalence. If X is the directed union of a family of compact subspaces X_α and if $f:X \to X$ is a self-map such that $fX_\alpha \subseteq X_\alpha$ and $f|X_\alpha$ is a homology equivalence for each α , then f is certainly a finitary self-homotopy equivalence of the nilpotent space X. It would surely be interesting to consider other conditions on f ensuring that it is a finitary self-homotopy equivalence. We note that it would certainly suffice that the spaces X_α above have finitely generated homology in each dimension.

Bibliography

1. J. P. Serre, Homologie singulière des espaces fibrés, Ann. of Math. 54(1951), 425-505.

2. J. M. Cohen, A spectral sequence automorphism theorem, Topology 7(1968), 173-177.

3. J. M. Cohen, Clarification to a result, Topology 9(1970), 299-300.

4. P. J. Hilton and J. Roitberg, On pseudo-identities I, Archiv der Mathematik, 41(1983), 204-214.

5. M. Castellet, P. J. Hilton, and J. Roitberg, On pseudo-identities II, Archiv der Mathematik. (to appear).

6. P. J. Hilton, On special group automorphisms and their composition, Can. Jour. Math. (to appear).

7. P. J. Hilton, A note on finitary automorphisms, Aspects of Mathematics and its Applications, North Holland (1984).

8. J. Roitberg, Finitary automorphisms and integral homology, Topological Topics, London Mathematical Society Lecture Notes 86(1983), 164-168.

RATIONAL HOMOTOPY AND TORUS ACTIONS

Stephen Halperin
Department of Mathematics, Scarborough College,
University of Toronto,
1265 Military Trail,
West Hill, Canada M1C 1A4

1. *INTRODUCTION*

The *toral rank* of a space M is the largest integer r such that an r-torus can act continuously on M with all its isotropy subgroups finite. In particular, the toral rank is zero if and only if every circle action on M has a fixed point.

This paper deals with the problem of bounding toral rank from above. For technical reasons we consider only spaces of finite cohomological dimension over $\mathbb{Q}$ - cf.[Q] - which are paracompact, Hausdorff, connected, locally path connected, semi locally one connected and satisfy $\varinjlim H^*(U(x);\mathbb{Q}) = \mathbb{Q}$, $x \in M$ - the limit being taken over all neighbourhoods of x. (Here, and throughout, cohomology is singular.) To simplify the theorems we shall refer to such spaces as *reasonable*.

Now for given M its toral rank may be bounded for reasons depending only on its local geometry. For instance, the wedge of two CW complexes always has toral rank zero, since the base point has to be fixed by any circle action. Here we want to find bounds which depend only on the global "algebraic topology" of M; more specifically which depend on its rational homotopy type. Then (for example) wedges can no longer be ruled out since if one pulls back the universal S^1-bundle to $S^2 \vee S^3$ via $S^2 \vee S^3 \xrightarrow{\text{collapse } S^3} S^2 \xrightarrow{\text{classify}} CP^\infty$ one gets a four dimensional CW complex admitting a free circle action, and having the rational homotopy type of $S^2 \vee S^3 \vee S^4$.

A positive result, conjectured by W.-Y. Hsiang and established in [A-H], asserts that for simple connected reasonable spaces, M, satisfying:

$$\dim H^*(M;\mathbb{Q}) < \infty \quad \text{and} \dim \pi_*(M) \otimes \mathbb{Q} < \infty$$

we have

$$\text{toral rank}(M) \le -\Sigma(-1)^i \dim \pi_i(M) \otimes \mathbb{Q}.$$

A small modification (as we shall see below) extends this to the non-simply connected case under (apparently) weaker hypotheses.

Indeed, let $Z_1(M)$ denote the subgroup in the centre of $\pi_1(M)$ of the elements which act trivially on each $\pi_i(M)$, $i\geq 2$. Then we prove

1.1 Theorem. Suppose M is reasonable and that for some m, $\pi_i(M)\otimes\mathbb{Q}=0$, $i>m$. Then $Z_1(M)\otimes\mathbb{Q}$ and each $\pi_i(M)\otimes\mathbb{Q}$ $(i\geq 2)$ are finite dimensional, and

$$\text{toral rank}(M) \leq \dim Z_1(M)\otimes\mathbb{Q} - \sum_{i\geq 2}(-1)^i \dim \pi_i(M)\otimes\mathbb{Q}.$$

When M is a homogeneous space we get equality in 1.1 ([A-H]). In general equality fails: we shall give (example 4.4) a rational homotopy type (containing a closed 57-dimensional manifold) such that every reasonable space in that rational homotopy type has toral rank zero. On the other hand, this homotopy type satisfies the second hypothesis of (1.1) and the bound predicted by the alternating sum is seven. This raises the

1.2 Problem. Is there a formula refining that of 1.1 which is sharp in each rational homotopy type; i.e. which always gives the least toral rank for M as M varies over the reasonable spaces (satisfying the hypotheses of (1.1)) in a given rational homotopy type?

In another direction one might ask the

1.3 Problem: Is there a good rational invariant i(M), defined for all reasonable spaces, majorizing the toral rank, and majorizing $\dim Z_1(M)\otimes\mathbb{Q} - \sum_{i\geq 2}\dim \pi_i(M)\otimes\mathbb{Q}$ when this is defined?

Of course $cd_{\mathbb{Q}}(M)$ itself is such an invariant; one would like to do better. One obvious suggestion is the rational Lusternik-Schnirelmann category (which does majorize the alternating sum - [F-H]); unfortunately it does not, in general, majorize the toral rank. (Consider the fibre of $\phi:N\to\mathbb{C}P^\infty\times\ldots\times\mathbb{C}P^\infty$ (k factors), where N is the connected sum of k copies of $\mathbb{C}P^2$, and ϕ is the classifying map.)

Another possible choice as a solution to 1.3 is raised by the next question (which first occurred to me in 1968):

1.4 Problem. Let r be the toral rank of a simply connected reasonable space, M. Is it true that

$$\dim H^*(M);\mathbb{Q}) \geq 2^r?$$

It is amusing to note that the answer to 1.4 is yes when M is a homogeneous space, and that this is a trivial consequence of [A-H] :

1.5 Proposition. If $K\subset G$ are compact connected Lie groups and

r is the toral rank of G/K then $\dim H^*(G/K;\mathbb{Q}) \geq 2^r$.

Proof: By [A-H] $r = \text{rank } G - \text{rank } K$ and so the Serre spectral sequence for $K \to G \to G/K$ gives : $2^r = \dim H^*(G;\mathbb{Q})/\dim H^*(K;\mathbb{Q}) \geq \dim H^*(G/K;\mathbb{Q})$.

□

Another interesting question (again due to W.-Y. Hsiang) is the

1.6 Problem. Must a circle acting on a non trivial connected sum of closed manifolds have a fixed point?

This question was recently answered in the negative by Aubry and Lemaire who construct a free circle action on $(S^2 \times S^3)\#(S^2 \times S^3)$. Under some additional hypotheses, however, the answer to 1.6 is yes, as shown by the main result of this article:

1.7 Theorem. Let M_i be a finite collection of closed simply connected manifolds whose rational homotopy is concentrated in finitely many degrees, all of which are odd. If at least two of the M_i satisfy $\dim \pi_*(M_i) \otimes \mathbb{Q} \geq 2$ then every circle action on the connected sum of the M_i has a fixed point.

1.8 Remark. Among homogeneous spaces satisfying the conditions on the M_i are Lie groups and Stiefel manifolds.

1.9 Example. On the 9-manifold $(S^3 \times S^3 \times S^3)\#(S^3 \times S^3 \times S^3)$ every circle action has a fixed point.

In the process of proving Theorem 1.7 we establish

1.10 Theorem. Let M be a simple connected reasonable space for which $\dim H^*(M;\mathbb{Q})$ is finite and the homotopy Lie algebra $L_M = \pi_*(\Omega M) \otimes \mathbb{Q}$ (Samelson product) is concentrated in even degrees. Assume L_M has no centre. Then every circle action on M has a fixed point.

2. *THE HOMOTOPY EULER CHARACTERISTIC*

Our principal objective here is Theorem 1.1. The following result and proof are due to Y. Felix (unpublished).

2.1 Theorem (Y. Felix). Let $\phi: X \to Y$ be a continuous map of simply connected CW complexes such that $\phi_\#: \pi_*(X) \otimes \mathbb{Q} \to \pi_*(Y) \otimes \mathbb{Q}$ is injective. Then the Lusternik-Schnirelmann category of the rationalizations satisfies

$$\text{cat}(X_{\mathbb{Q}}) \leq \text{cat}(Y_{\mathbb{Q}}).$$

2.2 Remark: When X and Y have finite type this is proved in [F-H].

Proof: We may assume $X = X_{\mathbb{Q}}$, $Y = Y_{\mathbb{Q}}$. Let F be the homotopy fibre of ϕ and consider the sequence $\Omega Y \xrightarrow{\psi} F \xrightarrow{i} X \xrightarrow{\phi} Y$. By hypothesis $\Omega Y \to F$ induces a surjection on homotopy groups.

On the other hand, the Cartan-Serre theorem implies that for each k, $\pi_k(\Omega Y) \to H_k(\Omega Y)$ is a linear injection (of rational vector spaces). A right inverse for this splitting is an element of $H^k(\Omega Y;\pi_k(\Omega Y))$ and so defines $f_k:\Omega Y \to K(\pi_k(\Omega Y);k)$. The product, $\prod_k f_k$ is a homotopy equivalence.

Now write $\pi_k(\Omega Y)=A_k \oplus B_k$ with B_k mapped isomorphically to $\pi_k(F)$. (Recall all these groups are rational vector spaces.) Then $\Omega Y \simeq \prod_k (K(A_k;k)) \times \prod_k (K(B_k;k))$ and the restriction of ψ to the second factor is a homotopy equivalence. Thus we get $\sigma:F \to \Omega Y$ with $\psi\sigma \sim id$. Hence $i \sim (i\psi)\sigma$ is null homotopic.

It follows that when ϕ is made into a fibration the inclusion of the fibre is null homotopic. This implies the theorem.

□

2.3 *Corollary*: If M is a simply connected space with $cd_{\mathbb{Q}} M = n < \infty$ and $\pi_i(M) \otimes \mathbb{Q} = 0$, $i > m$ (some m) then $\dim \pi_i(M) \otimes \mathbb{Q} = 0$ for all i.

Proof: The singular complex M' of M satisfies $H^i(M';\mathbb{Q})=0$, $i>n$ and $\pi_i(M') \otimes \mathbb{Q}=0$, $i>m$. In particular we may replace M' by an n-dimensional, 1-connected CW complex, N, with the same rational homotopy groups.

Suppose by descending induction that $\dim \pi_i(N) \otimes \mathbb{Q}$ is finite for $i>k$. Via a Postnikov decomposition for N it is easy to construct $\phi:N' \to N$ such that N' is (k-1)-connected, $\phi_{\#}:\pi_i(N') \to \pi_i(N)$ is an isomorphism for $i>k$ and injective for $i=k$, and $\dim \pi_k(N') \otimes \mathbb{Q} = r$, r any finite integer $\le \dim \pi_k(N) \otimes \mathbb{Q}$.

But then $\dim \pi_*(N') \otimes \mathbb{Q} < \infty$. By Felix's theorem, $cat(N'_{\mathbb{Q}}) \le cat(N_{\mathbb{Q}}) \le n$; hence according to [F-H] and [Ha$_1$]

$$\dim \pi_{even}(N') \otimes \mathbb{Q} \le \dim \pi_{odd}(N') \otimes \mathbb{Q} \le n.$$

In particular we must have $r \le n$; i.e. $\dim \pi_k(N) \otimes \mathbb{Q} \le n$.

□

Proof of 1.1: Corollary 2.3, applied to the universal cover, $\tilde{M}$, of M, shows that $\dim \pi_i(M) \otimes \mathbb{Q}$ is finite for $i \ge 2$. Let $\alpha_1,\ldots,\alpha_s \in Z_1(M)$ be linearly independent in $Z_1(M) \otimes \mathbb{Q}$ and let M' be the cover of M with $\pi_1(M') = \bigoplus_i \mathbb{Z} \cdot \alpha_i$. Then $\tilde{M} \to M' \to T^s$ (s-torus) is an oriented fibration. If n_1 and n' are the maximum degrees of non vanishing rational homology for $\tilde{M}$ and M' we then have $n \ge n' = n_1 + s$. It follows that $s \le n$ and so $\dim Z_1(M) \otimes \mathbb{Q} \le n$.

Now consider the action of an r torus T^r on M with only finite isotropy subgroups. The image of $\pi_1(T^r) \to \pi_1(M)$ is contained in $Z_1(M)$ and is a subgroup of the form $\bigoplus_{i=1}^{s} \mathbb{Z} \cdot \alpha_i \oplus$ torsion - the α_i are linearly independent in $Z_1(M) \otimes \mathbb{Q}$ and so $s \le \dim Z_1(M) \otimes \mathbb{Q}$. On the other hand we can find an (r-s)-torus $T^{r-s} \subset T^r$ such that $\pi_1(T^{r-s}) \to \pi_1(M)$ has finite image. A finite cover of T^{r-s} then acts on $\tilde{M}$. Since $\dim \pi_*(\tilde{M}) \otimes \mathbb{Q} < \infty$ and $cd_{\mathbb{Q}}(\tilde{M}) < \infty$ it follows that $\dim H_*(\tilde{M};\mathbb{Q}) < \infty$. Now [A-H] shows that $r-s \leq -\sum_{i\geq 2} (-1)^i \dim \pi_i(M) \otimes \mathbb{Q}$.

□

3. MINIMAL MODELS

We work over $\mathbb{Q}$ as ground field; graded vector spaces are supposed concentrated in degrees ≥ 0; graded algebras are called *commutative* if $ab = (-1)^{(\deg a)(\deg b)} ba$. A commutative graded differential algebra (cgda) is a cga equipped with a derivation d (the *differential*) of degree 1 satisfying $d^2 = 0$. A *quism* is a morphism of cgda's inducing a cohomology isomorphism.

If X is a graded space then ΛX=exterior algebra $(X^{odd}) \otimes$ symmetric algebra (X^{even}) denotes the free cga over X. A *KS complex* is a cgda of the form $(\Lambda X, d)$ in which X admits a well ordered basis x_α such that dx_α is in the subalgebra generated by the x_β, $\beta < \alpha$. It is *minimal* if the x_α can be chosen so that $\beta < \alpha$ implies $\deg x_\beta \leq \deg x_\alpha$.

A quism $\Lambda X \xrightarrow{\cong} A$ from a KS complex is called a *Sullivan model;* If $H^0(A) = \mathbb{Q}$ then A admits a minimal Sullivan model, unique up to isomorphism ([B-G], [Ha_2], [Su]). Now Sullivan (same references) constructs a functor A_{PL}:spaces $\to$ cgda's such that $H(M;\mathbb{Q}) \cong H(A_{PL}(M))$; if M is path connected we can construct the minimal model of $A_{PL}(M)$, called the *minimal model of M.* We say M and N nave the *same rational homotopy type* if their minimal models are isomorphic: this coincides with the classical definition if M, N are simply connected with finite rational betti numbers - cf.[B-G].

Suppose next that $(\Lambda X, d)$ is a *simply connected* $(X^0 = X^1 = 0)$ minimal KS complex of *finite type* $(\dim(\Lambda X)^i < \infty$, each i). The algebra ΛX can be graded by the spaces $\Lambda^p X$ spanned by elements of the form $x_1 \wedge \ldots \wedge x_p$, $x_i \in X$. Then $d = \sum_{i\geq 2} d_i$ with d_i a derivation mapping $X \to \Lambda^i X$. A graded Lie algebra $L = \oplus_p L_p$ is then defined by $L_p = \mathrm{Hom}(X^{p+1};\mathbb{Q})$ and

$$\langle x, [\alpha,\beta]\rangle = \pm \langle d_2 x; \alpha, \beta\rangle, \quad x \in X, \alpha, \beta \in L. \tag{3.1}$$

(For the exact signs see [Q] or [Ta].)

When $(\Lambda X, d)$ is the minimal model of a simply connected space M and $\dim H^i(M;\mathbb{Q}) < \infty$, each i, then L is isomorphic ([A-A]) with the homotopy Lie algebra $L_M = \pi_*(\Omega M) \otimes \mathbb{Q}$ equipped with the Samelson product.

In particular, let M and N be simply connected closed n-manifolds. Filling in the "connected tube" of the connected sum M#N with a disk defines a continuous map $M\#N \to M \vee N$. From this we deduce a homomorphism of homotopy Lie algebras, and in [H-L] it is shown that this gives a short exact sequence ($\amalg$ denoting coproduct or "free product")

$$0 \to I \to L_{M\#N} \xrightarrow{\rho} L_M \amalg L_N \to 0. \tag{3.2}$$

3.3 Lemma. (i) If $L_{M\#N}$ is filtered by the I-adic filtration then the associated bigraded Lie algebra has the form

$$\mathbb{L}(\sigma) \amalg L_M \amalg L_N ,$$

$\mathbb{L}(\sigma)$ denoting the free graded Lie algebra on a single generator, σ, of degree n-2.

(ii) ρ is an isomorphism in degrees $<n-2$.

(iii) If $\alpha, \beta \in L_{M\#N}$ are linearly independent commuting elements of even degree in $L_M \amalg L_N$ and if $\rho\beta$ is a non zero element of L_M then $\rho\alpha$ is also a non zero element of L_M.

Proof: (i) is proved in [H-L] and implies (ii) and (iii) easily.

□

4. THE BOREL FIBRATION

If an r torus T acts on a space M the action is studied via the fibration $M \to M_T = E_T \times_T M \to B_T$ introduced by Borel. It leads ([A-H]) to a sequence

$$(\Lambda(a_1,\ldots,a_r),0) \to (\Lambda(a_1,\ldots,a_r) \otimes \Lambda X_M, D) \to (\Lambda X_M, d) \tag{4.1}$$

of morphisms of KS complexes in which the KS complexes are respectively a minimal Sullivan model for B_T, a Sullivan model for M_T and a minimal Sullivan model for M. (In particular, deg a_i=2.) Note that if $X_M^1=0$ then the model for M_T is also minimal.

4.2 Proposition. Suppose M is a simply connected reasonable space with $\dim H^*(M;\mathbb{Q})$ finite. If an r-torus acts on M with only finite isotropy subgroups then the model (4.1) of the Borel fibration satisfies $\dim H(\Lambda(a_1,\ldots,a_r) \otimes \Lambda X_M) < \infty$.

Conversely if there exists a sequence of KS complexes of the form (4.1), ending in the Sullivan model of M, then there is a simple connected finite CW complex M' admitting a free r-torus action and such that M' has the same rational homotopy type as M.

4.3 Remark. If we define the toral rank of a rational homotopy type to be the maximum toral rank of simply connected reasonable spaces in that rational homotopy type then (4.2) gives an algebraic means of computing it.

Proof of 4.2. The first assertion follows (see [A-H]) from Hsiang's theorem that $\dim H(\Lambda(a_1,\ldots,a_r)\otimes\Lambda X_M)=\dim H^*(M_T;\mathbb{Q})<\infty$. For the second assertion, let N be a simply connected finite CW complex with Sullivan model $(\Lambda(a_1,\ldots,a_r)\otimes\Lambda X_M,D)$: the existence of N follows from the fact that $\dim H(\Lambda(a_1,\ldots,a_r)\otimes\Lambda X_M)<\infty$ - cf. [Ta]. The cocycles a_i then define elements $\alpha_i\in H^2(N;\mathbb{Q})$. Multiply by integers to get $k_i\alpha_i\in H^2(N;\mathbb{Z})$ and use these classes to define $\phi:N\to K(\mathbb{Z}^r;2)$. Pull back the universal T^r fibration to N to get M'.

□

4.4 Example. Let $X=X^3\oplus X^8\oplus X^{13}\oplus X^{15}$ with bases: $X^3:u_i(1\le i\le 5)$; $X^8:b_1,b_2$; $X^{13}:x_1,x_2$; $X^{15}:y_1,y_2$. Define a differential d in ΛX by

$$du_i=0,\ db_1=u_1u_2u_5,\ db_2=u_3u_4u_5,\ dx_1=u_1u_2b_1,$$
$$dx_2=u_3u_4b_2,\ dy_1=b_1^2-2x_1u_5,\ dy_2=b_2^2-2x_2u_5.$$

A theorem of Barge-Sullivan [B] asserts that $(\Lambda X,d)$ is the minimal model of a closed 57-manifold.

On the other hand a brute force calculation shows that in any sequence of the form (4.1) with r=1 and $(\Lambda X_M,d)=(\Lambda X,d)$ we must have $H(\Lambda a_1\otimes\Lambda X_M,D)$ of infinite dimension. Thus every circle action on a reasonable space in the rational homotopy type $(\Lambda X,d)$ has a fixed point.

□

5. *ANALYSIS OF $(\Lambda a\otimes\Lambda X,D)$*

Let $(\Lambda a\otimes\Lambda X,D)$ be a KS complex with $|a|$(=dega) even and >0, and Da=0; suppose the KS complex $(\Lambda X,d)$ obtained by setting a=0 is minimal, and satisfies $X^0=0$. We have just seen that these are central to the study of circle actions.

Filtering by the degree of a gives a spectral sequence $(E_i^{p,q}, d_i)$ with E_2-term $\Lambda a\otimes H(\Lambda X)$. It corresponds to the Serre spectral sequence for the Borel fibration. In that context the next lemma is standard; it follows by an easy induction in any case.

5.1 Lemma. Multiplication by a : $E_i^{p,q} \to E_i^{p+|a|,q}$ is surjective, and an isomorphism for $p \geq i$.

□

Of course $E_i^{p,*}=0$ unless p is multiple of $|a|$. The kernel of multiplication by $a^k : E_i^{0,*} \to E_i^{k|a|,*}$ is a graded ideal I_i, independent of k for large k. Denote $E_i^{0,*}/I_i$ by H_i. Then

$$E_i^{\geq i,*} \cong \left(\bigoplus_{j \geq i/|a|} a^j\right) \otimes H_i \tag{5.2}$$

as bigraded Λa-algebras.

If i+1 is not a multiple of $|a|$, then $d_i=0$. If $i+1=k|a|$ then in $E_i^{\geq i,*}$, d_i has the form $a^k \otimes \delta_i$ where δ_i is a derivation in H_i of degree-i and square zero. In the context of Borel fibrations the next lemma is also standard.

5.3 Lemma. (i) $H_{i+1}=H(H_i,\delta_i)$ if i+1 is a multiple of $|a|$; otherwise $H_{i+1}=H_i$.

(ii) If $H_i \neq 0$ then $H_{i+1}=0$ if and only if $i+1=k|a|$ and $a^k \in \mathrm{Im}\, d_i$.

(iii) Suppose $H(\Lambda X,d)$ is a Poincaré duality algebra with fundamental class ω of degree n. If $H_i \neq 0$ then ω survives as a cocycle until E_i, and H_i is a Poincaré duality algebra with fundamental class ω.

Proof: (i) and (ii) are obvious. Suppose by induction that (iii) holds for i-1. It is sufficient to show that $\delta_{i-1}\omega=0$ if $H_i \neq 0$. But if $\delta_{i-1}\omega \neq 0$ choose α so $(\delta_{i-1}\omega)\alpha=\omega$ in H_{i-1}. Then $\omega \cdot \delta_{i-1}\alpha = \pm(\delta_{i-1}\omega)\cdot\alpha \pm \delta_{i-1}(\omega\alpha)$; since $\omega\alpha=0$ we find $\delta_{i-1}\omega=\pm 1$ and $H_i=0$.

□

The following lemma is proved in [F-H]:

5.4 Lemma. If $H(\Lambda a \otimes \Lambda X, D)$ is finite dimensional so is $H(\Lambda X,d)$. The maximum degrees n' and n in which these cohomology algebras are non zero are then related by $n'=n-1$. If one algebra satisfies Poincaré duality so does the other.

□

Suppose now that $H(\Lambda a \otimes \Lambda X, D)$ is finite dimensional. Some power of a is then a coboundary, and it follows that for some $x \in X$, $Dx=a^{k+1}+\Omega$, $\Omega \in \Lambda a \otimes (\Lambda X)^+$. Minimize k. By modifying a basis, x_i, of X to elements of the form $x_i + a\omega_i$ we can arrange that $X=(u) \oplus U$ with U generating a D-stable ideal in $\Lambda a \otimes \Lambda X$ and

$$Du = a^{k+1}+\Gamma, \quad \Gamma \in \Lambda a \otimes (\Lambda U)^+. \tag{5.5}$$

Now write $D=\sum_{i\geq 0} D_i$ with D_i a derivation mapping a to zero and X to $\Lambda a\otimes\Lambda^i X$. Thus $D_0(U)=0$ and $D_0(u)=a^{k+1}$. The equation $D_0D_1+D_1D_0=0$ shows that D_1 preserves $\Lambda a\otimes U$. Define $f:U\to\Lambda a\otimes U$ and $\phi:U\to\Lambda a\otimes\Lambda^2 U$ by

$$D_2x = u\wedge f(x)+\phi(x)\ ,\quad x\varepsilon U.$$

From $D_1^2+D_0D_2+D_2D_0=0$ we deduce

$$f(x) = -D_1^2x\ ,\quad x\varepsilon U. \tag{5.6}$$

On the other hand recall the graded Lie algebra L ($L_p=\mathrm{Hom}(X^{p+1};\mathbb{Q})$) defined in sec. 3. A vector $\alpha\varepsilon L_{k|a|}$ is given by $\langle u,\alpha\rangle=1$ and $\langle U;\alpha\rangle=0$. If $f(x)=\sum_{i\geq 0} a^i\otimes f_i(x)$ ($f_i(x)\varepsilon U$) then

$$\langle x;[\alpha,\beta]\rangle = \pm\langle d_2x;\alpha,\beta\rangle = \pm\langle f_0(x);\beta\rangle$$

and so f_0 is dual to ad α (up to sign).

We are now able to prove Theorem 1.10 and half of Theorem 1.7.

5.7 *Proof of 1.10*. We apply the remarks above to the complex $(\Lambda a\otimes\Lambda X_M,D)$ associated with the Borel fibration of a circle action ($|a|=2$). We have to show the cohomology can not be finite dimensional. If it were we would have $0\neq\alpha\varepsilon\pi_{2k}(\Omega M)\otimes\mathbb{Q}$ with ad α dual to f_0, as described above.

On the other hand, D_1 is of degree 1 and since X_M is by hypothesis concentrated in odd degrees, $D_1=0$. By (5.6) $f_0=0$ and α is in the centre of the homotopy Lie algebra - a contradiction.

□

5.8 *Proof of 1.7 when the M_i are even dimensional*. It follows from Lemma 3.3 that in this case the homotopy Lie algebra of the connected sum of the M_i is concentrated in even degrees, and has no centre. Apply 1.10.

□

6. *PROOF OF 1.7 FOR ODD DIMENSIONAL MANIFOLDS*

Let n be the dimension and denote the connected sum of the M_i by M. Let $(\Lambda X_M,d)$ be its minimal model and let $(\Lambda a\otimes\Lambda X_M,D)$ be the model associated with the Borel fibration of a circle action (sec. 4). If the circle action were fixed point free then $\dim H(\Lambda a\otimes\Lambda X_M,D)$ would be finite. We assume this and deduce a contradiction.

Because we suppose $H(\Lambda a\otimes\Lambda X_M)$ finite dimensional, the second half of sec. 5 applies. Let u,U,α,f, and the D_i be as defined there, and suppose $|u|=2k+1$. ($||$ denotes degree.)

Let r be the least integer such that $X_M^r\neq 0$. Then by Lemma 3.3 one of the M_i (say M_1) has a minimal model $(\Lambda Y,d)$ with $Y^r\neq 0$. Let N be the connected sum of the remaining M_i, with minimal model $(\Lambda X_N,d)$. The map $M=M_1\#N\to M_1\vee N$ defines (again by (3.3)) inclusions $(\Lambda Y,d)$, $(\Lambda X_N,d)\to(\Lambda X_M,d)$ which may be taken to map $Y\oplus X_N$ injectively into X_M. In particular we have $Y^{<r}=X_N^{<r}=0$, and if $y\in Y$, $z\in X_N$ are d-cocyles then there is an element $w\in\Lambda X_M$ with $dw=yz$.

Note also from Lemma 3.3 that $X_M^{\leq n-2}$ is concentrated in odd degrees.

6.1 Lemma. $Dy=0$, $y\in Y^r$. $\frac{r+1}{2}$

Proof: If $Dy\neq 0$ then $Dy=\lambda a^{\frac{r+1}{2}}$ $(\lambda\neq 0)$ and so $y=\lambda u+v$, $v\in U^r$. We may therefore suppose $u=(1/\lambda)y$. Let $0\neq z\in X_N$ have minimal degree. Then $dz=0$ and there is some $w\in\Lambda X_M$ with $dw=uz$. Hence

$$Dw = uz+a\Omega.$$

Write $\Omega=u(\sum_i\Phi(i)) + \sum_i\Psi(i)$ where $\Phi(i)$, $\Psi(i)\in\Lambda a\otimes\Lambda^i U$. From $D^2w=0$ we get

$$z+a\Phi(1) + D_1\Psi(1) = 0.$$

Now $|\Omega|=|u|+|z|-2\leq n-2$ and this implies that $\Psi(1)\in\Lambda a\otimes U^{\leq n-2}$. Thus it is zero, since $|\Psi(1)|$ is even. Now $z+a\Phi(1)=0$, which is impossible. □

6.2 Lemma. Let $x\in X_M$, $\Phi\in(\Lambda a\otimes\Lambda X_M)^+\cdot(\Lambda a\otimes\Lambda X_M)^+$ satisfy $D(x+\Phi)=0$. Then $|x|$ is odd, and

$$2k \leq n-|x|-r-1.$$

Proof: The projection $\Lambda a\otimes\Lambda X_M\to\Lambda X_M$ maps $x+\Phi$ to a non trivial cohomology class, since $\mathrm{Im}\,d\subset\Lambda^{\geq 2}X_M$. Thus $|x|=|x+\Phi|\leq n-2$. By Lemma 3.3, $X_M^{\leq n-2}$ is concentrated in odd degrees, so $|x|$ is odd.

Moreover $x+\Phi\notin\mathrm{Im}D$. Choose the largest p for which $a^p(x+\Phi)\notin\mathrm{Im}D$. By Lemma 5.4, $H(\Lambda a\otimes\Lambda X_M,D)$ satisfies Poincaré duality with fundamental class of *even* degree n-1. Since any odd dimensional class has degree at least r we conclude that

$$\deg(a^p(x+\Phi)) \le n-1-r. \tag{6.3}$$

On the other hand, $a^{p+1}(x+\Phi)=D\Omega$. Now $|\Omega|\le n-r$ and $|\Omega|$ is even. Since $X_M^{\le n-2}$ is concentrated in odd degrees we get $\Omega=\Sigma\Omega(2i)$, $\Omega(2i)\varepsilon\Lambda a\otimes\Lambda^{2i}X_M$. Write $\Omega(2)=u\omega+\omega'$ with $\omega\varepsilon\Lambda a\otimes U$, $\omega'\varepsilon\Lambda a\otimes\Lambda^2U$. Write $\Phi=\Sigma\Phi(i)$, $\Phi(i)\varepsilon\Lambda a\otimes\Lambda^i X_M$. Then $\Phi(1)=a\sigma$, $\sigma\varepsilon\Lambda a\otimes X_M$.

Now from $a^{p+1}(x+\Phi)=D\Omega$ we deduce

$$a^{p+1}(x+a\sigma) = D_0\Omega(2) = a^{k+1}\omega.$$

Hence $k\le p$. Substitution of this in (6.3) gives $2k\le n-1-r-|x|$.

□

Lemmas 6.1 and 6.2 imply

$$2k \le n-2r-1. \tag{6.4}$$

6.5 Lemma. The projection $\rho:L_M\to L_{M_1}\amalg L_N$ (cf sec. 3) maps α to a non zero element of L_{M_1}.

Proof. By Lemma 3.3 it is sufficient to prove $[\alpha,\beta]=0$ where $\beta\varepsilon L_{M_1}$ is a non zero element of degree r-1. (Lemma 6.1 implies $|\alpha|>|\beta|$.) Were this to fail we could find $w\varepsilon U$ and $y\varepsilon X_M^r$ with

$$d_2w - uy\varepsilon\Lambda^2U.$$

Then

$$D_2w = u(y+a\sigma)+\Omega\ ,\quad \sigma\varepsilon\Lambda a\otimes U,\ \Omega\varepsilon\Lambda a\otimes\Lambda^{\ge 2}U.$$

From $D_1^2=-D_0D_2$ in U we conclude

$$D_1^2w = -a^{k+1}(y+a\sigma).$$

On the other hand, $|w|=|u|+|y|-1=2k+r\le n-r-1$ by (6.4). Thus $|w|\le n-2$. Since $U^{\le n-2}$ is concentrated in odd degrees, D_1 is zero there; i.e. $y+a\sigma=0$ - a contradiction.

□

6.6 Lemma. Let s be the least integer such that $X_N^s \neq 0$. Then

$$2k \geq n-s.$$

Proof: Choose $\gamma \in L_M$ projecting to a non-zero element of L_N^{s-1}. From Lemma 6.5 we deduce $[\alpha,\gamma] \neq 0$, and the lemma follows from the same argument as that of 6.5.

□

Now recall that $\pi_*(M_1)\otimes\mathbb{Q}$ is finite dimensional and concentrated in odd degrees. The same assertion applies, then, to Y. By Lemma 6.5 we may suppose that $u \in Y$; complete it to a homogeneous basis $y_1,\ldots,y_\ell,u$. From (6.6) we deduce

$$\sum_i^\ell |y_i| = n-|u| = n-2k-1 < s.$$

Since s is the least degree where $H^+(N;\mathbb{Q}) \neq 0$ and since $H^+(N;\mathbb{Q})$ is not just the fundamental class (cf. hypotheses of 1.7) and since $H(N;\mathbb{Q})$ satisfies Poincaré duality and since n is odd we deduce

$$\sum_1^\ell |y_i| < s < n-s < |u|. \qquad (6.7)$$

Let $Z \subset Y$ be the span of the y_i. From (6.7) we deduce

6.8 Lemma (i) $du=0$ and $H(\Lambda Y,d) = H(\Lambda Z,d)\otimes\Lambda u$.

(ii) Except in degrees 0 and n, $H(\Lambda X_N,d)$ is concentrated in degrees in the interval $[s,n-s]$ while $H(\Lambda Y,d)$ is concentrated in degrees in the intervals $[3,s-1]\cup[n-s+1,n-3]$.

(iii) $H^i(\Lambda X_M) = \begin{cases} H^i(\Lambda Z) & i \leq s-1 \\ H^i(\Lambda X_N) & s \leq i \leq n-s. \end{cases}$

In particular $\omega = y_1\wedge\ldots\wedge y_\ell$ is not in $d(\Lambda X_M)$.

□

It also follows from (6.7) that $Z = X_M^{<s}$ and that it generates a D-stable ideal. Thus $D\omega=0$. Applying Lemma 5.3 to the KS complex $(\Lambda a\otimes\Lambda Z,D)$ we deduce

6.9 Lemma. Let $\Phi_i \in \Lambda Z$ satisfy $D(\sum_{i\geq 0} a^i\Phi_i)=0$ and suppose for $p\geq 0$ $\Sigma a^{p+i}\Phi_i \notin D(\Lambda a\otimes\Lambda Z)+\Lambda^{\geq p+1}a\otimes\Lambda Z$. Then there are $\Psi_i \in \Lambda Z$ such that $D(\Sigma a^i\Psi_i)=0$ and $\Phi_0\Psi_0=\omega$.

□

Choose linearly independent $z,z' \in X_N$ so that $s=|z|\leq|z'|\leq|w|$ for any homogeneous $w \epsilon X_N$ independent of z,z'. Then $dz=0=dz'$. If we could solve $D(z+a\sigma)=0$ for $\sigma \epsilon \Lambda a \otimes \Lambda X_M$ we would get from Lemma 6.2 that $2k \le n-s-r-1$, in contradiction to (6.6). The same remark applies to z'. Thus there are maximum integers m,m' such that we can solve

$$D(z+a\sigma) = a^m \Gamma, \quad D(z'+a\sigma') = a^{m'}\Gamma'$$

for $\sigma,\sigma',\Gamma,\Gamma' \epsilon \Lambda a \otimes \Lambda X_M$. (Clearly $m,m' \ge 1$.)

Then $\Gamma = \Sigma a^i \Gamma_i$, $\Gamma' = \Sigma a^i \Gamma_i'$ where $d\Gamma_0 = 0 = d\Gamma_0'$ and Γ_0, Γ_0' represent non trivial cohomology classes of even degree $\le |z'|$ in $H(\Lambda X_M)$. From Lemma 6.8 we deduce $|\Gamma_0|$, $|\Gamma_0'| \le |\omega| < s$ and so $\Gamma,\Gamma' \epsilon \Lambda a \otimes \Lambda Z$. We wish to apply Lemma 6.9.

Suppose that for some $p \ge 0$, $a^p \Gamma \epsilon D(\Lambda a \otimes \Lambda Z) + \Lambda^{\ge p+1} a \otimes \Lambda Z$. Minimize p. Then $a^p \Gamma = D\Omega + a^{p+1}\Psi$ and $\Omega = \Sigma a^i \Omega_i$ with $\Omega_i \epsilon \Lambda Z$ and $\Omega_0 \ne 0$. In particular $|\Omega| \le |\omega|$ and so $2p + |\Gamma| \le |\omega| + 1$. Since $2m + |\Gamma| = |z| + 1 \ge s+1$ we deduce $2m + |\Gamma| \ge s+1 > |\omega| + 1 \ge 2p + |\Gamma|$; i.e., $m > p$. Hence $D(z + a\sigma - a^{m-p}\Omega) = a^{m+1}\Psi$, contrary to our supposition that m was maximized. Thus $a^p \Gamma \notin D(\Lambda a \otimes \Lambda Z) + \Lambda^{\ge p+1} a \otimes \Lambda Z$.

Now we can apply 6.9 to Γ (and to Γ') to obtain $\Phi_i, \Phi_i' \epsilon \Lambda Z$ with $D(\Sigma a^i \Phi_i) = 0 = D(\Sigma a^i \Phi_i')$ and $\Gamma_0 \Phi_0 = \Gamma_0' \Phi_0' = \omega$. Our next step is to show $|\Phi_0| = 0 = |\Phi_0'|$.

In fact, it follows from Lemma 6.8 that every d-cocycle in $\Lambda^{\ge 2} X_M$ whose degree $p \epsilon [|\omega|+1, |z|+|z'|-1]$ is a d-coboundary. Suppose $|\Phi_0| > 0$. We may clearly assume then that $|\Phi_i| > 0$, $i \ge 1$. Put $\Phi = \Sigma a^i \Phi_i$, and put $\Omega = (z + a\sigma)\Phi$. Then $\Omega \epsilon \Lambda a \otimes \Lambda^{\ge 2} X_M$ and

$$D\Omega = a^m(\omega + a\omega').$$

Now in particular writing $\Omega = \Sigma a^i \Omega_i$ we have that $\Omega_0 \epsilon \Lambda^{\ge 2} X_M$ is a cocycle of degree $|z| + |\Phi_0|$ which is $> |\omega|$ and $< |z| + |z'|$. Hence $\Omega_0 = d\Psi_0$, $\Psi_0 \epsilon \Lambda X_M^{<|z|+|z'|}$. In this subalgebra $D_0 = D_1 = 0$ and so $D\Psi_0 \epsilon \Lambda a \otimes \Lambda^{\ge 2} X_M$. Thus

$$D(\Omega - D\Omega) = a^m(\omega + a\omega')$$

and $\Omega - D\Psi \epsilon \Lambda^+ a \otimes \Lambda^{\ge 2} X_M$. Divide by a to get a new element $\tilde{\Omega} \epsilon \Lambda a \otimes \Lambda^{\ge 2} X_M$ such that $D\tilde{\Omega} = a^{m-1}(\omega + a\omega')$.

If $m-1 \ge 1$ we will have $|\tilde{\Omega}_0| > |\omega|$ and we can repeat the process. Eventually we will solve $D\tilde{\Omega} = \omega + a\omega'$ which will imply $\tilde{\Omega} \epsilon \Lambda a \otimes \Lambda Z$ and hence that ω is a d-coboundary in ΛZ. This impossibility followed from the supposition $|\Phi_0| > 0$. Hence $|\Phi_0| = 0$ and, similarly, $|\Phi_0'| = 0$.

This shows that Γ_0 and Γ_0' are scalar multiples of ω and hence that $2m=-|\omega|+|z|$, $2m'=-|\omega|+|z'|\geq 2m$. Clearly we may (after multiplying z,z') suppose $\Gamma_0=\Gamma'_0=\omega$. Then

$$D(z'-a^{m'-m}z+a\sigma') = a^{m'+1}\Gamma''.$$

If $m'>m$ this shows that m' was not maximum. If $m'=m$ put $z''=z'-z$. We could have used z'' instead of z' and so deduced the maximum m'' to be $|z''|-|\omega|=|z'|-|\omega|=m'$, whereas it clearly is at least $m'+1$.

This final contradiction follows from our original supposition $\dim H(\Lambda a\otimes\Lambda X_M)<\infty$. This is therefore false - and the theorem is proved.

References

[A-A] P. Andrews and M. Arkowitz, Sullivan's minimal models and higher order Whitehead products, Canad. J. Math. 30 (1978) 961-982.

[A-H] C. Allday and S. Halperin, Lie group actions on spaces of finite rank, Quart. J. Math. Oxford 28 (1978) 63-76.

[B] J. Barge, Structures différentiables sur les types d'homotopie simplement connexes, Ann. scient. Ec. Norm Sup. 9 (1976).

[B-G] A.K. Bousfield and V.K.A.M. Gugenheim, P.L. de Rham theory and rational homotopy type, Memoirs of the A.M.S. 179, 1976.

[F-H] Y. Felix and S. Halperin, Rational L.-S. category and its applications, Trans. Amer. Math. Soc. 273 (1982) 1-37.

[Ha_1] S. Halperin, Finiteness in the minimal models of Sullivan, Trans. Amer. Math. Soc. 230 (1977) 173-199.

[Ha_2] S. Halperin, Lectures on minimal models, Memoires de la Soc. Math. Math. France, 9/10, 1984, 1-261.

[H-L] S. Halperin and J.-M. Lemaire, Suites inertes dans les algebres de Lie graduees, preprint.

[Q] D.G. Quillen, The spectrum of an equivariant cohomology ring I, II, Ann. of Math. 94 (1971) 549-602.

[Su] D. Sullivan, Infinitesimal computations in topology, Publ. Math. I.H.E.S. 47 (1978) 269-331.

[Ta] D. Tenré, Homotopie rationnelle: Modèles de Chen, Quillen, Sullivan. Lecture Notes in Mathematics 1025, Springer Verlag 1983.

REMARKS ON STARS AND INDEPENDENT SETS

P. Erdös and J. Pach
Mathematical Institute of the Hungarian Academy of Sciences

1 INTRODUCTION

Let G be a graph with vertex set and edge set V(G) and E(G), respectively. A subset $X \subseteq V(G)$ is said to be *independent* if there are no two elements of X connected by an edge of G. We say that G has *property I* if for every independent set X there is a common neighbour in G, i.e. we can find a point y_X such that $y_X x \in E(G)$ for every $x \in X$. The neighbourhood $\{v \in V(G) | yv \in E(G)\}$ of a vertex y is usually called a *star* of the graph. In this terminology, property I means that every independent set of G is covered by a star.

A couple of years ago Erdös and Fajtolowicz [2] conjectured that there exists a positive constant ε such that every *triangle-free* graph G on n vertices having property I has a vertex of degree at least εn. Note that such a graph can be characterized by saying that the system of maximal (non-increasable) independent sets of G coincides with the system of stars. The above conjecture has been settled by Pach [6] who proved the following.

Theorem 1. Every triangle-free graph on n vertices having property I contains a vertex degree at least $\frac{n+1}{3}$. This result is best possible, provided $3|n + 1$.

Pach completely characterized all the graphs having these properties. He ended his paper with the question whether the same assertion holds if we drop the condition that the graphs are triangle-free. Our Theorem 2 answers this question in the negative. We can construct a graph of property I with maximum degree $(1+o(1))n \log\log n/\log n$, and we shall show that this bound cannot be improved.

What happens now if, instead of the condition that our graphs are triangle-free, we assume only that they do not contain K_r (a complete subgraph on r vertices), for some $r > 3$? In particular, is it true that every K_4-free graph of n vertices having property I contains a point of degree at least εn? Unfortunately, we can throw very little light on this simple question. Our only result in this direction (Theorem 3) is that, if $w(n)$ is any function tending to infinity as slowly as we please, then there exist graphs on n vertices containing no $K_{w(n)\log n}$, having property I, but whose maximum degree is $o(n)$.

In what follows, we need the following

Definition 1. Given any natural number k, a graph G is said to have property I_k if every independent set of cardinality k has a common neighbour in G.

Obviously, a graph has property I if and only if it has property I_k for all k. Property I_2 means that G has diameter 2.

We can sharpen Theorem 1 by proving

Theorem 4. Every triangle-free graph on n vertices and with property $I_{\lceil \log n \rceil}$ has a vertex of degree at least $\frac{n+1}{3}$.

Erdös and Fajtlowicz suspected that this result remains true (apart from the constant factor beside n) even if $I_{\lceil \log n \rceil}$ is replaced by the (essentially weaker) property I_3, but this was disproved in [6]. However, at the moment we have no idea how to attack the next question to arise naturally: Does there exist an $\varepsilon > 0$ such that every triangle-free graph with property I_4, having n vertices, contains a point of degree $\geq \varepsilon n$?

Let $f_{I_k}(r,n)$ denote the maximum integer f such that every K_r-free graph of n vertices having property I_k has a vertex of degree at least f.

It is not difficult to show (cf. [6]) that, for any fixed k and r ($k \geq 2$, $r \geq 3$), we have

$$f_{I_k}(r,n) \geq (1 - o(1))n^{1 - (1/k)}.$$

The great weakness of our results is that we cannot improve on this lower bound for any pair (k,r). The simplest case $k = 2$, $r = 3$ was considered by Erdös and Fajtlowicz [2], who proved

$$\sqrt{n}\log n \geq f_{I_2}(3,n) \geq \sqrt{n-1}\ .$$

We conjecture that here the *upper* bound is not far from the truth. On the other hand, a well-known construction of Erdös-Rényi [4,5] and Brown [1], using finite projective planes, shows that, in case $k = 2$, $r > 3$, the *lower* bound is asymptotically sharp. An extension of their construction (see [6]) proves that, in general,

$$f_{I_k}(k+2,n) = (1+o(1))n^{1-(1/k)}$$

holds for every $k \geq 2$.

Next we show that $f_{I_k}(k+1,n) = o(n)$. Let X be a $((k+1)m-1)$ - element set, and define a graph G whose vertices are the m-element subsets of X, two of them being joined by an edge if and only if their intersection is empty. It is now clear that G has property I_k and does not contain a complete subgraph on $k+1$ vertices.

Setting $n = |v(G)| = \binom{(k+1)\cdot m-1}{m}$, easy calculation shows that the degree $\binom{k\cdot m-1}{m}$ is $o(n)$. More exactly,

$$f_{I_k}(k+1,n) \leq (1+o(1))n^{\frac{k\log k-(k-1)\log(k-1)}{(k+1)\log(k+1)-k\log k}}$$

is valid, for every $k \geq 2$.

We almost certainly have that, if k is odd, then $f_{I_k}(k,n) = o(n)$, but we have not yet worked out the details. (The parity condition seems to be merely a technical requirement.)

Definition 2. ([3].) Given any natural number k, a graph is said to have property J_k if every set of k vertices has a common neighbour in G.

Property J_k is evidently stronger than I_k. It is also clear that every graph with property J_k contains a complete subgraph on $k+1$ vertices. On the other hand, a Kneser-type construction similar to the above one, shows that $f_{J_k}(k+2,n) = o(n)$. That is, there exist K_{k+2} - free graphs with property J_k with small maximum degrees.

It is easily seen that, given a triangle-free graph G, by the addition of some new *edges* creating no triangles we can obtain a graph with property I_2 (i.e. with diameter 2). This statement does not remain true if we require that our graph had property I_3. However, we are unable to decide whether or not the following assertion holds true: every triangle-free graph can be embedded as a subgraph into a triangle-free graph with property I_3. (Here addition of new *vertices* is also permitted.) If the answer to this question is in the affirmative, one can of course then ask how many extra vertices are needed for the embedding. Note that no 4-chromatic graph can be embedded into any triangle-free graph with property I, showing that property I_3 in the above problem cannot be replaced by I.

Another problem which can be raised is the following: given any natural numbers k,r determine the smallest value $g_{I_k}(r) = g$ such that there exists a K_r-free graph with g vertices and property I_k (and containing k independent points). The proof of our Theorem 4 gives that $g_{I_k}(3) \geq 2^{k+1}$.

2 RESULTS AND PROOFS

Theorem 2. Let $f_I(n)$ denote the maximum integer f such that every graph on n vertices having property I has a vertex of degree at least f. Then we have

$$f_I(n) = (1 + o(1)) \frac{n \text{ loglog } n}{\log n} .$$

Proof. The upper bound can be established by a routine calculation, as follows. Let G be a graph with property I and suppose that the maximum degree D is less than n loglog n/log n. Setting k = [log n/log log n], the number m_k of independent k-tuples can be estimated by

$$m_k \geq \frac{n(n-D)(n-2D)\dots(n-(k-1)D)}{k!} > \frac{(n-kD)^k}{k!} .$$

Using the fact that each independent k-tuple has a common neighbour in G, we obtain

$$\frac{n\cdot D}{2} \geq |E(G)| \geq \frac{m_k \cdot k}{2\binom{D-1}{k-1}} > (\frac{n}{D} - k)^k \frac{D}{2} ,$$

which yields the desired upper bound.

To prove that our result is sharp we take n points and divide them into t equal classes, where t will be specified later. Each class induces a complete subgraph, whereas every pair of points belonging to different classes will be joined by an edge independently with probability $p = \varepsilon/t$ ($\varepsilon > 0$ is an arbitrarily fixed small constant). Choosing now t points, one from each class, the probability that we cannot find a common neighbour for them is $(1 - p^t)^{n-t}$. Thus the probability that our graph has not got property I is at most

$$(\frac{n}{t})^t (1 - p^t)^{n-t} < e^{t\cdot \log(n/t) - (n-t)(\varepsilon/t)^t}$$

which tends to zero if $t = [(1-\varepsilon)\log n/\text{loglog } n]$. On the other hand, we almost certainly have that the maximum degree of the vertices does not exceed $(n/t) + (2\varepsilon n/t) < (1+3\varepsilon)n \log\log n/\log n$. □

Carrying out the same random construction a little more carefully, we can also ensure that no large independent sets appear in our graph. In this way we can obtain

Theorem 3. Let $f_I(r,n)$ denote the maximum integer f such that every K_r-free graph on n vertices having property I has a vertex of degree at least f. Then we have

$$f_I(w(n)\log n, n) \le (2 + o(1)) \frac{n \log\log w(n)}{\log w(n)},$$

where w(n) is an arbitrary function tending to infinity.

Proof. Let $\varepsilon > 0$ be a fixed small constant. Take n distinct points and divide them into t equal classes, where the value of t will be specified later. Define a random graph on these vertices, as follows. Any two points belonging to the same class are joined by an edge independently with probability $q = 1 - (4/w(n))$, while any edge running between different classes will be drawn in with probability $p = \varepsilon/t$. The probability that our graph contains a complete subgraph on $w(n)\log n$ vertices is at most

$$\binom{n}{w(n)\log n} \cdot q^{(w(n)\log n)^2/2} < e^{-w(n)\log^2 n}$$

which tends to zero as $n \to \infty$. On the other hand, it is almost certain that our graph will not contain any independent set S of size $\ge t(2+\varepsilon)\log n/\log w(n)$. Otherwise at least $s = [(2+\varepsilon)\log n/\log w(n)]$ elements of S would belong to the same class, and the probability of this event can be bounded (from above) by

$$t\binom{n/t}{s}\left(\frac{4}{w(n)}\right)^{s^2/2} < e^{\frac{s}{2}(2\log n - s\log(w(n)/4))}$$

which tends to zero again. Further, the probability that our graph does not have property I can be estimated by

$$\le st \binom{n}{st} (1 - p^{st})^{n-st}$$

and this approaches zero, if $t = [\log w(n)/(2+2\varepsilon)\log\log w(n)]$ and n is large. Finally, it is obvious that the maximum degree of the vertices is almost certainly smaller than $(1+2\varepsilon)n/t$. □

Figure 1

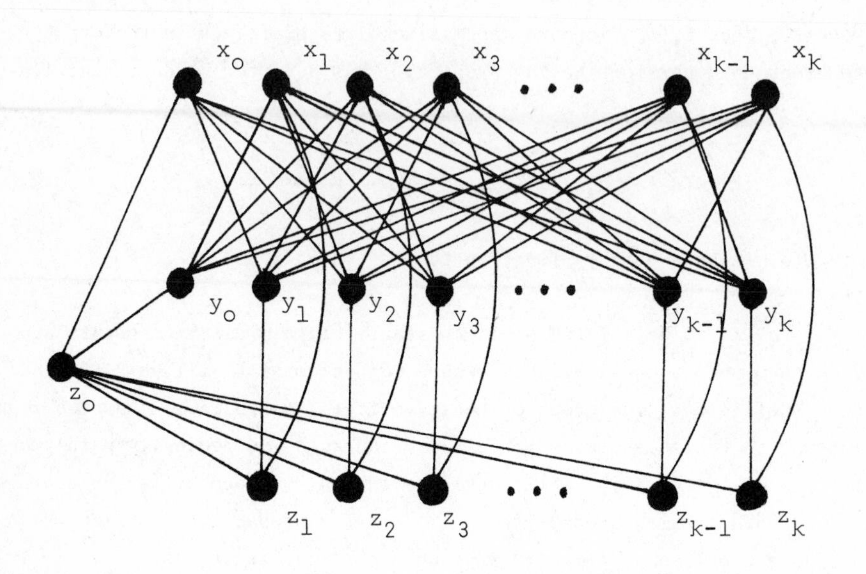

Theorem 4 (stated in §1) is an immediate consequence of Theorem 1 and the following

Lemma. Let G be a triangle-free graph on n vertices. If every set of $\lceil \log n \rceil$ independent vertices has a common neighbour in G, then G has property I.

Proof. Assume, indirectly, that there exists a maximum integer $k < n$ such that G has property I_k but one can find a $k+1$ element independent set $X = \{x_0, x_1, \ldots \ldots, x_k\} \subseteq V(G)$ with no common neighbour in G. Let y_i denote a vertex connected to all points in $X - x_i$; $i = 0,1,\ldots,k$. We obviously have that $x_iy_i \notin E(G)$. Further, let z_o be any point joined to both x_o and y_o and choose a common neighbour z_i to each triple $\{z_o, x_i, y_i\}$; $i = 1, 2,\ldots,k$. (See Fig. 1.) Using the fact that G is triangle-free it follows that all the above defined vertices are different from each other, and

$y_1, \ldots, y_k$, $z_1, \ldots, z_k$ induce a one-factor in G. Thus, for every subset $A \subseteq \{1,2,\ldots,k\}$,

$$H_A : = \{y_1 | i \in A\} \cup \{z_i | i \notin A\}$$

is an independent k-tuple and, by property I_k, we can find a vertex v_A connected to all elements of H_A. All the 2^k v_A's defined in this way are different, showing that $k < \lceil \log n \rceil$, a contradiction. □

References

[1] W.G. Brown, On graphs that do not contain a Thomsen graph, Can. Math. Bull. 9(1966), 281-285.

[2] P. Erdös and S. Fajtlowicz, Domination in graphs of diameter 2, preprint.

[3] P. Erdös and L. Moser, An extremal problem in graph theory, J. Austral. Math. Soc. 11(1970), 42-47.

[4] P. Erdös and A. Rényi, On a problem in the theory of graphs, Publ. Math. Inst. Hung. Acad. Sci. 7/A (1962), 623-641 (in Hungarian, with English and Russian summaries).

[5] P. Erdös, A. Rényi and V.T. Sós, On a problem of graph theory, Studia Sci. Math. Hung. 1(1966), 215-235.

[6] J. Pach, Graphs whose every independent set has a common neighbour, Discr. Math. 37(1981), 217-228.

COMPACT AND COMPACT HAUSDORFF

A.H. Stone
University of Rochester, Rochester, New York, USA

0 INTRODUCTION In this paper, separation axioms are not assumed without explicit mention; thus 'compact' means 'quasi-compact' in the sense of Bourbaki (every open cover has a finite subcover). Non-Hausdorff compact spaces arise, for example, in algebraic geometry (via the Zariski topology). In the summer of 1980, S. Eilenberg raised (orally) a question equivalent to the following: Is every compact space a quotient of some compact Hausdorff space? An affirmative answer, he said, would have interesting consequences. Unfortunately the answer is negative, as C.H. Dowker soon showed with the elegant example (hitherto unpublished) that follows. The present paper discusses various relations (some known, some new) between 'compact' and 'compact Hausdorff', suggested by Eilenberg's question and Dowker's answer to it.

1 EXAMPLE (Dowker) **There exists a countable compact T_1 space D, satisfying the second axiom of countability, that is not a quotient of any compact Hausdorff space.**

The space D consists of points a, b, c_{ij} (where $i, j \in \mathbf{N} = \{1,2,3,\ldots\}$), all distinct, topologized so that each c_{ij} is isolated, a neighbourhood base at a consists of the sets $U_n = \{c_{ij} : j \geq n\} \cup \{a\}$, $n \in \mathbf{N}$, and a neighbourhood base at b consists of the sets $V_n = \{c_{ij} : i \geq n\} \cup \{b\}$, $n \in \mathbf{N}$. Clearly this produces a compact T_1 topology; and D, being countable and first-countable, is also second-countable.

Suppose there is a quotient map $f : S \to D$ of some compact Hausdorff space S onto D. Write $A = f^{-1}(a)$, $B = f^{-1}(b)$; these are disjoint closed sets in the normal space S, so there is an open set $G \supset A$ in S with closure G^- disjoint from B. Put $P = f(G^-)$; then (since $b \notin P$)

$$f^{-1}(P) = A \cup \{f^{-1}(c_{ij}) : c_{ij} \in P\} = G \cup \{f^{-1}(c_{ij}) : c_{ij} \in P\},$$

because $A \subset G \subset f^{-1}(P)$. This shows that $f^{-1}(P)$ is open; and, since f is quotient, P must be open in D. Since $a \in P$, there is some $U_n \subset P$. Then U_{n+1} and the singletons $\{c_{ij}\}$, for c_{ij} in P, provide an open cover of P having no finite subcover (all the points c_{in}, $i \in \mathbf{N}$, are in $P \setminus U_{n+1}$), so P is not compact. But P is the continuous image of the compact set G^-, so P must be compact, giving the desired contradiction.

Example 1 shows that the following positive results, weak as they are, cannot easily be improved.

2 THEOREM Every compact space X is a quotient of a compact T_1 space Y.

Proof Let K denote the subspace $\{0\} \cup \{1/n : n \in \mathbf{N}\}$ of the real line — that is, a convergent sequence — let X be a compact space, and consider the space $X \times K$, with the usual product topology $\mathcal{P}$. Define a new topology $\mathcal{T}$ on $X \times K$ by taking, as a base for the open sets, all sets of the form $P \setminus F$, where $P \in \mathcal{P}$ and F is finite. (It is easily verified that this is a base for a topology.) Let Y denote $X \times K$ with topology $\mathcal{T}$; Y is clearly a T_1 space. Also Y is compact. For an arbitrary $\mathcal{T}$-open cover of $X \times K$ can be refined by one of the form $\mathcal{V} = \{P_\lambda \setminus F_\lambda : \lambda \in \Lambda\}$, where each P_λ is $\mathcal{P}$-open and each F_λ is finite. Then $\{P_\lambda : \lambda \in \Lambda\}$ is a $\mathcal{P}$-open cover of the compact space $X \times K$, so it has a finite subcover: say $X \times K = \bigcup\{P_{\lambda_i} : 1 \le i \le n\}$. For each point y of the finite set $\bigcup\{F_{\lambda_i} : 1 \le i \le n\}$ choose a member $V(y)$ of $\mathcal{V}$ that contains y; then $\{P_{\lambda_i} \setminus F_{\lambda_i} : 1 \le i \le n\} \cup \{V(y) : y \in \bigcup_i F_{\lambda_i}\}$ is a finite subcover of $\mathcal{V}$, proving Y compact.

The projection $f : Y = X \times K \to X$, given by $f(x, t) = x$, is continuous, because $\mathcal{T}$ is finer than $\mathcal{P}$. To see that it is a quotient map, suppose $E \subset X$ is such that $f^{-1}(E) \in \mathcal{T}$. Given $x_0 \in E$, the point $(x_0, 0)$ of $f^{-1}(E)$ has a $\mathcal{T}$-neighbourhood $P \setminus F$ contained in $f^{-1}(E)$, where F is finite and $P \in \mathcal{P}$; and then P contains a neighbourhood $G \times H$ of $(x_0, 0)$, where G is a neighbourhood of x_0 in X and H is a neighbourhood of 0 in K. Let $F = \{(x_i, t_i) : 1 \le i \le n\}$, and choose $t^* \in H$ different from all of $t_1, t_2, \ldots, t_n$. Then $G \times \{t^*\} \subset P \setminus F \subset f^{-1}(E)$, proving

$G \subset E$. This shows that each $x_o \in E$ has a neighbourhood contained in E , so that E is open, as required.

Remark As the construction shows, Y can be made countable if X is countable, and second-countable if X is countable and second-countable.

3 THEOREM Every finite space is a quotient of the compact metrizable space $K \times K$.

We prove rather more: **If X has n points, X is a quotient of $F \times K$ where F is a discrete space with n points.** Since $F \times K$ is trivially a quotient of $K \times K$, the theorem will follow.

Proof Put $F = \{1, 2, \ldots, n\}$ and let $X = \{x_i : i \in F\}$. We write the closure of the singleton x_i as x_i^- . For each $i \in F$ define $J_i = \{j \in F : x_i \in x_j^-\}$, and take a homeomorphic copy K_i of K , with (say) $a_i \in K_i$ corresponding to $0 \in K$. Trivially, $i \in J_i$. For each i such that $J_i \neq \{i\}$, partition $\mathbf{N}$ into $|J_i| - 1$ infinite subsets; these induce (by taking reciprocals) a partition of $K_i \setminus \{a_i\}$ into $|J_i| - 1$ infinite subsets L_{ih} , $h \in J_i \setminus \{i\}$, each of which forms a sequence converging (in K_i) to a_i . (If $J_i = \{i\}$, L_{ih} is not defined.) We may suppose the K_i's to be pairwise disjoint.

Let Y denote the topological sum (discrete union, coproduct) $\Sigma\{K_i : i \in F\}$; thus Y is homeomorphic to $F \times K$. Define a map $f : Y \to X$ as follows. Given $y \in Y$, consider the (unique) K_i to which y belongs. If $y = a_i$, or if $J_i = \{i\}$, define $f(y) = x_i$. Otherwise, y is in a unique L_{ih} , where $h \in J_i \setminus \{i\}$; define $f(y) = x_h$. It is clear that f maps Y onto X . To prove f continuous, say at $y \in K_i$, note that, if $J_i = \{i\}$, then f is constant on the neighbourhood K_i of y . So we may assume $J_i \neq \{i\}$. If now $y \neq a_i$, then y is an isolated point of Y , and again f is constant on a neighbourhood of y . So it suffices to consider $y = a_i$. Then $f(y) = x_i$. Let U be an arbitrary neighbourhood of x_i in X ; then (because $x_i \in x_j^-$ for all $j \in J_i$) we have $U \supset \{x_j : j \in J_i\}$. Now K_i is a neighbourhood of y in Y , and we have $f(K_i) \subset \{x_j : j \in J_i\} \subset U$, as required.

To complete the proof that f is a quotient, let $G \subset X$ be such that $f^{-1}(G)$ is open in Y ; we must show that G is open in X .

Suppose not; then there is some $x_i \in G \cap (X \setminus G)^-$; and, since $X \setminus G$ is finite, this shows $x_i \in x_j^-$ for some $x_j \in X \setminus G$. In particular, $j \neq i$, so $j \in J_i \setminus \{i\}$, and L_{ij} is defined. Now $a_i \in f^{-1}(x_i) \subset f^{-1}(G)$, an open set in Y . Thus $f^{-1}(G)$ meets L_{ij} , say in y . We then have $x_j = f(y) \in G$, giving the desired contradiction.

4 REMARK The preceding result is sharp, in that n is the <u>smallest</u> cardinal of a Hausdorff space F such that every n-point space X is a quotient of $F \times K$. In fact, it is easy to see that, if X is a quotient space (not necessarily finite) of $F \times K$, where F is discrete, then X can have at most $|F|$ non-isolated points. Hence, if an n-point space X having (for instance) the trivial topology is a quotient of $F \times K$, where F is any Hausdorff space, then F must have at least n points.

5 These results suggest a dual question: Is every compact space a continuous bijective image of a compact Hausdorff space? — that is, does it always have a finer compact T_2 topology? Here too the answer is 'no', as follows at once from the existence (known since 1948) of 'maximal compact' spaces that fail to be Hausdorff. A topological space is said to be 'maximal compact' (or 'minimal compact' by many authors) if it is compact and has no strictly finer compact topology. Evidently a maximal compact non-Hausdorff space can have no finer compact Hausdorff topology.

It is very well known that every compact Hausdorff space is maximal compact. The problem of finding non-Hausdorff maximal compact spaces goes back to Vaidyanathaswamy [8] in 1947. Examples of such spaces have been produced by Ramanathan [5], Balachandran [1], Hing Tong [7] and Smythe & Wilkens [6]. Balachandran's example is the simplest of these and will be considered in Example 11 below. Another particularly simple example is as follows.

6 EXAMPLE There is a countable, compact T_1 space X that is maximal compact but not Hausdorff; further, X is first-countable except ay one point.

Let **Q** denote the set of rational numbers, in their usual topology, and let $\mathbf{Q}^*$ denote the usual Aleksandrov (one-point) compactification of Q ; thus $\mathbf{Q}^* = \mathbf{Q} \cup \{p\}$, where points of **Q** have their usual neighbourhoods and the neighbourhoods of p are the complements of the (closed) compact subsets of $\mathbf{Q}$. $\mathbf{Q}^*$ is not Hausdorff, because $\mathbf{Q}$ is not locally compact. To see that $\mathbf{Q}^*$ is maximal compact, one uses the following neat characterization, due to Ramanathan [5] (see also [6]): _a space is maximal compact if and only if its compact subsets coincide with its closed subsets_. It is not hard to check that all compact subsets of $\mathbf{Q}^*$ are closed, and thence that this criterion applies.

7 EXAMPLE There is a compact T_1 space X that is not maximal compact but that has no finer compact T_2 topology; moreover X is countable and first- (hence second-) countable.

The underlying set of the space X is again $\mathbf{Q}^* = \mathbf{Q} \cup \{p\}$, and again the points of $\mathbf{Q}$ will have their usual neighbourhoods. To define a neighbourhood base at p, choose for each $q \in \mathbf{Q}$ a sequence $S(q)$ of distinct points of $\mathbf{Q} \setminus \{q\}$, converging to q (in the usual sense), and put $T(q) = S(q) \cup \{q\}$. A neighbourhood base at p consists of the sets of the form

$$X \setminus (F_1 \cup \bigcup\{T(q) : q \in F_2\}),$$

where F_1 and F_2 are finite subsets of $\mathbf{Q}$. It is easily checked that the neighbourhood axioms are satisfied, and that the resulting topology, say $\mathcal{T}$, is T_1. It is compact, for the complement of each basic neighbourhood of p is compact in the usual topology of $\mathbf{Q}$.

Suppose there is a compact T_2 topology $\mathcal{T}' \supset \mathcal{T}$ on X, and let $\mathcal{T}''$ denote the relative topology on $\mathbf{Q}$ induced by $\mathcal{T}'$. Then $\mathbf{Q} = X \setminus \{p\}$ is locally compact and Hausdorff in the topology $\mathcal{T}''$; and $\mathcal{T}''$ is finer than the usual topology (which is induced by $\mathcal{T}$). Now $\mathbf{Q}$ must have a $\mathcal{T}''$-isolated point, since otherwise a familiar construction (imitating the construction of a Cantor set) would produce a subset of $\mathbf{Q}$ with cardinal c. Let q_o be a $\mathcal{T}''$-isolated point of $\mathbf{Q}$. The infinite sequence $S(q_o)$ can have no cluster point in $(X, \mathcal{T}')$, because (a) p has the neighbourhood $X \setminus T(q_o)$ disjoint from $S(q_o)$, (b) q_o has the $\mathcal{T}'$-neighbourhood q_o disjoint from $S(q_o)$, and (c) each other $q \in \mathbf{Q}$ has a $\mathcal{T}$-neighbourhood having at most one point in common with $S(q_o)$. This contradicts the compactness of $\mathcal{T}'$.

To see that $(X, \mathcal{T})$ is not maximal compact, it is enough to note that the space Q^* of Example 6 provides a strictly finer compact topology on X. Alternatively one can invoke the following theorem.

8 THEOREM A maximal compact, first-countable T_1 space is necessarily Hausdorff.

Proof Suppose $(X, \mathcal{T})$ is a compact, first-countable T_1 but not T_2 space; we construct a strictly finer compact topology. By assumption there are points a, b of X with no disjoint neighbourhoods. Also there is an open neighbourhood base $\{U_n : n \in \mathbf{N}\}$ at a such that $X \setminus \{b\} \supset U_1 \supset U_2 \supset \ldots$, and an open neighbourhood base $\{V_n : n \in \mathbf{N}\}$ at b such that $X \setminus \{a\} \supset V_1 \supset V_2 \supset \ldots$. For each $n \in \mathbf{N}$, pick $c_n \in U_n \cap V_n$, and put $C = \{c_n : n \in \mathbf{N}\}$; note that C contains neither a nor b. Define a topology $\mathcal{T}'$ on X as follows.

A neighbourhood base at a will be $\{U_n : n \in \mathbf{N}\}$, as before. A neighbourhood base at each other point $x\ (\neq a)$ of X will consist of the sets $(W \setminus C) \cup \{x\}$, where W is an open $\mathcal{T}$-neighbourhood of x that does not contain a. The axioms for neighbourhoods are easily checked, and the resulting topology $\mathcal{T}'$ is clearly finer than $\mathcal{T}$, and in fact strictly finer because no $\mathcal{T}'$-neighbourhood of b can contain any V_n (because $c_n \in C \cap V_n$). To see that $\mathcal{T}'$ is compact, let $\mathcal{U}'$ be a $\mathcal{T}'$-open cover of X. Then $a \in$ some U_{n_0} $\subset$ some $U' \in \mathcal{U}'$, and (for each $x \in X \setminus \{a\}$) $x \in$ some $(W(x) \setminus C) \cup \{x\} \subset$ some $U'(x) \in \mathcal{U}'$, where $W(x)$ is $\mathcal{T}$-open. Then $\{U_{n_0}\} \cup \{W(x) : x \in X \setminus \{a\}\}$ is a $\mathcal{T}$-open cover of X; so it has a finite subcover of the form $\{U_{n_0}, W(x_1), \ldots, W(x_m)\}$. Now U_{n_0} contains all the points c_n with $n \geq n_0$. For each $n < n_0$ choose $U'_n \in \mathcal{U}'$ containing c_n; then $\{U', U'(x_1), \ldots, U'(x_m), U'_1, \ldots, U'_{n_0-1}\}$ is a finite subcollection of $\mathcal{U}'$ that covers X, as required.

9 Another closely related question raised by Vaidyanathaswamy [8] should also be mentioned. A topological space is said to be 'minimal Hausdorff' (or, of course, 'maximal Hausdorff' by some authors) if it is Hausdorff but has no strictly coarser Hausdorff topology. It is very well known that every compact Hausdorff space is minimal Hausdorff;

the question was whether the converse holds. Examples showing that the answer is 'no' have been given by Ramanathan [3], and by Smythe & Wilkens [6]. In fact [6] produces two topologies on the extended real line (both inducing the usual topology on $\mathbf{R}$), constituting a maximal compact non-Hausdorff space with a (strictly) finer minimal Hausdorff non-compact topology. An interesting characterization of 'minimal Hausdorff' was given by Katětov [2] and independently by Ramanathan [4]: a space is minimal Hausdorff if and only if it is Hausdorff and semi-regular and has a base of open sets with absolutely closed complements. For this and other properties of minimal Hausdorff spaces, see [2, 4, 6, 9, 10].

Examples 6 and 7 show that the following positive result, though weak, is not easily improved. It is implicitly contained in [7].

10 THEOREM Every compact space has a finer compact T_1 topology.

Proof Let $(X, \mathcal{T})$ be compact. Consider the topology $\mathcal{T}'$ having as a base (of open sets) the sets $U \setminus F$, where $U \in \mathcal{T}$ and F is finite. It is easily verified that $\mathcal{T}'$ is compact; and of course $\mathcal{T}' \supset \mathcal{T}$ (equality not excluded).

It follows that every maximal compact space is T_1.

Though not every compact space can be obtained from a compact Hausdorff space by taking a quotient or an injection, one might speculate that general continuous maps might suffice — that is, that every compact space X is a continuous image of some compact Hausdorff space. (By Theorem 10 it would suffice to deal with compact T_1 spaces X.) However, this too is false.

11 EXAMPLE There is a countable compact T_1 space X that is not a continuous image of any compact T_2 space.

We use Balachandran's example [1] of a non-Hausdorff maximal compact space; however, the property here derived for it appears to be new.

The underlying set $X = \{a, b\} \cup \{c_{ij} : i, j \in \mathbf{N}\}$ is the same as in Example 1, but there will be more complicated neighbourhoods at the point b. As in Example 1, the points c_{ij} are isolated and a neighbourhood base at a is $\{U_n : n \in \mathbf{N}\}$, where $U_n = \{a\} \cup \{c_{ij} : j \geq n\}$. A neighbourhood base at b is to be

$\{V_f : f \in \mathbf{N}^{\mathbf{N}}\}$, where $V_f = \{b\} \cup \{c_{ij} : i \geq f(j)\}$. One readily checks that this determines a T_1 topology, in which a and b do not have disjoint neighbourhoods (since $(n, f(n)) \in U_n \cap V_f$). Further, for all $n \in \mathbf{N}$ and $f \in \mathbf{N}^{\mathbf{N}}$, $X \setminus (U_n \cup V_f)$ is finite, since it consists only of points c_{ij} with $j < n$ and $i < f(j)$. Thus X is compact.

Suppose $F : S \to X$ is a continuous surjection, where S is a compact Hausdorff space. Write $F^{-1}(a) = A$, $F^{-1}(b) = B$, $F^{-1}(c_{ij}) = C_{ij}$; these sets partition S into (pairwise disjoint, non-empty) closed sets, and the C_{ij}'s are also open.

We first show:

(*) Let G be an open set (in S) containing A ; then $G \supset F^{-1}(U_n)$ for some $n \in \mathbf{N}$.

Suppose not; then for each n there is $j(n) \geq n$ for which not every $C_{ij}(n)$ is a subset of G . Take a subsequence $j_1 < j_2 < \dots < j_n < \dots$ of the sequence $\langle j(n) : n \in \mathbf{N} \rangle$, and for each n choose $i_n \in \mathbf{N}$ so that $C_{i_n j_n} \setminus G \neq \emptyset$ and choose $p_n \in C_{i_n j_n} \setminus G$. Define $f : \mathbf{N} \to \mathbf{N}$ so that $f(j_n) > i_n$ for all n ; the values of f for other values of j are arbitrary. Thus $p_n \notin F^{-1}(V_f)$, an open set containing B .

The infinite sequence $\langle p_n : n \in \mathbf{N} \rangle$ must have a cluster point p^* in S . But $p^* \notin A$, for $p_n \notin G$; and $p^* \notin B$, for $p_n \notin F^{-1}(V_f)$; so $p^* \in$ some C_{ij} . However, C_{ij} is then a neighbourhood of p^* that contains p_n for at most one n , giving a contradiction.

Next we prove:

(**) Let H be an open set (in S) containing B ; then $H \supset F^{-1}(V_f)$ for some $f \in \mathbf{N}^{\mathbf{N}}$.

For each $j \in \mathbf{N}$, $F^{-1}(U_{j+1}) \cup H$ is an open set containing $A \cup B$; so it must contain all but finitely many of the sets C_{ij} (otherwise we pick an infinite sequence of points $p_1, p_2, \dots$ from different sets $C_{i_n j} \setminus (F^{-1}(U_{j+1}) \cup H)$, and have as before that this sequence can have no cluster point in S) . But c_{ij} is not in U_{j+1} . Hence (still keeping j fixed) $C_{ij} \subset H$ for all but finitely many values of i . Define $f(j)$ to be greater than all these i's (if any); then $C_{ij} \subset H$ for all pairs (i, j) with $i \geq f(j)$. Thus $F^{-1}(V_f) \subset H$.

Now the disjoint closed sets A, B of the normal space S must have disjoint open neighbourhoods, say G, H. The foregoing shows $G \cap H \supset F^{-1}(U_n \cap V_f) \neq \emptyset$, a contradiction.

In the light of this example, the following positive result is not trivially improvable.

12 THEOREM Every compact space is a θ-continuous image of a compact Hausdorff space.

(A map $f : S \to X$ is called 'θ-continuous' if, for each $s \in S$ and neighbourhood V of f(s) in X, there is a neighbourhood U of s such that $f(U^-) \subset V^-$. If X is regular, θ-continuity coincides with ordinary continuity.)

<u>Proof</u> Let X be a compact space, with topology $\mathcal{T}$. Let X_o be X with the discrete topology, and consider its Stone-Čech compactification βX_o, consisting of all ultrafilters on X_o with the usual (compact Hausdorff) topology. Define a map $f : \beta X_o \to X$ as follows.

For each $p \in \beta X_o$, p is an ultrafilter on X, so it $\mathcal{T}$-converges to at least one point. In particular, if p is a 'fixed' ultrafilter, consisting of all sets containing a certain point of X, then p converges to that point (and perhaps others), and we define f(p) to be that point. In the remaining case, p is a 'free' ultrafilter, and we take f(p) to be an arbitrarily chosen point to which p $\mathcal{T}$-converges. Clearly f maps βX_o onto X (the fixed ultrafilters suffice). To see that f is θ-continuous, suppose V is $\mathcal{T}$-open and $f(p) \in V$. The ultrafilter p $\mathcal{T}$-converges to $f(p) \in V$, so $V \in p$. The set U of all ultrafilters q such that $V \in q$ is an open-closed neighbourhood of p in βX_o; and clearly $f(U) \subset V^-$.

13 FURTHER QUESTIONS One natural question left open by the preceding examples is this: Is every countable, compact space, with a countable base, necessarily a continuous image of a compact Hausdorff space (perhaps even under a 1-1 map)? Example 1 shows that it need not be a quotient of a compact Hausdorff space. As Theorem 10 shows, one can restrict attention to T_1 spaces.

It would also be interesting to have characterizations of those (compact) spaces that are images of compact Hausdorff spaces under continuous or quotient maps, or that have finer compact Hausdorff topologies.

REFERENCES

[1] V.K. Balachandran, Minimal bicompact spaces, J.Indian Math.Soc. (N.S.) 12 (1948) 47-48.

[2] M. Katětov, Über H-abgeschlossene und bikompakte Räume, Časopis Pěst.Mat.Fys. 69 (1940) 36-49.

[3] A. Ramanathan, Maximal Hausdorff spaces, Proc.Indian Acad.Sci., Ser.A, 26 (1947) 31-42.

[4] A. Ramanathan, A characterization of maximal Hausdorff spaces, J.Indian Math.Soc. (N.S.) 11 (1947) 73-80.

[5] A. Ramanathan, Minimal bicompact spaces, J.Indian Math.Soc. (N.S.) 12 (1948) 40-46.

[6] N. Smythe & C.A. Wilkens, Minimal Hausdorff and maximal compact spaces, J.Austral.Math.Soc. 3 (1963) 167-177.

[7] Hing Tong, Minimal bicompact spaces, Bull.Amer.Math.Soc. 54 (1948) 478-479.

[8] R. Vaidyanathaswamy, Set Topology (Madras, 1947; 2nd edition, New York, 1960).

[9] J. Vermeer, Minimal Hausdorff and compactlike spaces, Topological Structures II (Math.Centre Tracts 116, Math.Centrum, Amsterdam 1979) 271-283.

[10] J. Vermeer & E. Wattel, Projective elements in categories with perfect θ-continuous maps, Canad.J.Math. 33 (1981) 872-884.

T_1 - AND T_2 - AXIOMS FOR FRAMES

INTRODUCTION

C.H. Dowker and D. Strauss

A *frame* is a complete lattice in which the infinite distributivity condition $x \wedge \bigvee_{\alpha\in A} x = \bigvee_{\alpha\in A} (x \wedge x_\alpha)$ holds for every element X and every subset $(x_\alpha)_{\alpha\in A}$. A *frame map* is a map from one frame to another which preserves the joins of arbitrary sets and the meets of finite sets.

One of the motivations for the study of frames is the fact that the topology tX of a topological space X provides an example of a frame. If $f\colon X \to Y$ is a map of topological spaces, $tf=f^{-1}\colon tY \to tX$ is a frame map. We thus define a contravariant functor t from the category of topological spaces into that of frames, and many of the concepts of topology can be extended to this wider category.

The purpose of this paper is to suggest separation axioms for frames. While it is obvious what the definitions of regularity and normality ought to be, the T_1 and T_2 separation axioms are more elusive. These axioms, at first sight, seem to refer in an essential way to the *points* of a topological space, and in the category of frames we do not wish points to play a basic role.

We propose T_1 - and T_2 - axioms for frames which seem to be interesting conditions in the category of frames. We feel that proposition 4 below shows that our formulation of the T_2 - axiom is the right one for this category. Whether our formulation of the T_1 - axiom is the right one or not, it is at least an interesting condition to consider.

The condition that tX should be a T_1 - frame (T_2 - frame) is strictly stronger than the condition that a T_0 - space X should be a T_1 - space (T_2 - space). Indeed, there is no possibility of formulating a condition on the frame tX which would be equivalent to X being a T_1 - space, because it is possible to have two spaces with isomorphic topologies, with one being T_1 and the other failing to be. The corresponding situation cannot occur with T_2 - spaces, because any T_2 - space is sober. Nevertheless, while it would in principle be possible to have a frame condition on tX which would be equivalent to X being T_2, we do not feel that such a condition would be the right formulation of the T_2 - axiom for the category of frames.

Separation properties for frames have also been considered by other authors. A different definition of the T_1 - axiom has been proposed in [1], and the same definition as ours occurs in [6]. The class of frames satisfying our T_1 - condition has also been studied in [8] and [9], without being called "T_1". We do not claim that our results about T_1 - frames are new, but we include them here for the sake of completeness.

DEFINITIONS

If x is an element of a frame L, the element $\bigvee\{y \in L : y \wedge x\}=o$ will be denoted by x^*.

We shall say that a surjective frame map with domain L is a *subspace* of L. We give the subspaces of L a quasi - ordering by stating that $f \leq g$ if $f(x) = f(y)$ whenever x, y are elements of L for which $g(x) = g(y)$. If we identify subspaces f and g which satisfy $f \leq g$ and $g \leq f$, we find that the subspaces of L form a complete latticewhich is a dual frame. For each $x \in L$, the mapping $u \mapsto u \wedge x$ from L to $[o, x]$ will be called an *open* subspace, and the mapping $u \mapsto u \vee x$ from L to $[x, 1]$ will be called a *closed* subspace. A subspace of L whose codomain is the two-point frame $\{o, 1\}$ will be called a *point* of L.

A frame map h with domain L will be said to be dense if $h(x) \neq o$ for every $x \in L \setminus \{o\}$.

(If $L = tX$, where X is a topological space, the open (closed) subspaces of L are the maps of the form tf, where f is the inclusion map of an open (closed) subspace of X into X. If f is the inclusion map of a singleton subspace of X, tf will be a point of L. If $f\colon Y \to X$ is a map of topological spaces, tf will be dense if $f(Y)$ is dense in X).

An element p of L will be called *prime* if $p \neq 1$ and if, for any $x, y \in L$, $p = x \wedge y$ implies that $p = x$ or $p = y$. Note that the points of L correspond to primes. Each prime p defines a point $h : L \to \{0,1\}$ for which $h(x) = o$ if and only if $x \leq p$.

A frame will be called *spatial* if it is isomorphic to a frame of the form tX. This is the case if and only if every element of L is a meet of primes.

REMARKS

It is obvious that a T_2 - frame is T_1, and easy to see that a regular frame is T_2 (cf. proposition 3). The property of normality does not seem to have any simple relationship to the others. For example, we do not know whether every normal T_2 - frame is regular.

If X is a T_0 - topological space, X will be a T_1 - space if tX is a T_1 - frame, and X will be a T_2 - space if tX is a T_2 - frame. These facts are easy to see, and follow from propositions 1 and 2 below. However, we give an example to show that it is possible that tX may fail to be a T_1 - frame, even though X is a Hausdorff topological space.

A frame L will be called *compact* if every subset of L whose join is 1, has a finite subset with join 1.

A subspace $h : L \to K$ of a frame L will be said to have a frame property, such as compactness, if K has this property.

SEPARATION AXIOMS

A frame L is said to be $\underline{T_1}$ if, for every parallel pair, f, g, of frame maps with domain L, the condition $f(x) \leq g(x)$ for every $x \in L$ implies that $f=g$.

L is said to be $\underline{T_2}$ if, for every parallel pair, f, g, of frame maps with domain L, the condition $f(x) \wedge g(y) = o$ whenever $x \wedge y = o$ in L, implies that $f = g$.

L is said to be *regular* if, for every $x \in L$, $x = \bigvee \{u \in L : u^* \vee x = 1\}$.

L is said to be *normal* if, for every x, $y \in L$, satisfying $x \vee y = 1$, there exists $u \in L$ such that $x \vee u = y \vee u^* = 1$.

PROPOSITIONS

Proposition 1 - A point in a T_1 - frame is closed. Thus a prime in a T_1 - frame is a maximal element.

Proof - Let p be a prime element of a T_1 - frame L. Consider the two following mappings f, g from L into the frame $[p,1]$

$$f(x) = x \vee p; \quad g(x) = \begin{cases} p & \text{if } x \leq p \\ 1 & \text{otherwise} \end{cases}$$

For every $x \in L$, $f(x) \leq g(x)$. So $f = g$.

Proposition 2 Points in a T_2 - frame are separated in the following sense: If f_1, f_2 are distinct points, there are open subspaces g_1, g_2 such that $f_1 \leq g_1$, $f_2 \leq g_2$ and $g_1 \wedge g_2 = o$.

Proof Let f_1, f_2 be distinct points in a T_2 frame L. Then there are elements u_1, u_2 of L such that $u_1 \wedge u_2 = 0$, $f_1(u_1) \wedge f_2(u_2) \neq 0$. So $f_1(u_1) = f_2(u_2) = 1$. We define the open subspaces $g_i : L \to [o, u_i]$ by $g_i(x) = x \wedge u_i$ $(i = 1, 2)$. Then $f_i \leq g_i$ for $i = 1,2$, because $g_i(x) = g_i(y)$ implies that $x \wedge u_i = y \wedge u_i$ and hence that $f_i(x) = f_i(y)$.

Proposition 3 Subspaces of T_1 (T_2, regular) frames are T_1 (T_2, regular). Products and co-products of T_1 (T_2, regular) frames are T_1 (T_2, regular).

Proof We shall show that a product of T_1 - frames is T_1. The argument for T_2 - frames is very similar, and the other statements are entirely obvious.

Suppose then that (L_a) is a family of T_1 - frames and that $L = \Pi L_a$. Suppose that $f, g : L \to K$ are frame maps satisfying $f(x) \le g(x)$ for every $x \in L$. For each α, we define $j_a : L_a \to L$ by $\Pi_\beta\, j_a\,(\omega) = \omega$ if $\beta = a$ and $\Pi_\beta\, j_a\,(\omega) = 0$ if $\beta \neq \alpha$ Now $j_a(1_a)$, where 1_a denotes the unit element of L_a, has a complement v_α in L. Since $f\, j_a(1_a)$, $gj_\alpha(1_\alpha)$ are complements of $f\,(v_\alpha)$, g (v_α) respectively, and since $f(v_\alpha) \le g(v_\alpha)$, $f\, j_\alpha(1_\alpha) \ge g\, j_\alpha\,(1_\alpha)$. Thus $f\, j_\alpha\,(1_\alpha) = gj_\alpha\,(1_\alpha)$. It follows that the frame maps $f\, j_\alpha$, $g\, j_\alpha$ mapping L_α into $[o, fj_\alpha(1_\alpha)]$ are equal. Since any $x \in L$ satisfies $x = V_\alpha j_\alpha\, \Pi_\alpha(x)$, $f = g$.

Proposition 3 A regular frame is T_2.

Proof Suppose that L is regular, and that f, g are parallel frame maps with domain L such that $f(x) \wedge g(y) = o$ whenever $x \wedge y = o$. Given $x \in L$, $x = V\,\{u \in L : u^* \vee x = 1\}$. Now $u^* \vee x = 1 \Rightarrow g(u)^* \vee f(x) = 1$ (because $g(u) \wedge f(u^*) = o$ and so $g(u)^* \ge f(u^*)) \Rightarrow g(u) \le f(x)$.
So $g(x) \le f(x)$.
Similarly, $f(x) \le g(x)$.

Proposition 4 For any frame L, the following three statements are equivalent:

i) L is T_2;

ii) The coequaliser of any parallel pair of frame maps with domain L is closed;

iii) The diagonal of $L + L$ is closed.

<u>Proof</u> i) $\Rightarrow$ ii)

Suppose that L is T_2. Let h_1, h_2 : $L \to K$ be frame maps and let g : $K \to M$ be their coequaliser. Put $u = \bigvee \{x \in K : g(x) = o\}$. Consider the closed frame map k from K to $[u,1]$ for which $k(x) = x \vee u$. We shall show that k and g are equivalent.

Clearly, $k(x) = k(y) \Rightarrow g(x) = g(y)$. We shall show that $kh_1 = kh_2$. It will then follow that $g(x) = g(y) \Rightarrow k(x) = k(y)$, because k can be expressed as g followed by some frame map.

Now, if $x \wedge y = o$ in L, $g\,(h_1(x) \wedge h_2\,(y)) = o$ because $gh_1 = gh_2$. So $h_1(x) \wedge h_2\,(y) \leq u$, and therefore $k\,(h_1(x) \wedge h_2\,(y)) = o$. Since L is T_2, $kh_1 = kh_2$.

<u>ii) $\Rightarrow$ iii)</u>

The diagonal of $L + L$ is the unique map d for which the following diagram commutes, where q_1, q_2: $L \to L + L$ are the canonical maps:

<u>Diagram 1</u>

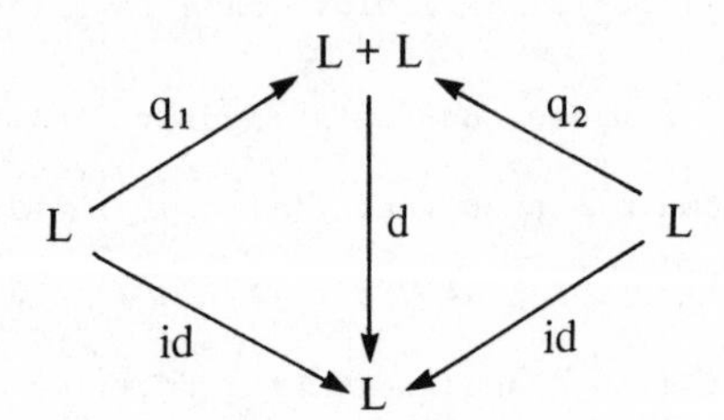

It is readily seen that d is the coequaliser of q_1 and q_2. To see this suppose that h : $L + L \to K$ is a frame map for which $h\,q_1 = h\,q_2 = k$, say. It then follows that $kd = h$, because $kdq_1 = k = hq_1$ and $k\,d\,q_2 = k = hq_2$.

iii) $\Rightarrow$ i)

Assume that d is closed. Then, if $\xi \in L + L$ is defined to be $V\{q_1(x) \wedge q_2(y) : x \wedge y = o\}$, the condition that $d(\eta) = d(\zeta)$ implies that $\eta \vee \xi = \zeta \vee \xi$ in $L + L$.

Consider frame maps $h_1, h_2 : L \to K$ for which $h_1(u) \wedge h_2(v) = o$ whenever $u \wedge v = o$ in L.

Diagram 2

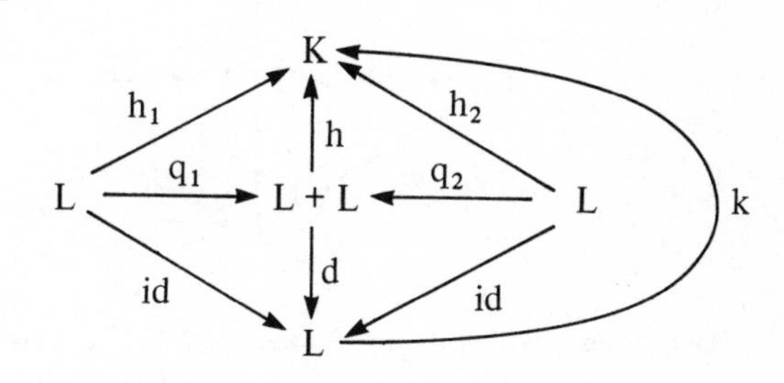

Then $h(\xi) = o$, where $h : L + L \to K$ is the frame map for which $hq_1 = h_1$, $hq_2 = h_2$; for, if $u \wedge v = o$ in L, $h(q_1(u) \wedge q_2(v)) = h_1(u) \wedge h_2(v) = o$.

Thus $d(\eta) = d(\zeta)$ implies that $h(\eta) = h(\zeta)$. It follows that there is a frame map $k : L \to K$ for which $kd = h$. Hence $h_i = hq_i = kdq_i = k$ for $i = 1,2$. So $h_1 = h_2$.

Proposition 5 A frame L is T_2 if and only if the following statement holds:

Two frame maps $h_1, h_2 : L \to K$ are equal if $gh_1 = gh_2$ for some dense frame map with domain K. (Thus the dense maps are the monic maps in the category of T_2 frames).

Proof

i) Suppose first that L is T_2, and that $h_1, h_2 : L \to K$ are frame maps for which $gh_1 = gh_2$, where g is a dense frame map with domain K. Then if $u \wedge v = o$ in L, $g(h_1(u) \wedge h_2(v)) = o$ and so $h_1(u) \wedge h_2(v) = o$. Thus $h_1 = h_2$.

ii) Suppose that two frame maps h_1, h_2 . $L \to K$ are equal if $gh_1 = gh_2$ for some dense frame map g with domain K.

Let d be the diagonal of $L + L$.

Diagram 3

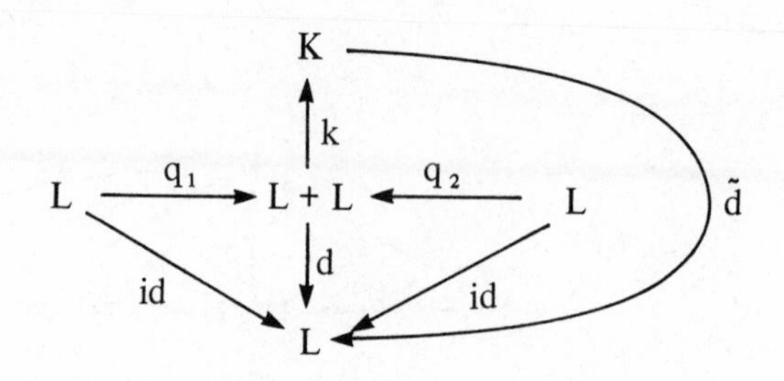

Let ξ be the maximum element of $L + L$ for which $d\ (\xi) = o$. Let $K = [\xi, 1]$ and let $k : L + L \to K$ be the closed subspace of $L + L$ for which $k\ (\eta) = \eta \vee \xi$.

We define a dense frame map $\tilde{d} : K \to L$ by $\tilde{d}\ (\eta) = d\ (\eta)$. Since $\tilde{d}kq_1 = \tilde{d}kq_2$, $kq_1 = kq_2$. Hence k factorises through d, the coequaliser of q_1 and q_2. It follows that d is equivalent to k. Thus d is closed, and so L is T_2.

The proof of the following proposition uses the axion of choice.

Proposition 6 A compact T_2 frame is spatial.

Proof Let L be a compact T_2 frame. Note that, for any $x \neq 1$ in L, there is a prime p of L satisfying $p \geq x$. This prime can be obtained as the join of a maximal chain of elements in $[x, 1)$.

We first use proposition 2 to prove the following statement: for any $x \in L$ and any prime p of L satisfying $x \nleq p$, there is an element u of L for which $u \vee x = 1 = u^* \vee p$.

Let Q_x denote the set of primes in $[x, 1]$. By proposition 2, for each $q \in Q_x$ there exists $u_q \in L$ such that $u_q \vee q = 1 = u_q^* \vee$ p. Now $\bigvee_{q \in Q_x} u_q \vee x = 1$, as there cannot be a prime in $\bigcap_{q \in Q_x} [u_q \vee x, 1]$.

Thus there is a finite subset F of Q_x for which $\bigvee_{q\in F} (u_q \vee x) = 1$. Put $u = \bigvee_{q\in F} u_q$. Then u has the required properties.

We next show that L must be normal. Suppose that $x \vee y = 1$ in L. For each $q \in Q_x$, there exists $v_q \in L$ such that $v_q \vee q = 1 = v_q^* \vee y$. Now $\bigvee_{q\in Q_x} (v_q \vee x) = 1$. So $\bigvee_{q\in K} (v_q \vee x) = 1$ for some finite subset K of Q_x. Put $v = \bigvee_{q\in K} v_q$. Then $v \vee x = 1 = v^* \vee y$.

We finally show that L must be regular. It will follow that L must be spatial, as it is well known (if one assumes a.c.) that any compact regular frame is spatial.

For any $x \in L$, put $\hat{x} = \bigvee \{y \in L : y^* \vee x = 1\}$. Since $y_1^* \vee x_1 = 1 = y_2^* \vee x_2$ implies that $(y_1 \wedge y_2)^* \vee (x_1 \wedge x_2) = 1$, the mapping $x \to \hat{x}$ preserves finite meets. We shall show that it preserves arbitrary joins.

We first show that $(x_1 \vee x_2)\hat{} = \hat{x}_1 \vee \hat{x}_2$ for every $x_1, x_2 \in L$.

Suppose that $y \in L$ satisfies $y^* \vee x_1 \vee x_2 = 1$. Then there exists $u \in L$ such that $y^* \vee u \vee x_1 = 1 = y^* \vee u^* \vee x_2$. There also exists $w \in L$ such that $w^* \vee u^* = w \vee y^* \vee x_2 = 1$. We have $(y \wedge u^*)^* \vee x_1 = 1$ and hence $y \wedge u^* \le \hat{x}_1$. We also have $(y \wedge w^*)^* \vee x_2 = 1$ and hence $y \wedge w^* \le \hat{x}_2$. Since $y = (y \wedge u^*) \vee (y \wedge w^*)$, $y \le \hat{x}_1 \vee \hat{x}_2$. So $(x_1 \vee x_2)\hat{} = \hat{x}_1 \vee \hat{x}_2$.

Now suppose that $(x_\alpha)_{\alpha\in A}$ is a family of elements of L and that $y \in L$ satisfies $y^* \vee \bigvee_{\alpha\in A} x_\alpha = 1$. Then $y^* \vee \bigvee_{\alpha\in F} x_\alpha = 1$ for some finite subset F of A.
Thus $y \le (\bigvee_{\alpha\in F} x_\alpha)^\wedge = \bigvee_{\alpha\in F} \hat{x}_\alpha$. This shows that $(\bigvee_{\alpha\in A} x_\alpha)^\wedge = \bigvee_{\alpha\in A} \hat{x}_\alpha$.

We have proved that the map $x \to \hat{x}$ is a frame map of L to itself. Since $\hat{x} \leqq x$ for every $x \in L$ and since L is T_1, we must have $\hat{x} = x$. So L is regular.

<u>Remark</u> The method used in the above proof shows that any compact normal T_1 frame must be spatial.

<u>Example</u> We give an example of a Hausdorff topological space X for which tX fails to be a T_1 frame.

Let X denote the unit interval $[0,1]$, with the topology in which the open sets are those of the form $U \cup (Q \cap V)$, where U and V are open subsets of $[0,1]$ in the usual topology.

Let B denote the complete Boolean algebra formed by the subsets of $[0,1]$ which are regular open sets in the usual topology.

We define two maps $f, g : tX \to B$ as follows:

$f(U \cup (V \cap Q)) = \overline{U}^{\,o}$, $g\,(U \cup (V \cap Q)) = \overline{U \cup V}^{\,o}$, the operations of forming closures and interiors being carried out in the usual topology. These maps are well-defined, for if $U_1 \cup (V_1 \cap Q) = U_2 \cup (V_2 \cap Q)$, the fact that $U_1 \cap Q' = U_2 \cap Q'$ implies that $\overline{U}_1 = \overline{U}_2$, and the fact that $(U_1 \cup V_1) \cap Q = (U_2 \cup V_2) \cap Q$ implies that $\overline{U_1 \cup V_1} = \overline{U_2 \cup V_2}$. It is easy to check that f and g are frame maps. Now $f(W) \leq g(W)$ for every $W \in t\,X$; but $f \neq g$, because $f\,(Q) = 0$ and $g\,(Q) = 1$.

References

[1] B. Banaschewski and R. Harting. Lattice aspects of radical ideals and choice principles. Proc. London Math. Soc. (to appear).

[2] C.H. Dowker and D. Papert, Quotient frames and subspaces, Proc. London Math. Soc. (3), 16 (1966), 275-296.

[3] C.H. Dowker and D. Papert, On Urysohns' Lemma, General Topology and its relations to modern analysis and algebra II, Prague, 1966, 111-114.

[4] C.H. Dowker and D. Strauss, Sums in the Category of Frames, Houston J. of Math., (3) No. 1, 1976, 17-32.

[5] C.H. Dowker and D.P. Strauss. Separation axioms for frames. Colloq. Math. Soc. Janos Bolyai 8 (1974), 223-240.

[6] M.P. Fourman. T_1 spaces over topological sites. J. Pure Appl. Algebra 27 (1983), 223-224.

[7] J.R. Isbell. Atomless parts of spaces. Math. Scand. 31 (1972) 5-32.

[8] J.R. Isbell. Function spaces and adjoints. Math. Scand. 36 (1975) 317-339.

[9] P.T. Johnstone. Stone Spaces. Cambridge Studies in Advanced Math. no. 3, Cambridge University Press 1982.

[10] P.T. Johnstone. Wallman compactification of locales. Houston J. Math. (to appear).

[11] H. Simmons. A framework for topology. In Logic Colloquium 77, Studies in Logic vol. 96, North-Holland 1978, 239-251.

[12] H. Simmons. The lattice theoretic part of topological separation properties. Proc. Edinburgh Math. Soc. (2) 21 (1978), 41-48.